THE COMPLETE GUIDE TO

SAVING SEEDS

THE COMPLETE GUIDE TO
SAVING SEEDS

322 VEGETABLES, HERBS, FLOWERS, FRUITS, TREES, AND SHRUBS

ROBERT GOUGH and
CHERYL MOORE-GOUGH

Storey Publishing

The mission of Storey Publishing is to serve our customers by publishing practical information that encourages personal independence in harmony with the environment.

Edited by Gwen Steege and Fern Marshall Bradley
Art direction and book design by Cynthia N. McFarland

Cover photography by © Rosemary Kautzky: back cover and spine; © GAP Photos, Ltd., Fiona McLeod: front cover top; © GAP Photos, Ltd., Jo Whitworth: front cover top
Interior photography by © Rosemary Kautzky, except for © Cheryl Moore-Gough: 1 bottom, 15 bottom left and right, 23 right; Mars Viluabi: 20; © Joe De Sciose: 105; H. Zell/Wikimedia Commons: 109; © Ada Roth/agefotostock.com: 127; © J. Kottmann/agefotostock.com: 140; © Philip Nealey/agefotostock.com: 142; © StockFood LBRF/agefotostock.com: 148; Enrico Blasutto/Wikimedia Commons: 155; © Kristina DeWees/agefotostock.com: 183; © Alfred Osterloh/agefotostock.com: 193; © Creativ Studio Heinem/agefotostock.com: 268; © Maria Kemp/agefotostock.com: 271; © Siepmann/agefotostock.com: 276
Illustrations by Beverley Duncan: 13, 14, 16, 17, 21, 22, 29, 63, 75, 83, 129, and 131, and Alison Kolesar: 28, 29, 30, 41, 46, 47, 72, and 79

Indexed by Christine Lindemer, Boston Road Communications

© 2011 by Bob Gough and Cheryl Moore-Gough

Storey Publishing
210 MASS MoCA Way
North Adams, MA 01247
www.storey.com

Printed in the United States by Quad/Graphics
10 9 8 7 6 5 4 3 2 1

LIBRARY OF CONGRESS CATALOGING-IN-PUBLICATION DATA

Gough, Robert E. (Robert Edward)
 The complete guide to saving seeds / by Robert Gough and Cheryl Moore-Gough.
  p. cm.
 Includes index.
 ISBN 978-1-60342-574-2 (pbk. : alk. paper)
 1. Seeds. 2. Seeds—Harvesting. 3. Seeds—Viability. 4. Plant propagation.
 I. Moore-Gough, Cheryl. II. Title.
SB118.3.G68 2011
631.5'21—dc22
                                    2010051179

contents

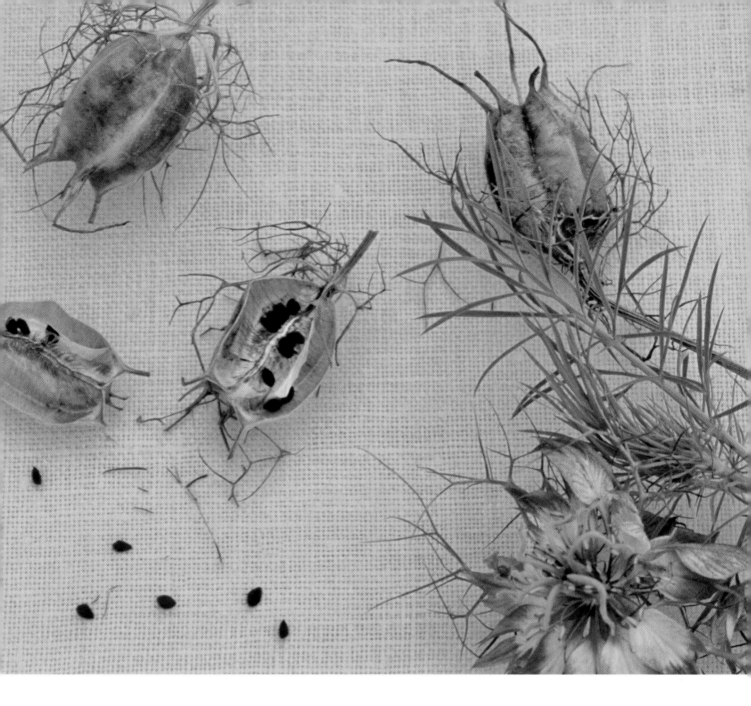

To our parents and grandparents, who inspired us,
and to our children and grandchildren,
who will benefit from their wisdom

Bad seed is a robbery of the worst kind: for your pocket-book not only suffers by it, but your preparations are lost and a season passes away unimproved.

— George Washington

THE SEED-SAVING REVIVAL

ELCOME TO THE DYNAMIC and wonderful world of seeds. Nature has excelled at saving seeds for millions of years, but improving upon natural systems is part of human nature. We have done well by probing the mysteries of seeds, learning how to harvest and store them and coax them to germinate and grow abundantly to suit our needs.

A bit more than 10,000 years ago, humankind learned that saving seeds and planting them in specially prepared gardens meant that we no longer had to wander through muddy slime in warm Mediterranean swamps to gather celery, nor did we have to stalk wild cabbages along the cold and windy Baltic coast. Thanks to sojourning pioneers who plied ancient trade routes and returned to Europe with seeds of a strange new tree, we no longer had to live in Kazakhstan to enjoy wild apples.

Seed saving and agriculture birthed civilizations. By planting seeds, humans could more easily enjoy more types of food in greater quantity, which left them more leisure time for developing art and astronomy, music and mathematics. Early gardeners quickly began to understand that if they sowed seeds of, for example, spinach, each plant that sprouted displayed a slight variation in the traits of its parent, while still conserving the overall attributes of spinach. The gardeners observed variations on a theme, so to speak, noticing a few plants with a more tender or larger leaf, while some other plants were more or less robust than the rest. These first generations of seed savers kept the seeds only from the best plants and destroyed plants with less desirable traits (a process called *rogueing*). Through the hundreds of generations of gardeners and horticulturists that followed, modern seed-saving techniques were refined and ambitious plant-breeding programs flourished. Thus, in the course of only a few thousand years, our collective seed-saving endeavors have accomplished what nature, working far more slowly, would have taken millions of years to accomplish through evolution.

▼ Sowing seeds that you've collected from your own plants offers exciting new possibilities for your gardening endeavors.

Those Amazing Seeds

Seeds are amazing entities. Those of witchweed (*Striga asiatica*), which is a parasitic weed, are less than one-hundredth of an inch long — as small as dust motes on the wind. A teaspoon will hold millions. About 350 spinach seeds, moderately sized as seeds go, fill a teaspoon. In contrast, consider the coconut: each seed is as large as a cannonball.

In every viable seed, nature has packed an embryo and a small food supply, a sort of box lunch, to nourish it during its seed life. The nucleus of every cell in that seed contains in its DNA the complete instructions for making a plant like itself. If we could translate this genetic information into instructions spelled out in English, the written blueprint for spinach might require several hundred encyclopedic volumes, each about 800 pages long. How marvelous it is to ponder that the spinach seed, in its every cell, contains the entire blueprint for making another spinach plant! In this cellular blueprint, there are minor variations caused by minute differences in gene expression or some maverick mutation, which may result in a spinach plant with unexpected traits. That's one of the things that make seed saving so fascinating. When you plant seeds you've collected and saved, you never know precisely what characteristics the new group of seedlings will show. Nature promises no absolutes, only probabilities.

Why Bother Saving Seeds?

Through the 1940s, home seed saving was fairly common. Bob helped his grandfather save vegetable seeds. Gramps wouldn't think of buying seeds for ten cents a packet if he could save them for nothing. So Bob, along with others of his generation, learned firsthand how to save seeds — and money — and found seed saving to be easy, fun, and rewarding. Each harvest was an adventure.

Following World War II, however, interest in home seed saving declined, and a process that we call *social rogueing* became an important force in changing the dynamics of plant breeding and seed saving. Social rogueing involves the interaction of gardeners and seed sellers. If gardeners fail to purchase seeds of a certain variety, the seller replaces that variety with a more popular one. Many gardeners, rightly perhaps, bemoan the passing of older varieties such as the 'General Grant' tomato and point the finger of blame at the seed sellers, frequently accusing them of deliberately and demonically diminishing the selection of varieties for some sinister purpose. That many of the replacement varieties are hybrids and sometimes proprietary adds fuel to the controversy. But it's our view that we gardeners must accept some of the blame. If we

◀ Heirloom varieties have found new popularity with today's gardeners. Once you learn how to save seeds, you can even experiment with developing your own "heirloom" vegetables and flowers.

▶ Whether your collection is large or small, your seed-saving efforts help to preserve genetic diversity in the plant world.

found the 'General Grant' tomato so wonderful, why didn't we continue to purchase its seeds? If we found Grandma's rosebush so enticing with its sweet-scented flowers, why didn't we propagate it ourselves?

Select for Specific Traits

These days, plant breeders often work to develop garden plants for better shipping quality, because fewer Americans garden than those who demand fresh produce year-round.

And varieties are selected for greater pest resistance because fewer of us use pesticides, and for greater vigor because gardeners now shy away from using commercial fertilizer. The consuming public has, in a sense, made demands, and the seed companies have delivered. Unfortunately, many new selections do not have the flavor of the older varieties, but in our opinion, many new hybrids, especially of sweet corn, have flavor and quality far superior to that of many older varieties.

Our message is this: It's much more exciting to look at the wonderful possibilities that seed saving presents today than to bemoan what has changed from the "good old days." Take matters into your own hands! Make your own plant selections for taste or color or fragrance, and save the seeds from them to increase your genetically diverse stock. Save your own seeds and custom-select the best plants for your garden. We've written this book especially to guide you

THE STORY IN A NAME

MANY OLD-TIME VARIETIES carry colorful place-names, which often suggest the specific locality where a variety was developed by foresighted gardeners through years of purposeful selection. Others bear the names of famous figures from history. A few of these varieties, which may date back as far as the midnineteenth century, are still jealously guarded by wise gardeners, but many have disappeared. 'Rhode Island Asylum' sweet corn was developed on the state-run market garden for supplying the state's prison, mental hospitals, and such. It must have been well suited for the purpose but is now long gone. 'Mandan' corn, selected in North Dakota and named for an Indian tribe, is mighty hard to come by. 'President Grant' tomato, like its namesake, passed from the scene many years ago.

We found ourselves so fascinated by historical details such as these while researching this book that we gathered them together in a history of seed saving and selling, which begins on page 279.

through the twists and turns along the seed-saving trail.

Preserve Diversity

If we put all our eggs in one basket, we lose genetic diversity, leaving our plant populations without resistance to many pests. In the early 1970s, much of the corn crop in the United States suffered a disastrous infection of bacterial southern corn leaf blight. Researchers found that so many of the varieties planted at that time were bred from the same or similar parents that they all had low resistance to the disease; they lacked genetic diversity! Fortunately, scientists overcame the problem quickly by breeding resistance to the pathogen into new corn varieties.

At about this time, realizing that the potential for catastrophe was mounting as genetic diversity was lost, some foresighted gardeners took matters into their own hands. The waning of genetic diversity germinated a new seed-saving movement. Why let seed companies do what you could do yourself for a fraction of the cost and for the pure enjoyment of discovery? Echoes of Gramps! So these crusading pioneer seed savers began saving and trading among themselves seeds of non-hybrid, open-pollinated varieties. The nascent preservation effort grew rapidly. The rediscovered passion for open-pollinated heirloom plants in turn spawned small businesses that sold seeds of the best old-time varieties, and seed-swapping groups encouraged gardeners to share their seeds with other gardeners throughout the world.

The humble task of saving seeds has come full circle, and gardeners are becoming more self-reliant. Now, in the twenty-first century, as people have done for hundreds of generations, many of us are again saving our own seeds for future generations.

Save Money

We plant more than 30 vegetable and flower varieties each year. Packets of seeds that once cost a dime now sell for $3 apiece. Our yearly seed bill would be about $100 if we didn't save our own seeds. It could be argued that investing $100 to produce a year's supply of vegetables for two people is a good value, but why spend the money if you can save your specially selected seeds easily and for nothing but a few minutes of your time?

Besides, seed packets may cost $3 now, but what might they cost 10 years from now? There are good reasons to suspect that the relative cost of food and of seeds will increase in the next few decades, so be ready for it and hone your seed-saving skills now.

Create Superb Plants

The improvement in your gardening that results when you save your own seed from year to year is something that takes time to appreciate, but it's actually one of the key benefits of saving seeds. You see, you will be able to better care for your plants than could any commercial grower, and you'll pay more attention to the selection of the very best plants for seed saving, too. Because of this, over time your seed-saving efforts will pay off by providing you with plants that are uniquely adapted to succeed in your environment. And rather than selecting for the standard characteristics that seed-industry breeders look for — such as suitability for shipment and storage — you can focus on saving seeds that produce the *best-tasting* results instead.

Even when you're growing old-time varieties, you'll still find that some do better than others in your garden, and you'll be able to save seeds selectively from varieties that are your personal top performers. You may even come up with a strain of an heirloom crop that is significantly superior in your region to the original.

A 'DELICIOUS' TASTE OF TIMES GONE BY

BOB WAS LUCKY ENOUGH to have tasted the original strain of 'Delicious' apple, and it was terrific . . . it was crisp, juicy, and tangy, much different from the fruit of that name that we buy at the grocery store. But that original strain probably doesn't exist anymore, and the original strain of the 'McIntosh' apple is gone, too. In the case of apples, varieties are propagated by grafting, not by seed, but even so, there are genetic mutations that occur to the plants over time, such as brighter red color and better shipping quality, which are selected for by breeders, sometimes at the cost of flavor.

▶ Saving seeds from your family's heirloom crops and flowers is a wonderful way to introduce your children to their great-great-grandparents.

Go on an Adventure

Many seeds available today are hybrid varieties. If you collect seeds from a hybrid plant, those seeds won't produce plants that are "true to type." In other words, if you save seeds from a 'Big Boy' hybrid tomato plant and sow them in a seed flat or in your garden, you'll get a fine crop of tomato seedlings. But those seedlings won't grow up to look like 'Big Boy' tomato plants, and they won't produce 'Big Boy' fruits. Most likely they will be off-type fruits that are much less desirable than 'Big Boy'. We'll explain the reasons for this in detail in chapter 1, but our point here is that sometimes the goal of a seed-saving project is simply to have fun seeing what comes up!

Saving seeds from a backyard fruit tree will be an adventure, too, because apples, peaches, and other tree fruits, by their nature, don't produce seedlings that are true to type. Every seedling will be unique in its growth and in the quality, appearance, and taste of its fruit. For gardeners who enjoy experimentation and surprises, therein lies the fun. If you have a 'McIntosh' apple tree growing in your yard, why not try planting a few seeds from it and see what you get? The chances of getting a fine-tasting apple from one of the seedlings are mighty slim, but you just might be the gardener who plants the lucky seed! After all, 'McIntosh' itself was a chance seedling. And even if your seed-produced trees don't put forth fine fruit, you can still enjoy the fragrant blossoms and let the animals enjoy the fruit. They're not as fussy as we are.

The art of saving seeds may so intrigue you that you'll branch out to your own plant-breeding adventures with such long-term seed-saving projects as growing oaks from acorns. We'll show you how and tell you what works and what does not to ease your journey through these fascinating ventures.

Ensure a Vital Link

Plants and animals are interdependent cohabitants of Earth. Plants absorb the carbon dioxide that animals exhale, blend it into a chemical soup with water, chlorophyll, sunlight, and some minerals, and produce the food that we animals eat. In the process, plants release the oxygen that we breathe. In a sense, each of us lives on the others' waste, but we are the more dependent beings in this delicate and delightful dance. Higher plants thrived for a couple of hundred million years without us; our lifespan without them would wither to perhaps a month or, with luck, a few months, until we devoured the last morsel of food and breathed the last, long lungful of precious oxygen. It is imperative that we, as stewards of Earth, ensure a safe and varied seed supply to pass along to future generations.

▲ When you save your own seeds, you can experiment with a wide range of varieties or specialize in just one or two. You're in control of the variations.

How to Use This Book

This book is divided into two sections. Part 1 is called Saving Seeds: The Basics and Beyond. Unless you're already an experienced seed saver, start at the beginning of part 1. In the opening chapter we present an overview of how plants produce seeds, including explanations of many important terms regarding flowers, fruits, pollination, and seed formation that we'll continue to use throughout the book. Understanding the physiology of how plants produce seeds is fundamental to seed-saving success, and it might just make you a better gardener overall, too.

Once we've laid out the process of seed formation and defined terms, in chapter 2 we'll present an overview of some of the special plant-care techniques, such as hand-pollination and overwintering biennial crops, that you'll need to master to produce high-quality seeds, especially of vegetable crops. Chapters 3 and 4 cover the practical details of how to collect, clean, and store seeds for maximum vigor and viability. In chapters 5 and 6, we complete the cycle, discussing various types of special pregermination treatments, sowing homegrown seed, and caring for seedlings until they're ready to be transplanted to the garden. As a bonus, in chapter 7 we'll discuss some of the principles and practices of plant breeding on the home-garden scale for those advanced gardeners who'd like to experiment with creating new varieties through controlled cross-pollination.

Part 2, called The Handbook: From Vegetables to Nuts, is a guide to how to collect, clean, store, and germinate seeds from specific food crops and ornamentals. We have divided this part of the book into four chapters: vegetables; herbs; flowers; and nuts, fruits, and woody ornamentals. In these chapters you'll find plant-by-plant entries with detailed information on all aspects of saving seeds of specific plants, including flowering and pollination, ensuring seed purity by controlling pollination, determining when fruits or seeds are ready for collecting, extracting seeds from fruits, cleaning and drying seeds, overcoming dormancy, providing ideal germination conditions, and raising healthy seedlings. We also tell you how to harvest and clean seeds and how to store them properly.

In the appendix, which follows part 2, we offer a history of seed saving in North America — fascinating stories of early seed-swapping groups in colonial America, the formation of the first seed companies (including the birth of the famous Burpee Seeds), seed imports versus seed exports, a government seed-give-away program in the nineteenth century, and more.

On a more practical level, we've also included a glossary at the back of the book, so that if you can't quite remember what a *silique* is or what *dehiscence* means, you can quickly look it up in the glossary and get back to the details of how to save seeds of your beans or dill or sunflowers.

Learning Plant Lingo

Let's take a few minutes here to look at some plant vocabulary. We all like to call plants by their common names rather than botanical names in our everyday gardening conversations. Common names are much more familiar and easier to pronounce, and most of the time we have no problem communicating what plant we are talking about. After all, we all know what apples and alyssum and hollyhocks are.

VARIETY VERSUS CULTIVAR

THE TERM *VARIETAS* refers to a subdivision of a species and is part of the botanical name for the plant, but not all species have a varietas. And botanically speaking, a variety is a subdivision of a varietas. The word *cultivar* derives from "[culti]vated [var]iety," and for 50 years it has been the official term for referring to varieties of cultivated plants: in other words, varieties that have resulted from deliberate plant-breeding work. 'Bloomsdale Long Standing' spinach and 'Big Boy' tomato are examples of well-known cultivars.

Technically speaking, the term *variety* is outdated, having been replaced by *cultivar* more than 50 years ago. However, many gardeners still use the older term, and so we do in this book too. In this book we offer seed-saving instructions for some true varieties of many wild plants, as well as discuss cultivars of domesticated plants. For simplicity's sake, we use the term *variety* throughout instead of *cultivar*.

▲ This planting of multicolored lettuce combines four strikingly different cultivars.

When it comes to giving directions about harvesting and propagating seeds, it's not enough simply to know that a plant is a hollyhock; seeds of different species of hollyhocks have different requirements for breaking dormancy. If you don't know precisely what kind of hollyhock you have, you won't know which protocol to follow to propagate it from seed most efficiently. Thus, in this book, it serves us well to use botanical names, because for the most part, there is only one currently accepted botanical name for each kind of plant. (Plant classification is always a work in progress, and scientists continue to reclassify plant relationships as new information becomes available.) You'll notice that we use the term *botanical name* rather than *Latin name*. That's because only about a third of plant names are in Latin. The rest are Greek or latinized English.

Taxonomists lump plants into groups based upon shared characteristics. For example, all seed-bearing plants that have needles instead of leaves and bear cones are called conifers. Among the conifers, pines (*Pinus* spp.) bear their needles in bundles (fascicles). Firs (*Abies* spp.) and spruces (*Picea* spp.) bear their needles singly attached to the stems. The cones of fir stand upright on the branch; cones of the others do not. So you see, we continue to subdivide plants into smaller and smaller groups that have certain traits in common.

The largest groups of seed-bearing plants that we deal with in this book are the divisions Gymnospermae (gymnosperms) and Angiospermae (angiosperms). These divisions are subdivided into classes, the classes into orders, the orders into families, the families into genera (singular = genus), the genera into species, and, finally, the species into varietas and the varietas into varieties (also referred to as cultivars). Because this provides a systematic way of classifying plants, this branch of botany is called systematics. Thus, to a taxonomist, or systematist, a 'Golden Acre' cabbage looks like this:

DIVISION	Angiospermae
CLASS	Dicotyledoneae
ORDER	Cruciferales
FAMILY	Brassicaceae
GENUS	Brassica
SPECIES	*oleracea*
VARIETAS	*capitata*
VARIETY	(cultivar) Golden Acre

In most cases, you don't need to worry about which groupings a plant belongs to above the family level. In chapter 7, we give some general characteristics to look for in each plant family. For example, members of Brassicaceae (the cabbage family) almost always have four-petaled flowers and bear seeds in a capsule; members of Fabaceae (the legume family) almost always show hardseededness and bear their seeds in a pod. Knowing these general characteristics will help you understand better how to collect and process the seeds.

One final point about botanical names: Scientists occasionally come up with new botanical names for plants or plant families as new genetic information about the plants comes to light. We won't bog you down with the details of why particular names have been changed, but we want you to be aware that for many plant families, there are now two names in popular use. For example, the grass family used to be called Gramineae, but taxonomists have changed the name to Poaceae. Thus, while older books may still use Gramineae, newer ones such as ours use Poaceae to refer to the same family. The same is true for Leguminosae, which is now Fabaceae; Cruciferae, which is now Brassicaceae; and a few others.

PART 1

Saving Seeds:
The Basics & Beyond

The entire fruit
is already present
in the seed.

— Tertullian

CHAPTER 1

SEED BIOLOGY 101

SAVING SEEDS is a fascinating and delightful pastime. Seeds are sometimes the "poor stepchild" in the garden, overlooked by gardeners in their pursuit of a bountiful harvest of fruits and vegetables. But we've discovered that collecting seeds has been a satisfying extension of our other gardening efforts, and being seed savers has made us better gardeners overall.

Plants do a fine job of producing their own seeds year after year, but when we save seeds for replanting, we must manage the process so that we end up with the healthiest, most vigorous, and most reliable seeds possible. Once our plants set seed, we must know how to collect and store those seeds so their seedlings grow well the next year.

Most gardeners know the basic process by which plants make seeds, but it's very helpful to learn more about the physiological processes leading to seed production. Among

these are how plants produce flowers, how flowers are pollinated, a bit about plant breeding, and how seeds form and ripen. Along the way, we'll familiarize you with some of the terminology that botanists use to distinguish different types of fruits and seeds. So put on your student cap, and we'll get started right away with our review of seed biology.

Before we delve into the physiology of how plants produce seeds, let's distinguish higher plants (plants that produce seeds) from lower plants, such as ferns and mosses, that reproduce by spores or other means. We discuss only higher plants in this book. The two main divisions of higher plants are *gymnosperms* (which have naked seeds) and *angiosperms* (which bear seeds in a fruit).

ANGIOSPERMS. Angiosperms are classified as monocotyledons

(monocots) or dicotyledons (dicots), according to how many *cotyledons* (seed leaves) their seeds contain. Woody monocots are not very common in North America, but yucca is one. Important herbaceous monocots are asparagus, onion, and corn; among the flowers, iris and gladiolus. Most of the vegetables and flowers we grow are dicots, and so are many common woody plants, such as maples, apples, hollies, and roses.

▲ Dicot seedlings produce two leaf-like cotyledons before they form true leaves.

GYMNOSPERMS. For home gardeners, the two main groups of interest among the gymnosperms are ginkgo (*Ginkgo biloba*) and conifers, such as pines, spruces, and yews.

Here's a little twist in the terminology. From a botanist's perspective, an angiosperm produces true flowers but a gymnosperm produces a *strobilus*, which we would call a "cone." For our purposes here, when we talk about flowering, we're referring both to true flowers and to strobili (plural of *strobilus*).

Flowers Come First

Sexual reproduction in all higher plants begins with the formation of flower buds (the precise terms for this are *flower bud induction* and *differentiation*). The process ends with seed maturation and dispersal. Annual plants accomplish this cycle in 1 year; biennials, in 2 years or growing seasons. Perennials repeat the cycle year after year once they have reached reproductive maturity. In most trees and shrubs, a 2-year cycle is common. In the first year, flower buds form anywhere from the middle to near the end of the growing season. In the second year, flowers bloom, and pollination and fertilization occur. The embryo grows rapidly, and seeds mature by summer or early fall. Most pines follow a 3-year cycle and some other plants yet another cycle. We'll explain the unusual cycles in a bit more detail in part 2.

FIGURING OUT THE FANCY TERMS

WHEN WE TALK about the parts of a plant, sometimes our only option is to use a scientific term, because it's the only word that exists to name a particular plant part. For example, the scar on a seed where the seed detached from the ovary is called a *hilum*. There's no other name for it. The same is true for *endosperm*, which is the tissue inside a seed that nourishes the embryonic plant until it can establish a root after it sprouts.

From a botanical point of view, there are special names for various types of fruits, too, such as *capsule*, *pod*, *samara*, and *achene*. Each of these names evokes a different fruit with a particular form and characteristics, and it's important to become familiar with these different types of fruits when your goal is to save seeds.

There are some cases where we can use common terms rather than fancy scientific ones. For example, a *testa* is simply a seed coat, so we talk about the seed coat of a seed, not its testa. We use the common term *husk* but also the more technical term *involucre* when talking about that leathery "rind" on a walnut.

Don't let the terminology discourage you. Before you know it, you'll find that you breeze right along when you encounter terms like *monoecious* and *stratification*. And if you come across a form that's unfamiliar, just turn to the glossary, page 285.

The Juvenile Period

All plants propagated from seeds pass through a period of *juvenility* before they become sexually mature and can start producing flowers and seeds. How long does a plant's juvenile period last? It can take as little as a few weeks for some annuals. Among the herbaceous perennials, such as hostas, the juvenile period can be 5 to 10 years. For some trees it lasts half a century. It takes a very patient gardener to grow such trees from seed! Most temperate-zone trees, though, reach sexual maturity in about a dozen years, and shrubs generally flower and fruit in 4 to 8 years from seed. (Stress, environment, and genetic differences can affect the length of the juvenile period of plants.)

Plants in the juvenile phase may look different from what you're used to. For example, a juvenile English ivy plant is a vine with a characteristic leaf shape, while mature English ivy is a shrub that has a different leaf shape. And juvenile apple, citrus, and pear trees have long spines, which make them look like hawthorns.

As a plant enters adulthood, it sheds its juvenile characteristics and assumes adult characteristics, the most prominent of which, and the most important to us, is flowering. During the transition phase, the bottom of a plant may remain juvenile while the top becomes a reproductive adult or vice versa.

Flower Structure and Development

As mentioned above, flowers develop from flower buds, and you can distinguish flower buds from vegetative buds (which produce leaves and

PARTS OF A COMPLETE FLOWER

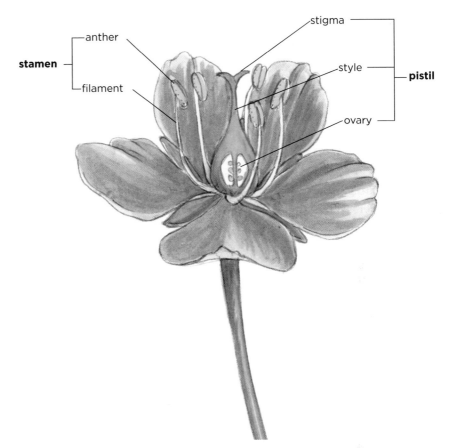

stems) by their appearance and location. Flower buds are usually larger than vegetative buds and in some species, such as flowering dogwood, have distinctive shapes. Watching for flower-bud formation in annual crops is important because it puts you on alert that your plants will be flowering soon. And locating the flower buds on woody plants helps you confirm that a plant is sexually mature and also lets you know where you'll be able to find fruit and seeds to collect later in the season.

Flowers come in an amazing variety of forms and sizes. It's important to recognize what type of flowers a plant has when you want to save seed from it, especially if you need or want to hand-pollinate it (we explain more about hand-pollination in chapter 2).

There are some basic structures that are the building blocks of all flowers. The typical flower of an angiosperm has the following parts.

- *Stigma*, *style*, and *ovary*: the female parts of a flower
- *Pistil*: the collective term for the stigma, style, and ovary. The word *carpel* is also used to describe the stigma, style, and ovary. Some flowers may have multiple pistils.
- *Anther*: a sac that contains pollen
- *Filament*: a stemlike structure upon which the anthers are attached
- *Stamen*: the collective term for the anther and filament; the male sexual organ of a flower

The sexual parts of the flower are usually surrounded by the petals, collectively called the *corolla*. Outside the corolla are the sepals (petal-like structures), collectively called the *calyx*. Both the corolla and the calyx are nonsexual parts of a flower. Flowers are usually borne on a stalk called a *pedicel* or *peduncle*.

A flower that contains all these parts is called a *complete flower* (see illustration above). Any type of flower that is lacking one or more of the basic parts is called an *incomplete flower*. And as you might have guessed, botanists have come up with a specific term to describe each variation in flower structure. On the next page, you'll find some of those descriptive terms.

PERFECT FLOWER

STAMINATE (MALE) FLOWER

PERFECT. A perfect flower has both male and female parts but may not have all the nonsexual parts. Such flowers are also called *bisexual* or *hermaphroditic*. Perfect flowers are not necessarily self-fertile; i.e., they may not be able to fertilize themselves and produce seed. Many plants with perfect flowers are self-sterile and thus require pollen from another plant in order for fertilization to occur.

IMPERFECT. An imperfect flower has only one type of sexual organ, or, if both types are present, only one is functional. These flowers are also called *unisexual*.

▲ These star-of-Bethlehem (*Ornithogalum* spp.) blossoms are complete flowers, with six stamens encircling a central pistil.

STAMINATE. Also called a male flower, this type contains stamens but not pistils.

PISTILLATE (FEMALE) FLOWER

PISTILLATE. Also called a female flower, this type contains pistils but not stamens.

Many vegetable plants, including tomato, pepper, and bean, produce complete, and therefore perfect, flowers. Others, such as spinach, asparagus, and squash, bear imperfect, and therefore incomplete, flowers. The willow flower is odd indeed and lacks both a calyx and a corolla but has both sexual parts. It's perfect but incomplete. Let's just say it gets right down to business.

MALE PLANTS, FEMALE PLANTS, AND MORE

To further complicate things, it's not just flowers that can be male or female; plants may have a "gender," too. Two fundamental terms that relate to this concept are *monoecious* and *dioecious*. Monoecious ("one house") plants have both sexes on one plant. Squash, birch, and striped maple are monoecious: they bear separate male and female flowers on the same plant. Dioecious ("two houses") plants bear male flowers and female flowers on separate plants. Asparagus, spinach, and hollies are dioecious. Most conifers are monoecious, but juniper is dioecious.

The plants with male flowers are termed *androecious* while those with female flowers are *gynoecious*. We can also call them male plants and female plants, and thus we talk about male and female asparagus and holly plants. Only female plants produce fruits and seeds (and only when male plants are growing within pollinating range, of course), so not all dioecious plants produce fruit.

But wait, there's more! Some species of plants can have all these flowering habits. Cucumber is one example. And beyond that there is such a thing as an *andromonoecious* species, which means that it may produce male flowers and perfect flowers on the same plant. Some plants, such as hackberry and bittersweet, are *polygamous* and produce both bisexual (perfect) and unisexual (staminate or pistillate) flowers on the same plant. A *polygamo-monoecious* plant is one that is functionally monoecious with imperfect flowers but produces some bisexual flowers as well. A *polygamo-dioecious* plant produces some

MORE EXAMPLES OF FLOWER TERMINOLOGY

▲ In the blueberry, following fertilization the ovary in each blossom develops into a blueberry fruit. The five-lobed pair of "lips" at the top of each berry make up the calyx of the flower.

▲ Each urn-shaped flower on a blueberry bush has a small stem called a pedicel, which joins the main stem, or peduncle, of the inflorescence.

◄ The strobili (cones) of gymnosperms, such as those on this blue spruce, can be staminate (male) or pistillate (female). Staminate strobili are often bright yellow, red, or purple when fully developed. Pistillate strobili develop into woody, durable cones that bear seeds.

▶ The strobili (flowers) on this mugo pine are not yet mature.

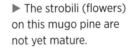

male flower

female flower

◄ ▶ Squash plants bear separate male and female flowers on the same plant. As a female flower matures, the ovary at the base of the flower swells. Male flowers appear first, and although plants usually produce more male flowers than female, female flowers are more numerous than male at the end of the season.

bisexual flowers but is functionally dioecious. In a few genera, flowering patterns vary from species to species. For example, Norway maple is monoecious but silver maple can be monoecious or dioecious.

The Architecture of Flower Stalks

Now that we've covered all the permutations of the form of an individual flower, let's talk about how flowers are arranged on a plant stem. The technical term for this is *inflorescence*, and there are many possible variations, including the number of flowers per stem and the order of bloom within a cluster of flowers. An inflorescence may hold a single, solitary flower or many flowers in a cluster, with or without leaves, bracts, or even a flower stalk. (*Sessile* flowers do not have a flower stalk and are attached flat to the stem.) Inflorescences form either along plant stems in the leaf axils — angles where leaf stalks join the plant stem — or at the tip of the axis (the main stem of the plant).

If an inflorescence located at the tip of the axis prevents elongation of the shoot, this is called a *determinate growth habit*. If shoots continue to elongate and flower as inflorescences form at the leaf axils, then the plant has an *indeterminate growth habit*.

Type of inflorescence is important to us as seed savers because it dictates the method of collecting the seeds. For example, carrot flowers (and therefore seeds) are held in a type of inflorescence called an *umbel*. The flowers within an umbel open and are pollinated over time, which means that the seeds will ripen over

time, too, and that's something we need to take into account when it's time to harvest the seeds. Spinach inflorescences are different, though. They are called *panicles*, and they form at the leaf axils. The seed heads they produce ripen all at once.

There are three main categories of inflorescences: solitary flowers, racemose inflorescences, and cymose inflorescences. There are also mixed types, with racemose clusters borne on cymose inflorescences and the opposite as well.

Solitary flowers are borne singly. They may be held at the growing tip, within the axils of leaves, or on a stalk or stem. Examples are pumpkin, tulip, and narcissus.

RACEMOSE INFLORESCENCES

Generally speaking, the inflorescences in this category are indeterminate. This means that the flowers at the base of the inflorescence are older than the flowers at the tip of the inflorescence. Seeds may be formed at the base of the inflorescence at the same time that young flowers are just opening at the tip.

CYMOSE INFLORESCENCES

Cymose inflorescences are determinate, developing from a flower bud at the tip of the meristem. As opposed to racemose inflorescences, the oldest flowers occur at the center of the inflorescence and the youngest at the base.

CYMOSE INFLORESCENCES

FASCICLE

CYME

CYME. A flowering cluster that is flat-topped or domelike. Examples are tomato and baby's-breath (*Gypsophila*).

FASCICLE. A cyme with flowers borne on short peduncles in a dense cluster, with much longer and about equal-length pedicels all attached at about the same point. Examples are sweet William, cherry, and plum.

RACEMOSE INFLORESCENCES

CATKIN

CORYMB

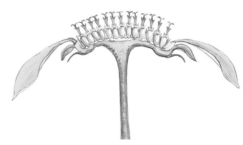

HEAD

CATKIN. A deciduous raceme or spike with petal-less unisexual flowers. Examples are aspen and walnut flowers.

CORYMB. A flowering cluster that is flat-topped or domelike, because the lower pedicels are longer than the upper. The outer flowers open first, so the youngest flowers are in the middle of the cluster. Examples are yarrow, highbush blueberry, and hawthorn.

HEAD. A dense cluster of short, sessile or almost sessile flowers on a flattened receptacle. Examples are sunflower, coneflower, and marigold.

PANICLE. A branched raceme or corymb with a loose flower cluster, usually longer than wide. Examples are beet, chard, and sorrel.

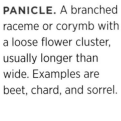

PANICLE

RACEME. Flowers borne on pedicels of approximately equal length along an elongated stalk. Examples are chokecherries and some blueberries.

RACEME

SPIKE

SPIKE. Flowers held on a stalk, sessile or almost so. An example is delphinium.

UMBEL

UMBEL. A flowering cluster with pedicels of approximately equal length attached at the same point and spreading like an umbrella. Examples are onions, carrots, and dill.

Factors That Affect Flowering

As gardeners, we strive to have something in flower in our yards and gardens throughout the growing season, and because plants are so diverse in their growing habits, it's an achievable goal. But when our goal is to save seeds, we need to think about the timing of flowering from a different perspective.

In general, annual vegetables, herbs, and flowers initiate flower buds in late spring or early summer and bloom during the same growing season. Biennials, however, form their flower buds in late summer or early fall. The buds must survive over winter in order for a plant to flower and complete its reproductive cycle the following year. Some plants, such as Swiss chard and many "annual" flowers, may be either annual or biennial, depending upon environmental conditions. We treat many vegetable crops that we grow for food as annuals, but, in fact, they may be biennial or perennial plants.

Trees and shrubs usually initiate flower buds in summer or fall that must persist through the winter before they can open and set fruit. Most temperate tree and shrub species bloom in the spring, though witch hazel flowers in fall or winter and the September elm — you guessed it — in September.

You probably know the usual time when the peonies or purple coneflowers bloom in your area, but there are factors that can cause plants to bloom unusually early or late, or that affect flowering in other ways. Some of the most important influences on flowering are temperature, photoperiod (day length), light intensity, and nutrition.

TEMPERATURE. High temperatures and/or mild drought during summer and lack of adequate cold during winter both can affect flowering. For example, in cucurbits, high temperature causes a shift to maleness; that is, it encourages the production of male flowers over female flowers. So if you have a very hot summer, your zucchini plants may form mostly male flowers, which means you'll end up harvesting fewer fruits. Heat and mild drought both also have the potential to stimulate earlier flower-bud formation as well as increased flowering in many woody species.

Most temperate perennial species need a period of cold to stimulate adequate flowering. For example, daffodils and tulips must fulfill a chilling requirement of between 400 and 1,500 hours below 45°F (7°C), depending upon species, in order to bloom properly. Many fruiting plants and woody ornamentals require a chilling period as well.

PHOTOPERIOD. We all know that plants need light in order to grow, because light supplies the energy that plants use to make food. But light also has other fundamental effects on plant growth, including flowering. Plants contain a pigment called *phytochrome* that is a photoreceptor, which means that its chemical structure changes in response to light and/or day length. The chemical changes in phytochrome in turn can promote the physiological changes in a plant that lead to flowering. (For more details about phytochrome, see page 64.) Flowering in most annuals is strongly affected by day length, but flowering in most woody species is not. For example, long days favor formation of male flowers on squash plants while short days favor formation of female flowers. Most bush fruits form flower buds during the short days of late summer and autumn, whereas most tree fruits form flower buds during the long days of midsummer.

LIGHT INTENSITY. Light intensity is at least as important as photoperiod. Plants growing in direct sun normally

<aside>

SHORT-DAY PLANTS AND LONG-DAY PLANTS

A LONG-DAY RESPONSE to photoperiod occurs when a plant is exposed to a certain number of days that have a light period longer than the critical day length for the species. A short-day response occurs when a plant is exposed to a certain number of days that have a light period shorter than the critical day length for the species.

The actual number of hours that corresponds to a plant's critical day length varies from species to species. For example, a plant that forms flower buds when given photoperiods of 14 hours or less, such as cocklebur, is still called a short-day plant, even though it forms flowers on a day that we might consider a "long day" (i.e., a day that has more hours of daylight than darkness).

</aside>

produce more flowers than do plants in shade, and in the northern parts of the country, more flowers form on the south and west sides of a plant and on the top of a plant than on the north and east sides or inside the plant canopy. Where the light is, there the flowers will be.

NUTRITION. Plants under slight moisture stress during flower-bud formation generally set more flower buds. You may have noticed that the stress created by root pruning, which temporarily interferes with moisture uptake and nutrient flow, increases fruit and seed production. But watch it! Too much stress will kill a plant.

Plants that are given adequate fertilizer will set more flowers and seeds than plants that are undernourished, but supplying too much nitrogen will keep a plant vegetative and reduce flower production. If your goal is to save seeds from a plant, it's generally a good idea to use a fertilizer that is higher in phosphorus than in nitrogen, but take care that your plants don't suffer a nitrogen deficiency.

Pollination

Pollination is the transfer of pollen from stamens to pistils or from staminate strobili to ovulate strobili. Without pollination, there can be no union of the male and female gametes, and therefore fertilization of the ovule cannot occur. If fertilization does not occur, there will be no fruit and, thus, no seeds.

A *self-pollinated* plant is one that is pollinated by its own pollen due to flower structure or *isolation* (physical separation from the pollen of other varieties or species of plants). *Cross-pollinated* plants are those that are pollinated by pollen from other varieties of the same kind of plant or, in some cases, by other species of plants in the same genus.

Seed savers need to be mindful of the sources of pollen that their seed-stock plants are exposed to because of the potential for different varieties of a plant to cross-pollinate. Cross-pollination can lead to unpredictable results because it "mixes up" the genetic composition of the seeds. For example, pollen from a red cabbage plant carries the genes for red leaf color. If pollen from a red cabbage plant lands on the pistil of a green cabbage plant, those red color genes will become intermixed with the green color genes in the seeds that form. If you plant those seeds, you will end up with "off-type" plants. Most will still have green leaves, but they will not necessarily look like the green cabbage parent.

Plant breeders have come up with a variety of methods for preventing unwanted cross-pollination. Some of these methods aren't practical for home gardeners, but others work very well. We'll explain the nuts and bolts of these methods, which are referred to as *isolation techniques*, in chapter 2.

How Pollen Gets Around

Pollen is dispersed from one flower to another mainly by wind (botanists call this *anemophily*) and by insects (*entomophily*). Wind pollination is common in conifers and in some angiosperms that lack showy flowers, such as spinach, corn, and hickory

▲ Purple coneflowers and other perennial plants usually bloom at the same time each year in a given location, but factors such as weather and soil conditions can cause a shift in bloom time.

▲ Attracting bees to your garden will improve pollination of many types of plants, but sometimes you'll need to protect certain plants from accidental cross-pollination by bees or other insects.

been accomplished. But it will matter to you, the seed saver, because of your wish to maintain genetic purity of the seeds produced.

A few plants are pollinated by bats (*chiropterphily*), wasps, butterflies, and other agents. *Magnolia*, a primitive genus, is pollinated by beetles. Yucca is pollinated by moths; the toad cactus (*Stapelia*), which smells like rotted flesh, by flies; and the saguaro cactus (*Carnegiea*) by bats. Water pollination (*hydrophily*) is rare but occurs in eelgrass (*Vallisneria*). Almost all flowers pollinated by birds (*ornithophily*) are red.

Fertilization

When pollen lands on a stigma (which is moist and sticky when receptive), the pollen grain germinates, and a pollen tube grows down the style and into the ovary, releasing, in most cases, two sperm cells. One sperm cell unites with the egg in the ovule to form the embryo, and the other unites with two polar bodies to form the *endosperm*, a tissue that will nurture the embryo while it's in the

trees. It's the most common type of pollination.

Species with brightly colored and/or scented flowers, such as roses, squash, and apple, often have heavy or sticky pollen grains that cannot float on the wind. These plants are insect-pollinated, primarily by bees. Quite a few shrubs and understory plants such as rhododendron depend on insect pollination because the overstory plants block free wind movement. These plants would have died out long ago were it not for insects. Some species, notably some maples, willow, and mulberry, are pollinated both by wind and insects.

There are plants that are self-pollinated; these require no assistance from wind, insects, or other critters to set fruit. Tomatoes and legumes, such as beans and peas, fall into this category. Cross-pollination may still occur in these plants, though, due to opportunistic insects seeking pollen or to a brief gust of wind that blows

pollen from one flower to another. For instance, a bee may happen to visit a tomato blossom in your garden, pick up pollen, and transfer it to another variety of tomato. After all, the bee doesn't know that those flowers are supposed to be self-pollinated! In the long run, it doesn't matter much to nature; the mission has

INBREEDING DEPRESSION

SELF-POLLINATION increases *homozygosity* (sameness), decreases genetic diversity, and tends to produce poor, weak growth and diminished flowering and fruiting capacity over time. Enforced self-pollination in plants normally cross-pollinated by insects often yields poorly developed seeds, weak seedlings, small plants, and reduced yield; this loss of vigor is called *inbreeding depression*. Some plants show inbreeding depression after only a generation or two of self-pollination. With other types of plants, it may take several generations of selfing to notice a decline in vigor, and in still other species this phenomenon is seldom seen. Some vegetables are particularly susceptible to inbreeding depression. With these crops, seed should be saved from more than one plant, and in some cases (corn, for example) from as many plants as possible in each generation.

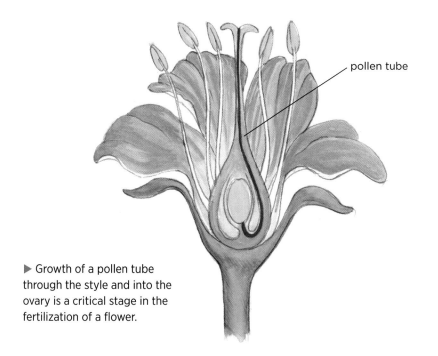

pollen tube

▶ Growth of a pollen tube through the style and into the ovary is a critical stage in the fertilization of a flower.

seed, just as an egg yolk nurtures an unborn chick. In this case, the flower undergoes double fertilization.

Pros and Cons of Cross-Pollination

Natural cross-pollination increases genetic diversity, vigor, fruitfulness, and pest resistance, but it also can result in the loss of desirable traits you want to perpetuate. Controlled, or intentional, cross-pollination by plant breeders (or by you, the home garden seed saver) can result in plants that have more desirable traits.

Most older varieties of vegetables and flowers are *open-pollinated*, which means that they are pollinated by some natural means, whether it's wind or insects or bats. They are different from F_1 *hybrid* varieties, which have been created by very specific and controlled cross-pollination between two parent plants that have particular characteristics.

Open-pollinated plants will produce seeds that are reasonably true to type only if planted in isolation, and this is a key point for seed savers. Let's study the example of spinach, which is a naturally cross-pollinated species. 'Bloomsdale Long Standing' is an open-pollinated variety of spinach. Commercial seed growers grow 'Bloomsdale Long Standing' in fields that are isolated from all other varieties of spinach. That way they can be sure that their female plants are being pollinated only by pollen from males of the same variety. Thus, the genetic traits incorporated into the seeds produced will be within the acceptable known characteristics of 'Bloomsdale Long Standing'.

If you plant only 'Bloomsdale Long Standing' spinach in your garden, you can be reasonably sure that the seed you save from your plants will produce a 'Bloomsdale Long Standing'–type plant from year to year. However, if you plant different varieties of spinach in your garden in the same year *and* allow them both to bloom, they will cross-pollinate.

Let's say, for example, that you planted not only 'Bloomsdale Long Standing', which has crinkly dark green leaves, but also 'Lombardia', which has smooth leaves that are not so deeply green. The seeds that result from cross-pollination of these varieties will carry a combination of traits from the two. Of course, if you allow only one of the varieties to flower, no cross-pollination can occur. Now, what if your neighbor grows spinach, too? Can you still save seeds from your spinach plants? You'll need to visit your neighbor's garden and check on the spinach plants. If none is in flower, there can be no cross-pollination. You're safe!

Isolating open-pollinated plants may be difficult or impossible if you are collecting seed in the wild. For example, you won't find a single oak in the woods, but instead a stand of oaks that likely represent a wide variety of genetic traits. Since they'll cross-pollinate, the acorns you save cannot reproduce precise replicas of the parent.

POLLINATION OF CONES

IN GYMNOSPERMS, the pollination process is a bit different from that of angiosperms. The scales of ovulate cones spread apart when the ovules are receptive, and small drops of liquid called *pollination drops* form. Pollen grains blown between the scales by wind are "captured" by the drops, and fertilization follows.

HYBRIDS AND POLLINATION

From a horticulturist's perspective, a *hybrid* plant is one that results from a cross between two inbred lines (plant lines that have been produced by a series of self-pollinations). Scientists control this process very closely; that is, they carefully choose two parent plants with certain characteristics and pollinate one with the other. Many varieties of vegetables and flowers, and some varieties of woody ornamentals, are what are known as F_1 hybrids. The designation F_1 means the "first filial generation," or the first generation of a plant after a controlled cross was made.

In the wild, though, hybridization takes place easily between naturally cross-pollinated species, but since we don't know to which generation a particular wild plant belongs and what the parent plants were, we don't call them F_1 hybrids. The results from this type of natural hybridization are much less predictable than those from controlled hybridization.

F_1 hybrids usually produce vigorous, high-yielding, pest-resistant plants with high-quality flowers, fruits, or roots. Or they may have been bred to produce fruit that ripens early, that will store longer, or that has redder fruit, more colorful foliage, disease resistance, or any number of other characteristics the breeder sought.

F_1 hybrid seeds produce plants true to type, but if you save seeds from an F_1 hybrid plant, the plants that sprout and grow from those seeds won't be, because they won't have the precise genetic mix that results from the original controlled cross. Such plants are called the F_2 generation. Many F_2 plants are likely to have fairly desirable qualities as well, but as you continue to save seeds, those qualities may be lost. Most likely, if you keep on saving the seeds, you'll wind up with plants far inferior to the original F_1 parents. If you want to experiment with saving seeds from hybrids or with making your own hybrid crosses, you can learn about that in chapter 7.

It can be lots of fun to experiment with saving seeds from the natural hybridization that's happening in your yard. Willows, for instance, are wind-pollinated, and pollen from many different willow trees may blow onto the stigmas of your tree, there to unite their genetic material with that of your tree. The result will be seeds that will produce plants with variable characteristics. Nature provides you with a variety of choices, as though to say, "Here are multiple seedlings with many characteristics. Choose what you want, and propagate it." If you want to reproduce a plant that is exactly like the parent, then take cuttings or try some other method of asexual propagation. But if you like the idea of exploring the potential for surprise, work with the seeds.

Fruit and Seed Development

Following fertilization, as both the embryo and the endosperm grow, the ovule tissues surrounding both harden into the testa, or seed coat, and the seed is formed. While it grows within the seed, the embryo forms a miniature plant with a *radicle* (rudimentary root), *hypocotyl* (stem), *plumule* (bud), and *cotyledons* (seed leaves). In many plants, the process of seed development takes only a matter of weeks. Annuals usually mature their seeds within 2 to 6 weeks after flowering. Most biennials flower in early to late spring of their second year and mature their seeds in mid- to late summer. Seeds of many perennial grasses and herbaceous perennials mature 2 to 3 weeks after bloom. Some trees, such as poplars, maples, and elms, produce mature seeds less than 2 months after bloom.

Other plants take a little longer for their seeds to mature. Plums, peaches, magnolias, and chestnuts require 3 to 4 months. Apples, dogwoods, and most tree nuts require about 5 months after flowering; pine trees mature their seeds 2 to 3 years after flowering. The double coconut (*Lodoicea maldivica*) requires 7 to 10 years!

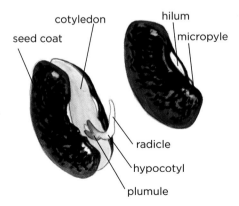

DEVELOPING SEED EMBRYO

seed coat

cotyledon

hilum

micropyle

radicle

hypocotyl

plumule

FRUIT TYPES

NATURE HAS GIVEN US a wonderful variety of fruit shapes and sizes. Knowing what type of fruit your plant produces is important both to identifying the plant and in knowing how to harvest its seeds. Botanically speaking, tomato fruits are *berries*, which means that their seeds are enshrouded in fleshy pulp. (You'll need to separate the seeds from the pulp in order to save and store them.) Apple fruits are *pomes*, with large, easy-to-extract seeds, while caragana fruits are *pods* that pop open when the seeds are ready for harvesting. As a seed saver, you'll need to pay special attention to any fruits that are classified as *dehiscent*. These fruits tend to split open or break apart and shed their seeds, often before you get around to harvesting them.

We've summarized the different types of angiosperm fruits here so that if you can't keep all the terms straight in your mind, you can refer to this list to refresh your memory. Gymnosperm fruit types (called *strobili*) are far fewer in number than those we find in the angiosperms.

Simple Fruit

These are just what their name implies: simple and easy to handle. Some are dry at maturity, some are fleshy. Still others are in-between; these are called dry-fleshy fruits.

DRY FRUITS

INDEHISCENT

These fruits do not split open when ripe.

ACHENE. A single seed attached to the ovary at only one point (buttercup, sunflower)

UTRICLE, OR SEEDBALL. A modified achene, such as in beet and chard, often containing more than one seed

CARYOPSIS OR GRAIN. A single seed (corn and all grasses)

CARYOPSIS (CORN)

SAMARA. One or two seeds with wings (maple, elm, ash)

SAMARA (MAPLE)

NUT. Hard fruit (hazelnut, acorn, pecan)

NUT (ACORN)

SCHIZOCARP. Two or more carpels that separate at maturity (carrots and many related crops)

DEHISCENT

These fruits split open (dehisce) when ripe.

LEGUME. One carpel, which splits along two lines (bean and caragana)

FOLLICLE. One carpel, which splits along one line (columbine, milkweed, peony, larkspur, euonymus)

FOLLICLE (EUONYMUS)

CAPSULE. Two or more carpels (azalea, okra, iris, poppy)

SILIQUE. Two carpels with a partition wall between them (cabbage and most related crops except radish)

STROBILE. A dry, conelike fruit developing from pistillate catkins (alder and birch)

HEAD. A multiple fruit that forms a compact cluster of simple fruits (sweet gum)

(continued on next page)

DRY-FLESHY FRUITS

DRUPE. One-seeded fruit with a bony endocarp, or "stone," surrounding the kernel (peach and viburnum)

DRUPE
(PEACH)

POME. Papery endocarp forming the "core" (apple, hawthorn, holly; individual fruits inside holly pomes are drupes)

POME
(APPLE)

HYP. A complex fruit made up of multiple drupes (pomegranate)

HYP (POMEGRANATE)

HIP. A complex fruit consisting of many achenes (rose)

HIP (ROSE)

FLESHY FRUITS

BERRY. A fruit in which the fleshy portion is the fruit wall (tomato, pepper, avocado)

HESPERIDIUM. A fruit that has a leathery rind (oranges and most other citrus fruits)

PEPO. A fruit type that has a shell surrounding the flesh (squash and most of its relatives)

Compound Fruits

These fruits result either from pollination of a single flower that contains multiple pistils or from pollination of many flowers that unite to form the fruit.

AGGREGATE

A fruit resulting from pollination of a flower that contains many pistils (raspberry and blackberry)

MULTIPLE

A fruit that results from pollination of many flowers that unite after pollination (pineapple, globe artichoke, hop)

SYNCONIUM. A type of pseudocarp in which achenes are actually borne on the inside of a hollow receptacle (fig)

SOROSIS. A fruit derived from the ovaries of several flowers (mulberry)

COENOCARP. A fruit incorporating ovaries, floral parts, and receptacles of many flowers (large-bracted dogwoods and breadfruit)

Dry Strobili

CONE

Woody structure that opens and releases seeds (fir, spruce, and most pines)

CONE (SPRUCE)

Fleshy Strobili

DRUPELIKE FRUIT

A fleshy fruit that encloses a single seed (ginkgo, yew, *Torreya*, and some junipers) or multiple seeds (some junipers)

DRUPELIKE FRUIT (YEW)

CHAPTER

2

GROWING PLANTS FOR SEEDS

Becoming a seed-saving gardener is an excellent exercise in the art and science of cultivating plants. You need to be in touch with your plants' development throughout the growing season, not just at season's end when you gather the seeds. The planting, watering, and feeding of plants for seed production is similar to that of plants you're growing for food, flowers, or foliage, but you must also monitor the flowering process to ensure that only the exchange of desired pollen takes place. In the case of biennials, you must ensure that the plants survive over the winter so they can produce seeds in their second year of growth. And with any type of plants you're growing for seed, you must monitor growth throughout the season to rogue inferior and disease-ridden plants and select the most robust, healthiest plants for their seed.

You can also select for desirable characteristics and slowly, over the years, improve the varieties you grow to best fit your needs. There is a range of characteristics you might want to select for. For instance, if one of the tomato plants in your backyard produces particularly delicious, solid fruit, saving and planting seed from that plant will most likely give

you tomato plants with similar fruit the following year. We suggest several characteristics you might want to consider, but the particular traits that matter most depend on whether you're saving seeds from tomatoes or lettuce or zinnias or oak trees. Whatever your goal, careful observation of your plants throughout the growing season will help you

BEST BETS FOR BEGINNERS

IF YOU'RE NEW TO GARDENING or trying your hand at saving seeds for the first time, you can boost your confidence by choosing plants from which it's easy to collect seeds and have them sprout and grow. Here are 25 plants that should give you good results even if you've never tried saving seeds before.

Vegetables: beans, leaf lettuce, peas, peppers, spinach, tomatoes

Herbs: borage, cilantro, dill, parsley, pot marigold, sweet cicely

Flowers: California poppies, pinks (*Dianthus*), snapdragons, sunflowers, sweet peas

Trees/shrubs/vines: apple, grape, horse chestnut, oak, Ohio buckeye, peach, pear, plum

▲ Beds placed near the house make it possible to carefully monitor your plants during the growing season for the best seed production.

determine which ones show the traits you want to select for.

Super Spacing for Seed Production

Most plants that will be producing seed stalks require more space than plants grown solely for the harvest of their leaves, stems, or roots. Thus, when you're planting a vegetable or herb that's not normally grown for its fruit or seeds, allow triple or even quadruple the normal spacing between plants and at least double the recommended spacing between rows. For example, you might normally space Swiss chard plants a few inches apart in the row to produce their succulent greens. But Swiss chard grown for seeds is an entirely different plant, with branched seed stalks that may grow to 5 feet tall. You'll want to be sure that Swiss chard plants intended for seed production are thinned to stand at least 18 inches apart. Crowding may de-

crease seed yields and encourage pests. Of course, this spacing rule does not apply to woody plants or herbaceous plants normally grown for their flowers or fruit. In part 2 we'll offer guidelines for plants that need to be spaced wider than usual when grown for seed production.

Preventing Unwanted Pollination

When your goal is to save and replant seed from open-pollinated varieties of vegetables, herbs, or flowers, make sure that you maintain genetic purity. In other words, you need to create some type of barrier to stop the "wrong" kinds of pollen from pollinating the flowers of your stock plants. The barrier can be an actual physical structure that blocks pollen-carrying insects or it can be the element of time or distance. These tactics are referred to as "isolating your plants."

Cross-pollinated plants such as those in the Chenopodiaceae, Asteraceae, Cucurbitaceae, Apiaceae, Rosaceae, and Brassicaceae families and in families of most woody species are the most numerous type. These plants have a diverse hereditary makeup, and must be isolated if you are to maintain any sort of control over their hereditary patterns.

Keep in mind that on a home-garden scale, isolating crops sometimes is not a concern. As we explained in chapter 1, if you're the only gardener in your neighborhood who is trying to produce spinach seed, then you may be the only gardener who has spinach plants in flower — because most gardeners simply uproot and compost any spinach plants in their gardens that start to go to seed. However, if your neighbor is saving spinach seed, too, or if you want to save seed of more than one variety, you need to isolate your plants.

Isolating Plants with Distance

At the commercial seed-production level, seed producers sometimes rely on physical distance to isolate their crop from others. The idea is to plant a crop far enough away from others so that pollen can't travel from one to the other either via insects or on the wind. There is quite a bit of information available on the isolation-distance requirements of vegetable crops but very little on the isolation of flower crops and almost nothing on the isolation requirements of woody species. Recommended isolation distances for cross-pollinated plants vary by authority and by mode of pollination. Often, the recommended minimum isolation distance is at least half a mile, and that's just not practical for home gardeners. If you have a relatively small garden, our advice is to separate all varieties of a crop that is cross-pollinated by insects by at least 200 yards to reduce the chances of crossing among varieties. As a general rule, plants of species cross-pollinated by wind should be separated from plants of the same species by at least a mile. In part 2, we include information on recommended isolation distances for specific vegetable crops and flowers for which they are a concern.

Less than 4 percent of all plant species are self-pollinated, but among those species are peas, beans, flax (*Linum* spp.), lima beans, flowering tobacco (*Nicotiana* spp.), lettuce, and tomatoes. These crops have limited genetic composition; you can be fairly sure of getting plants true to type from seeds saved from an earlier generation of nonhybrid parent plants. Chance factors such as a windy environment can allow some cross-pollination to occur. For example, research has shown that the degree of cross-pollination in "self-pollinated" tomatoes is very small, perhaps between 2 and 4 percent.

Nevertheless, just to make sure, if you plant more than one variety of these self-pollinated plants for seed-saving purposes, try to separate them by at least 150 feet or by a row or two of a tall, nonrelated species to help block movement of any pollen from one to the other. For example, a row or two of corn planted between varieties of beans will work just fine.

Eggplant, pepper, celery, and squash-family crops are partially cross-pollinated, with the amount of cross-pollination dependent upon the environment. If you plan to save the seeds of any of these crops, plant them in isolation just to be sure. Squash and pumpkins belonging to certain species also intercross and must be isolated from each other to remain reasonably true to type. For more details on this, refer to the Squash entry, page 130.

Isolating with Bags and Cages

If you're concerned that you can't rely on distance to prevent accidental cross-pollination of your seed-stock plants by wind or insects, you can use bags or cages to isolate them. Bags for this purpose are available from mail-order supply companies (see Resources). The bags come in

◄ Because peas are self-pollinated, nonhybrid plants usually come true from seed.

METHODS FOR ISOLATING SEED STOCK

◀ You can construct a simple frame for an isolation cage out of 1×2s.

▶ Staple row cover to the frame along all edges so that no pollinating insects can penetrate the cage.

◀ Use a piece of row cover to bag individual flowers or flower clusters to prevent unwanted pollination.

HERE, BEE!

CAGES MAY ALSO be used with introduced pollinators, but this technique is usually best left to the experienced seed saver or professional seed grower. If you want to try it, place a plate covered with honey in the garden to attract bees. When you have attracted 15 or so bees, pick up the plate and place it on the ground near the plants to be pollinated, then set the cage over the plants and plate.

different sizes and materials, or you can make your own from floating row-cover material such as Reemay. Either tie a piece of row cover around the flowers or sew a piece of row cover to make a bag. Be sure the bag is sealed and has no holes. The best bags are water resistant but not waterproof. For vegetable crops that usually self-pollinate but may also be visited by pollinators, bagging individual blossoms or inflorescences to exclude pollinators does the trick. Remove the bag once seed has set, and if needed, clearly label the fruit or plant so you don't accidentally pick and eat the fruit by mistake.

Using the same principle as bagging individual inflorescences, you can cage a group of self-pollinating plants to prevent any chance of pollen transfer by wind or insects. Build a simple wooden frame out of 1×2s, large enough to enclose several plants, and staple standard window screen, Reemay, or some other porous material onto the frame to make the cage. Or, you can cage individual plants by surrounding them with a wire tomato cage and wrapping it with row cover fabric. Or, set up wire hoops over a whole row of plants and cover it with row cover, as you would to keep out insect pests.

A variation is "alternate-day caging," in which two varieties that would otherwise cross are caged on alternate days to avoid cross-pollination by insects. Remove one cage in the morning, replace in the evening, then do the same the next day with the second cage. This allows you to plant two different varieties right next to each other. Keep in mind that you'll need to construct cages large enough to house the recommended minimum number of plants for seed saving for the crop you want to grow.

Isolating by Planting Time

Isolation can also be accomplished by planting different varieties of a plant at different times of the season or by planting two varieties with different periods of time to maturity. After all, if the two varieties don't flower at the same time, they can't cross-pollinate. For example, say you have two varieties of corn, one a 65-day corn and the other an 85-day corn. If you plant them both at the same time, there is little chance they will flower at the same time, so there will be little or no crossing. But if one variety is a 65-day corn and the other a 67-day corn, they are very likely to bloom at the same time if you plant them at the same time. Instead, plant them at least 10 days apart to stagger their bloom times.

Hand Pollination

Pollinating plants by hand is generally a simple process in which you collect pollen or entire male flowers and rub, brush, or pour the pollen onto stigmas of your seed-stock plants. It's a skill that all seed savers will want to develop, and it's easy and fun to do. The details of how to hand-pollinate vary from plant to plant. Three of the most common plants that home gardeners hand-pollinate are carrots, squash, and corn.

Hand-Pollinating Carrots

With carrot plants, keep in mind that the individual flowers in the umbels do not open all at once, and thus you will need to hand-pollinate daily for several weeks to ensure good seed set. Before flowers begin to open, decide on your seed-stock plants and bag those flowers to prevent unwanted pollination. Then each

day, uncover the flowers and use a camel-hair brush or the palm of your hand to stroke the flowers from several umbels back and forth to collect pollen. Then transfer the pollen to other flowers. Re-bag and repeat the process daily. Wait to remove the bags until all seeds have set. For more details on this method of hand-pollination, see page 30.

This technique works for all members of the carrot family and for other plants that have umbel-type inflorescences.

Hand-Pollinating Squash

It's best to hand-pollinate squash (and other squash-family crops) early in the season, because these crops often stop setting fruit after they have successfully set a few. To start with, you'll need to learn to tell the difference between male and female flowers. Female flowers have a small, squash-shaped ovary at their base, whereas male flowers do not. Also, examine the differences in the internal structure of the flowers.

Next, get to know what the blossoms look like when they're getting ready to open. They'll usually start to show a little color and the end of the blossom may loosen. Follow the steps at right to hand-pollinate.

▲ Use a camel-hair brush to hand-pollinate carrot flowers.

1) Bag or tape shut male and female flowers in the evening, and mark them well for ease of location in the morning.

2) Early the next morning, clip the male flowers from the plant and remove the corolla to reveal the anthers.

3) Remove the bag or tape from one female flower at a time and quickly "paint" pollen from the anthers to the stigma, distributing pollen evenly. Use more than one male to pollinate each female.

4) Re-bag or re-tape and flag the flower. If any bees land on the female flower while you're working, do not save seed from that flower.

Hand-Pollinating Corn

Hand-pollinating corn involves collecting pollen from the male plants that you've designated as pollen donors and sprinkling that pollen onto the silks of your designated seed-production plants before the silks become contaminated with undesired pollen. Here's how to do it (and see the illustration below):

BAG THE EARS. Covering developing ears before the silks begin to "show" (protrude from the husk) prevents any undesirable pollen from reaching the silks. Watch for husk leaves to develop on the stalks of the seed-production plants. Pull these leaves off, exposing the developing ears and husks. At this point you can remove the tip of a husk by cutting down with a knife or scissors within ½ to 1 inch of the tip of the cob (**1**).

Do not cut the cob. Place a bag over the husk and secure it to the plant with a twist tie or staples. Depending on your variety, you may be able to do this over just a few days. Bagging the bottom ear before the top ear sometimes gives better success, but be sure to bag the top ear as well, as it is the last to be aborted in dry conditions. The silks will grow out of the husks over the next few days (**2**).

COLLECT THE POLLEN. One method of pollination is to simply cut a tassel from one plant, remove the bag from the silks, and shake or rub the pollen from the tassel onto the silks, replacing the bag afterward. Or, to ensure genetic diversity, you can bag tassels on several plants and mix the pollen. When the anthers start to show on the tassels, gently place a bag over the top of each tassel. Make sure the stem of the plant is off-centered in the bag to help with sealing it. Fold the mouth of the bag tightly around the stalk and secure with string or tape, so you don't lose the pollen. Do not bag the top leaves, as they may have collected

HAND-POLLINATING CORN

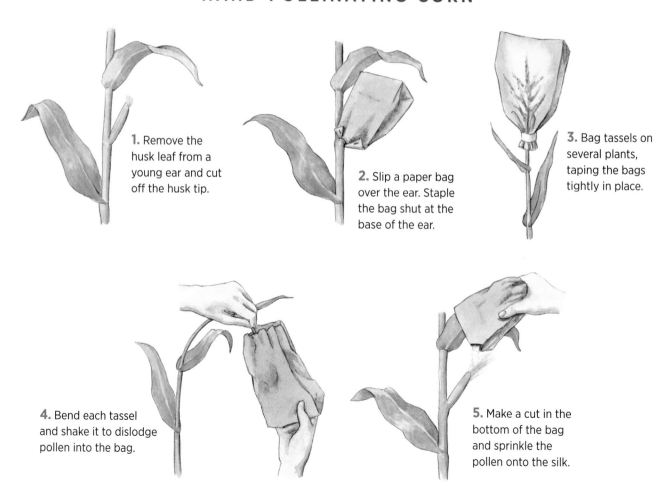

1. Remove the husk leaf from a young ear and cut off the husk tip.

2. Slip a paper bag over the ear. Staple the bag shut at the base of the ear.

3. Bag tassels on several plants, taping the bags tightly in place.

4. Bend each tassel and shake it to dislodge pollen into the bag.

5. Make a cut in the bottom of the bag and sprinkle the pollen onto the silk.

contaminated pollen. Bag at least as many tassels as you have bagged ears. Do this in either the evening or the early morning, because the heat that builds up in the bags from the sun can kill the pollen (**3**).

Return to the bags in late morning and start your collection by noon. Bend the top of a plant so the pollen will settle to the bottom of the bag while you shake the top of the plant to dislodge pollen. Remove the bag and continue to the next tassel (**4**).

Mix the pollen from all the tassel bags to encourage genetic variability by pouring all the pollen into one bag and shaking. Dead anthers are lighter than the pollen and may be floated out of the bag by tipping the bag and gently pouring them off. Shed pollen typically remains viable for 10 to 30 minutes in the field and may be stored for 24 hours or so in your refrigerator.

DISTRIBUTE THE POLLEN. Sprinkle the pollen from a cut made in the bottom of the bag. Remove one ear-shoot bag at a time and sprinkle the pollen mixture evenly onto all the silks, replacing the ear-shoot bag with a larger bag after pollinating to allow for growth of the ear (**5**).

Pollen does not need to contact the tips of the silk; the silks are receptive to pollen along most of their length. If the silks are long and difficult to work with, cut them back to an even brush, 2 inches long, prior to sprinkling with pollen. Fasten the bag over the ear loosely to allow for growth of the ear but tight enough so it doesn't come off in the weather. Bags may be left on until harvest or removed when the silks brown, but be sure to mark these special ears so they are not harvested for eating by mistake.

Selecting Your Seed-Stock Plants

As we explained earlier, saving seeds from your own garden plants offers you a great opportunity to select for superior plant performance in your local conditions. And surely there are aspects of the plants you grow that you'd like to improve: pest resistance, for example. But how do you go about making selections to get the best results?

Say you have a row of bean plants, all but five of which have powdery mildew. Those five plants may have some resistance to powdery mildew, so you collect the seeds from those five plants and sow them the following year. Each year from then on, you save seeds only from bean plants that don't show any signs of powdery mildew. Over the generations, you may develop a powdery mildew–resistant strain of beans that grows better for you than any other strain because it was developed under the very precise conditions of *your* garden. You can select for resistance to root maggots, root rots, downy mildew, wireworms, and a host of other pests and diseases by following the same approach.

Another characteristic that you can select for is soil adaptability. We all want plants as widely adapted to soil conditions as possible. Plants that do well only in rich soil won't do you much good if your soil is sandy. This is a characteristic that will take many years to develop through selection. For example, roses do very well in neutral to acidic soils. If you see one growing well in alkaline soil, propagate its seeds. Keep selecting for that trait over the generations, and you might wind up with a rose that will grow well in alkaline soils.

Once you decide upon the characteristic you want to select for — early ripening, slow bolting, brilliant flower color, good flavor, resistance to fruit cracking — then go ahead

ROGUEING AND TAGGING

THERE IS AN OLD ROMAN SAYING that the footsteps of the master make the best manure. And in truth, the gardener who spends the most time carefully tending her crops often produces the best crops. The same is true when it comes to growing plants for seed. If you notice a plant that is developing a disease problem or is being overwhelmed by insect pests, pull it out and get rid of it. If you see an off-type plant or one that bolted early, pull it out. This process of removal of undesirable types is called *rogueing*. It's a simple way to get inferior plants out of the way and out of the gene pool. On the other hand, if you see a strong, healthy plant with desirable characteristics, tag the plant with a bright ribbon or a string. It can be helpful to write a description of favorable traits on the tag to remind you later when it's time to save seed.

and do it. Keep in mind, though, that making good selections is sometimes a bit tricky. Read on to learn more about this.

Early Ripening

When it comes to the vegetable garden, everyone is eager to harvest that first tomato or pepper of the season, so selecting for early ripening is a popular choice. But think twice! Too many folks plant half a dozen tomato plants, harvest what they want for eating and preserving, and then, when their freezers are full, save the seeds from the last fruit of the season for planting next year. In doing that they are unconsciously selecting for lateness. If you do this over many years, conceivably you could end up with plants that will bear later and later until finally the fruit will not ripen at all, and you won't have seeds to plant next year.

The rule is this: If you are selecting for earliness, save the seeds from the first fruit that ripens, not the last. There is a slight catch-22 to this rule, though. The earliest fruit to ripen may not always be the fruit that has the best flavor, and it's usually the fruits that ripen later that have better keeping quality. And in the case of plums and other plants in the genus *Prunus*, most of the early-ripening varieties produce seeds that have nonviable embryos; you won't be able to propagate them from seeds anyway.

Fruit Appearance

Fruit shape (also referred to as *typiness*) and fruit color are two qualities you can select for. Many folks prefer an apple that has the conical shape of a 'Delicious' and the same five carpel

bumps on the bottom, so they select seeds only from trees that have fruit shaped like that. As for color, if you want to grow apples that will produce the very reddest fruits, save seeds only from the reddest fruits. Note, though, that fruit trees are heterozygous, so saving seeds from red fruit is no guarantee that your seedlings will produce red fruit.

Root Size

Selecting for size in the root crops can be a complicated matter. Of course we want the biggest roots, but sometimes large roots are also quite woody and poorly flavored. And there's no way to double-check all these characteristics, because if you dig up a root such as a carrot or parsnip to taste its flavor, you won't be able to then replant that root and have it produce seeds. So go ahead and save the seeds from the parsnip and carrot plants with the largest

roots, and see what results you get. With luck, those seeds will sprout and grow to form large roots that also have great flavor and texture.

Small Root Core

Again, it's tough to select for small cores in root crops, because you can't cut open the roots to evaluate the cores if you want those plants to produce seeds. But perhaps if you harvest 12 carrots in a row with small cores, the thirteenth plant in that row would have a small core, too. It's random probability here, not hard science, but save that lucky plant number 13, collect its seeds, and take a chance on growing them.

Crack Resistance

Some fruits have a tendency to crack; for example, some varieties of tomatoes readily crack, whereas some are resistant to cracking. If fruit cracking is a problem in your garden, select

▲ The fruits of this dual-purpose ornamental pepper (*Capsicum annuum* 'Nosegay') are not only beautiful, they're also edible.

seeds from fruits that seem to have some resistance.

High Heat

Some hot peppers are hot, some are hotter. If it's heat you're after, sample one or two fruits from each pepper plant you're growing. On whichever plants are producing the hottest fruits, allow some of those fruit to mature and save seeds from them for planting the following year. Keep in mind, however, that a pepper's degree of "heat" depends not only on its genes but also on your watering regimen, climate, and the type of fertilizer you use.

Slow Bolting

If you're like us, just about the time you figure out a delicious recipe that would use up all that lovely spinach your garden is producing, the plants send up seed stalks and the leaves turn bitter and tough. This phenomenon, called *bolting*, is all too familiar. Sometimes, though, among all those bolting plants, you find one that is slow to form a seed stalk. Wait for that one to flower and cross your fingers that it's female. If it is, save the seeds from that plant and turn the rest under. Those seeds should carry the genetic tendency for resistance to bolting. Continue this process over time, and you may develop a strain of spinach that gives you a significantly longer harvest period.

Hardiness

You may find an old abandoned orchard with many dead trees. But among them stands a tree that is still alive and bearing fruit. Take the seeds from those fruits, and continue to select for hardiness from among the seedlings. Keep in mind, however, that this is not a short-term project; you may be leaving it to your grandchildren to complete.

Flower and Foliage Color

Select seeds from plants with large, showy flowers. In dogwoods, magnolias, and crab apples, select for those with the prettiest and brightest-colored flowers or bracts or the brightest and most persistent fruit. Fall foliage color is a good trait to select for in the maples and cotoneasters and maintenance of green winter color in arborvitae and juniper. Select for desirable seasonal foliage color such as bright blue in Colorado blue spruce and blue Atlantic cedar, golden in some of the junipers, and bright red leaves in chokecherry, beech, and some maples. Unusual and desirable foliage texture is an attribute in cut-leaf birch and some *Chamaecyparis.* Some evergreens will show a dwarf character that you might select for. Other plants may display a weeping habit, as in beech, birch, and Norway spruce, or interesting bark characteristics, as in red osier dogwood and sycamore maple (*Acer pseudoplatanus*). You could take cuttings from these plants to propagate these traits, but saving seeds from these specimens will give you far greater genetic diversity among the seedlings. After you select among these seedlings for

▲ It might be worth saving seeds from this large, richly hued anemone flower.

the most desirable characteristics, you can take cuttings or try other asexual propagation techniques to reproduce these traits.

Special Handling for Biennials

Saving seeds from annuals and perennials is easy, but the seeds of biennials are borne in the second season of growth, following a cold period. Therefore, the plants must overwinter to produce flowers and fruit the following year. Learning how to overwinter biennials successfully is an important part of saving vegetable-crop seeds, because many popular leafy crops and root crops are biennials. There are two basic methods for overwintering these crops. Which one to use depends on whether the area you live in has cold winters or mild winters. In mild locales, it's often sufficient simply to trim back plant tops and cover the plants with mulch right in the garden (see Seed to Seed, next page). Those who live in colder climates must usually dig up biennial crops and store them in a cool, damp location over the winter, to replant the following spring (see Plant to Seed, next page).

LIFE CYCLE OF A BIENNIAL: CABBAGE

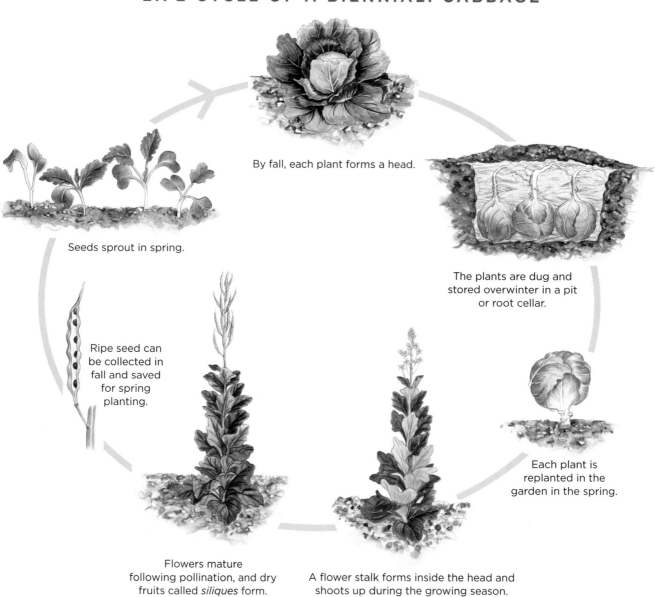

By fall, each plant forms a head.

Seeds sprout in spring.

The plants are dug and stored overwinter in a pit or root cellar.

Ripe seed can be collected in fall and saved for spring planting.

Each plant is replanted in the garden in the spring.

Flowers mature following pollination, and dry fruits called *siliques* form.

A flower stalk forms inside the head and shoots up during the growing season.

Seed to Seed

This is shorthand for planting the seeds (or moving the transplants) and letting the plants remain in place through the winter and into the second year. This saves you work, since you won't have to dig and store plants such as cabbage or roots of carrots over winter, but it may not always work if you live where winters are cold but there is little snow. If your winters are mild, and/or if there is plenty of snow cover that will insulate the plants, then this easier method might be for you.

When the plants die down in the fall, clean up any debris around them, remove the carrot tops and the dead wrapper leaves from the cabbages, and let the plants remain in the garden for the winter. For extra protection, mound up soil over the plants and mulch them with a thick layer of leaves or clean straw, then snow if you are fortunate enough to have it. Pull away the mulch and the extra soil when growth starts in the spring, and let the plants finish their cycles.

Plant to Seed

We'll use the same examples of cabbage and carrots to explain this technique, which is important to master if you live where winters are very cold. During the growing season, select the plants from which you plan to save seeds. In the fall, after the first light frosts, dig the cabbage plants, roots and all, and store them upside down in a covered pit or root cellar over the winter. In the spring, replant them in the garden to complete their life cycle.

With root crops like carrots, dig the roots as late in the fall as you

---

WHERE DOES THE ROOT GO?

IF YOU'VE EVER OVERWINTERED BEETS for a second season's growth and then uprooted one of the plants after harvesting the seeds, you may have been surprised by the puniness of the roots. Where did the beautiful round red beet root go?

For the answer, let's think about the biennial life cycle. A biennial plant forms a taproot and a rosette of leaves in its first year of growth. All through the growing season, the plant stores up carbohydrate reserves in its taproot. Its biological strategy is to use those reserves for early growth the following spring. Thus, when we harvest a biennial root crop at the end of the first growing season, we are taking advantage of the plants' strategy for our own benefit: we get the food value that the plants "intended" for their own regrowth.

If you overwinter or replant beets (or any biennial root crop) to grow for a second season, the plants use up the reserves in their taproots early in the season. After that, the plants rely on small lateral roots to collect moisture and nutrients. Thus, by the time your beautiful beet plant forms its seed stalk, the main taproot is dried up, shrunken, and not fit for consumption, unless you are a wireworm! The moral of the story is: You can't have your seeds and your roots, too.

---

can and check for desired characteristics. These first-year roots are called *stecklings*. Remove the tops, being careful not to cut into the crowns. Pack the roots in moist sand to overwinter in a root cellar. Replant them the following spring.

The storage area for overwintered biennials must be as close to 32°F (0°C) as you can keep it, and certainly below 41°F (5°C), but should never be allowed to drop below about 30°F (-1°C) or the roots might freeze.

Mice and other critters such as voles and chipmunks can wreak havoc on plants and roots stored in a pit or root cellar. A cat goes a long way in helping to minimize the problem, but for added protection, many folks store their plants inside a simple cage made of half-inch galvanized hardware cloth.

To make a cage, construct a frame of rot-resistant wood, such as black locust, cypress, or white oak. Staple or nail the hardware cloth to the frame (use galvanized poultry-wire nails). The top face of the frame should be a simple door that has galvanized hinges and a sturdy latch. Bury the entire cage so that the top lies several inches below the soil surface. After you fill the cage with plants, close the door, place a few inches of straw or leaves over it, and cover the whole with several inches of soil. A cage like this works well for storing produce over winter, too.

HARVESTING AND CLEANING SEEDS

▲ Harvest seeds from dead-ripe fruit only. Fruit color is often, but not always, a good indicator of ripeness.

ONE SEED-SAVING TASK THAT'S often satisfying and fun is harvesting. It's a sign of success! Actually, it's often the fruit you'll be collecting, and then you'll separate the seeds from the fruit by a variety of techniques, depending on the type of fruit or seed. Once you've separated the seeds from the fruit, you'll need to clean them up so they're ready for planting or storage.

Know What You're Harvesting

Let's take a minute here to consider a very basic but important point: Make sure you know the identity of the species and variety of the plant from which you are collecting seeds. After all, what good does it do to go through all the work of collecting and propagating seeds if you don't know what you're growing? Now is the time to verify the plant's identity and

to write it down! It's much easier to identify a plant than it is to identify seeds. The cleaned seeds of related species such as turnip and cabbage are nearly impossible to tell apart, for example. But more important, seeds of different species within a genus may have different requirements for germination (we'll talk about germination requirements in detail in chapter 5). Unless you know precisely what kind of seeds you have collected, you may not be able to get them to sprout.

Sometimes identification is easy. But even when you can tell at a glance that you're collecting seeds from a marigold or a pepper, for example, it's important to note the variety name for your records, too. Also, if you're collecting seeds from trees or shrubs, you may need to consult a field guide or take a sample of the plant to an expert to ID the plant to the species level. Pay close attention

▶ The ripe lily seeds are dry and separate easily from the pods.

to nomenclature and keep precise records of collection dates and locations. If you have any doubt about the identity of a plant, preserve a small branch from the plant along with the seeds for use in identifying it later.

Timing the Harvest Just Right

Timing of fruit and seed harvest is critical because maximum seed viability and seedling vigor occur only when fruits are physiologically mature, which often differs from their horticulturally mature stage. *Horticultural ripeness* refers to the stage at which people like to eat a fruit; that same fruit is *physiologically ripe* only after its seeds have ripened. If you harvest too early, you'll end up with lower-quality seed (due to embryo immaturity), as well as lower seed yield overall. If you harvest too late, however, you may lose seeds due to shattering, predation, and seed

rots. Save seeds from physiologically ripe fruit, and you won't go wrong.

Estimating seed maturity correctly is difficult because the rate at which seed matures is influenced by weather, plant characteristics, site quality and aspect, and location of the fruit on the plant. Here are just a few examples:

- Cucumbers planted on sandy, drier soil ripen faster than those planted on deep silt loam soils.
- Lupine seeds in lower pods ripen before the seeds in upper pods.
- Dill seeds are borne in umbels and do not all ripen at once.
- When drying winds or high temperatures occur in the autumn, seeds of several western conifers shed quickly, as do those of eastern white pine (*Pinus strobus*). Conversely, cold, rainy conditions can delay shedding.
- Cones and fruit on plants growing at lower elevations ripen

before those at higher elevations. The same is true for plants growing on south- and west-facing slopes compared to north- or east-facing slopes.

Another wrinkle is that in some species, fruits require 2 or 3 years for full development. You can mistakenly collect fruit too early, especially when the immature fruits closely mimic mature fruits in color and size. Sierra juniper (*Juniperus occidentalis*) is one example.

BIG SEEDS ARE BETTER

ONE GENERAL RULE to keep in mind at seed-harvest time is to save your biggest seeds for planting. Across species, the largest seeds generally produce the most vigorous seedlings.

Clues to Ripening

Judging maturity requires careful observation of the appearance of fruits, seeds, or cones. Immature green color changes to yellow-green, yellow, brown, reddish, or purple as fruits ripen and soften. Cone scales or bracts crack or flex. A few early-maturing individuals begin to drop their seeds. Luckily, for many species, a week or more elapses between seed maturity and seed shed or fruit drop.

Gauging ripeness by these characteristics alone doesn't always work, especially for carrots and other members of the carrot family. Also, color changes may not be uniform among individual plants. Grapes and eggplants may turn purple, cherries red, and blueberries blue long before they're ripe, especially when air temperatures are high. On the whole, however, these indicators are

BEING A SAGE SEED SAVER

NO, WE'RE NOT TALKING about saving sage seeds. We're advising you to be smart about the seeds you choose to save. Before you collect the seeds from any plant, you should make sure you know the answer to these two questions: 1) Is this plant considered invasive *in my area?* and 2) Is this plant toxic?

Invasive or Not?

From the beginning of agricultural practice, when people have migrated from one area to another, they've taken along seeds of their important food, fiber, and herb plants. Plus, for centuries, plant explorers have traveled to remote terrain in search of unique new plants to introduce into the horticultural trade in their home countries. The result has been a nearly unlimited range of beautiful plant choices for modern gardeners. But the downside of these plant introductions has been a rise in problems with invasive plants — plants that can escape into natural areas and outcompete the native vegetation.

As we researched this book, we pondered how best to write about this problem, because it's not a black-and-white issue. Most states have developed their own distinct lists of species that are invasive within their boundaries. However, these lists are far from identical. Just one example: Norway maple is on the invasive species list in Massachusetts, but where we live, in Montana, Norway maple is difficult to grow, it does not spread in the wild, and it's considered a valuable tree. More examples: Grape, wild parsnip, and wild carrot are labeled as noxious weeds in Ohio, but does that mean we should not include grapes, parsnips, and carrots in this book?

We concluded that trying to give you specific information on invasiveness for the more than 1,000 species covered in the book would be a task that would take more than a lifetime. What we have done instead is include brief "Alerts" in the plant entries in part 2 calling attention to plants that are labeled invasive in some areas. These alerts are by no means comprehensive, and we strongly advise that you do your own research before you save and germinate seed from any plant, especially nonnative species. You can find information about invasive plants by contacting your local weed board or by consulting the USDA website on noxious plants: http://plants.usda.gov/java/noxiousDriver.

Toxic to Touch?

Some very popular garden flowers, such as floxglove and lily of the valley, are toxic if ingested. Others, such as some types of iris, can cause dermatitis when handled. Because collecting and saving seeds from plants requires more prolonged contact with the plants than simply growing them does, it's wise to make sure you cover your skin when you collect fruit or seed heads from toxic plants, and wash your hands afterward. If you're storing seeds that are toxic if ingested, keep them in a secure place out of reach of children and pets.

As with invasiveness, including comprehensive listings of the toxic potential of all the plants in this book would be an impossible task. We have put in warnings for some popular garden plants under heads like Crop Alert and Flower Alert, but we strongly encourage you to do your own research regarding potential toxicity of any ornamental plants from which you plan to collect seeds.

practical and fairly accurate. When in doubt, it's better to shorten the collection period than to collect immature seeds. Early collection (that is, before the natural maturity date) will still provide viable seeds in some species, including some pines, firs, and spruces.

VEGETABLES AND FRUIT CROPS At physiological ripeness, many common vegetables and fruits will have turned yellow and become very soft or even begun to rot. The seeds of some crops will change color. For example, seeds of beets, watermelons, and beans turn dark brown when they're ripe. But those of other crops, such as cucumbers, peppers, and squash, may be white. A physiologically ripe cucumber is large, soft, and yellow, and its seeds have developed a tough, leathery seed coat. Ripe muskmelon seeds are pinkish white, and ripe pea seeds are green or yellow.

To save seed from squash or pumpkins, it's best to leave the fruits on the vine until 10 days to 2 weeks after they have reached prime harvest maturity for eating. Corn, which is horticulturally ripe at the milk stage, becomes physiologically ripe at the hard, dry "dent" stage.

Color often tells you when the fruit is ripe. For example, when the color of fleshy fruits turns from green to a bright red (*Berberis*), yellow, orange, or blue-black (*Cotoneaster*), they are ripe, or pretty close to it. Cherries and some other fruits actually become "dead" ripe about a week after they turn color, and you'll want to hold off harvesting such fruits as long as you can if your goal is saving seeds. Eggplant turns to its ripe color long before either the fruits or seeds are ripe. Wait to harvest eggplants for seed until the plant naturally sheds the fruits, which is long after they have become inedible.

▲ Eggplant fruits turn deep purple long before they are physiologically ripe. Wait until the fruit has dropped naturally from the plant before harvesting the seeds.

Hold off harvesting all fruits for as long as you can, as long as the critters do too. Most fruits should also separate easily from the stems, spurs, or branches. If you have to twist and tug hard to pick them, they're not ripe. Aim for fleshy fruits not only to have developed full ripe color but also to have softened. In fact, tomato seeds germinate better if the fruit is partly rotted before seed extraction, and pepper seeds will have better germination if the fruit is overripe at harvest.

For certain crops, such as good ripe peaches and muskmelons, horticulturally ripe fruits are also physiologically ripe.

▲ We eat cucumbers at horticultural ripeness. When they are physiologically ripe, they are large, soft, and yellow.

▲ An immature sunflower head (bottom) is covered with floral structures; ripe seeds are revealed when a head has matured (top).

FLOWERS/WOODY ORNAMENTALS.
Seeds of most flower species are best harvested when the flower heads or fruits are completely brown and dry. Ripe flower seeds may be black, gray, or brown. In some plant families, such as the sunflower family, seeds ripen over a long period of time. For these plants, bag the seed heads and wait as long as you can to harvest the seeds, until the heads are thoroughly dry and shattered.

Seeds of most woody plants will have turned dark brown or black at maturity. Cut open a few cones or fruits to see what you have. Cut cones lengthwise and count the number of plump, healthy seeds on a cut face. In firs, for example, half the seeds should look plump and brown in that longitudinal cross section; for white pines, it should be 75 percent. The cups of acorns release easily when the fruits are mature, and the scales on the cones of white pine will flex open when cones are bent double,

indicating they are ready for harvest. You can also easily estimate whether a cone is mature by measuring its specific gravity. This simple, fun method also lets you practice some arithmetic. (See The Specific Gravity Method on the facing page.)

The seeds of some plants, such as ash and ginkgo trees, do not mature until after natural seed fall. Thus, even if you wait for the plants to shed their seeds naturally, you may not be collecting mature seed. On the other hand, seeds of some species germinate better if harvested while still slightly immature. Some seeds are ready to germinate at once after fruit-shed, whereas some require an afterripening period or other treatment after shedding before they can germinate.

In part 2, we'll discuss the little idiosyncrasies of judging maturity and harvesting seed for individual species of vegetables, herbs, flowers, fruits, nuts, and woody ornamentals.

Seed-Gathering Techniques

Collecting ripe seed can be as simple as picking a tomato, but sometimes it's not quite so easy. Some plants do not release their fruits or cones readily even when they're ripe. We call this type of fruit or cone *persistent*. Either hard twisting and pulling or cutting with pruning shears is needed to sever the connection. However, twisting or pulling can damage the fruit-bearing spurs on some plants, such as pears, plums, and apples, ruining subsequent cropping.

As a general rule, we advise that you think about how the process works in nature and wait, if you can, for the plants to shed their fruit naturally. There are a couple of exceptions to this rule. One is fruit that is in danger of being eaten by wildlife such as raccoons — or bears! (In our neck of the woods, it isn't unusual in the fall to see bears climbing trees in search of apples and crab apples.) In

▲ An immature pinecone (left) is tightly closed. When mature, the cone's scales open (right) to enable seed shed.

THE SPECIFIC GRAVITY METHOD

THE SPECIFIC GRAVITY of a solid or liquid is the ratio of the weight of a given volume of that substance to that of an equal volume of water. You can use measurements of specific gravity to gauge the maturity of conifer cones and seeds of some plants in the genus *Prunus.*

As a cone matures, it loses water, and thus its specific gravity decreases.

Scientists have studied this for many types of plants and recorded the particular specific gravity of ripe cones. You can find lists of the specific gravity values for ripe cones of various species in part 2. To calculate the specific gravity of a cone, you can do a simple test using water in a graduated metric cylinder marked in increments of 1 milliliter (available through mail-order scientific equipment and supply companies). Use the smallest-size cylinder into which the cone will fit. (We like using plastic cylinders because they don't shatter if you accidentally drop them.) Make your estimates of specific gravity immediately after the cone is harvested from the tree, before moisture loss begins.

1. Partly fill a graduated cylinder with water, leaving enough room so that the entire cone can be submerged. Record the volume of water.

4. Subtract the original volume of water in the cylinder from the volume of water you noted in step 3. Because the specific gravity of water is 1 (1 milliliter of water weighs 1 gram), the volume of water displaced (measured in milliliters) is approximately equal to the weight of the cone or seed (measured in grams).

6. Divide the weight of the cone (from step 4) by the weight of the displaced water (from step 5) and you'll have the specific gravity of the cone.

• • •

Let's walk through an example. Say you have put 50 milliliters of water into the graduated cylinder. You place a cone of white fir (*Abies concolor*) into the cylinder and it floats. The water volume rises to 80 milliliters. Therefore, the volume of displaced water (with the cone floating) is

$$80 - 50 = 30 \text{ milliliters}$$

Next, you push on the cone so that it is completely submerged. The water volume then rises to 82 milliliters, for a new displacement of

$$82 - 50 = 32 \text{ milliliters}$$

This displacement of 32 milliliters equals the total volume of the cone. The first displacement divided by the second displacement gives the specific gravity:

$$30 \div 32 = 0.938$$

Thus, the specific gravity of the cone is 0.938. This is within the range (0.850–0.960) of specific gravities for ripe cones of white fir.

2. Place the cone or seed into the water. If it sinks, the specific gravity is greater than 1, which means the cone or seed is immature, and the test is over. If it floats, its specific gravity is less than 1, and thus it may be ripe. Continue to step 3.

3. Make a note of the new water level.

5. Next, push on the cone or seed until the water just fully covers it. Use a pencil point or a pin so that your finger doesn't confound the water displacement. Record the water level with the cone or seed submerged. This will indicate the weight of the volume of water displaced by the object.

cases like this, harvest when you have the opportunity.

The other exception to the "let it happen naturally" rule is plants with seed heads that tend to *shatter*, which means that the seed heads are brittle and are prone to losing their seeds. Shattering of seed heads is common in members of the carrot family and in some cabbage-family plants. Fruits of some plants, such as impatiens and hardy geraniums, actually explode to release their seeds, throwing them as much as 10 feet from their plants.

Gather seeds from plants that tend to shatter as follows: As the first fruits begin to brown, place paper bags over the seed clusters and fasten them to the stems. Batting or other soft filling may be used between a stem and a bag to ensure that no seed escapes. Every day or so, give the stems a shake to loosen the ripened seeds, which will fall into the bags. After a couple of weeks, snip the stems and place the bags with the seed clusters in them in a warm place to dry further, then rub the seeds from the clusters.

Alternatively, bag the clusters and snip them off the plants as soon as the stems begin to turn brown. Bring the bagged clusters inside to dry in a warm place. After the clusters are fully dry, rub them vigorously in your hands to remove the last few seeds that cling to them.

Another collection option is to wait until most of the seeds are brown and look as though they are going to shed, then harvest an entire head, place it in a bag, and put the bag away for the seeds to dry (we explain more about drying seeds on page 45). However you do it, you have to catch the seeds before they drop!

HOW MUCH TO COLLECT?

This chart provides ballpark estimates of how much seed you can expect to collect from a 10-square-foot planting of a crop. Compare that to how many ounces of seeds it takes to plant 100 row feet, and you'll see that a little seed saving goes a long way! Note that seed yields are highly variable depending upon location, weather conditions, plant spacing, variety, and cultural management practices.

CROP NAME	SEED YIELD FROM 10 SQ. FT. OF CROP (OZ.)	SEED QUANTITY OR NUMBER OF PLANTS TO SET OUT PER 100-FT. ROW
VEGETABLES		
Beans, lima (pole types)	6.0	6.0 oz.
Beans, snap (bush types)	4.8	8.0 oz.
Beet	4.4	0.5 to 0.75 oz.
Broccoli	4.4	0.12 to 0.25 oz.
Brussels sprouts	1.2	50 plants
Cabbage	2.0	0.12 to 0.25 oz.
Cabbage, Chinese	1.6	0.12 oz.
Carrot	2.0	0.18 oz.
Cauliflower	1.6	0.12 to 0.25 oz.
Celery	3.2	150 to 200 plants
Chard	4.4	2.0 oz.
Chicory	1.2	0.5 oz.
Corn, sweet (hybrid varieties)	8.0	4.0 oz.
Corn, sweet (nonhybrid varieties)	6.0	4.0 oz.
Cucumber	1.6	1.0 oz.
Eggplant	0.4	50 to 75 plants
Endive	2.0	0.5 oz.
Fennel, sweet	4.8	0.25 oz.
Kale	2.0	0.5 oz.
Kohlrabi	4.8	0.5 oz.
Lettuce (head)	7.2	0.12 oz.
Lettuce (leaf, romaine)	1.6	0.12 oz.
Muskmelon	1.0	0.25 oz.
Mustard	2.0	0.5 oz.
New Zealand spinach	8.0	0.5 oz.
Okra	4.8	2.0 oz.
Onion	1.6	0.5 oz.
Parsnip	3.2	0.5 oz.
Peas	4.0	16.0 oz.
Pepper	0.16	50 to 75 plants
Popcorn	4.8	4.0 oz.

HOW MUCH TO COLLECT? (CONTINUED)

CROP NAME	SEED YIELD FROM 10 SQ. FT. OF CROP (OZ.)	SEED QUANTITY OR NUMBER OF PLANTS TO SET OUT PER 100-FT. ROW
Pumpkin	1.2	1.0 oz.
Radish	1.6	1.5 oz.
Rutabaga	3.2	0.25 oz.
Salsify	2.8	1.0 oz.
Spinach	2.0	1.0 oz.
Squash	1.6	1.0 oz.
Tomato	0.18	0.25 oz. or 50 plants
Turnip	2.0	0.5 oz.
Watermelon	0.8	0.7 oz.
HERBS		
Anise	2.0	300 to 400 plants
Caraway	4.0	600 plants
Coriander (cilantro)	2.8	0.2 oz.
Dill	4.0	150 plants
Parsley	3.2	0.25 oz.

Adapted from Hawthorn and Pollard, 1954

FLOWER-SEED YIELDS

With flower seeds, it's easy to collect all you might need for your own gardening needs. You'll probably end up with extra seed to share with friends, too. As with vegetable and herb seeds, these quantities are estimates, and are affected by weather, choice of variety, and other factors.

FLOWER NAME	SEED YIELD (OZ.)/10 SQ. FT. OF PLANTING	FLOWER NAME	SEED YIELD (OZ.)/10 SQ. FT. OF PLANTING
Alyssum	1.0	Linaria	1.0
Antirrhinum (snapdragon)	0.4	Lobelia	0.4
Aster	1.2	Lupinus (lupine)	2.0
Calendula (pot marigold)	1.2	Malva (hollyhock mallow)	0.8
Centaurea (bellflower)	1.4	Matthiola (stock)	1.0
Cheiranthus (wallflower)	1.2	Papaver (poppy)	0.8
Chrysanthemum	1.2	Petunia	0.3
Clarkia	1.2	Phlox	0.6
Consolida (larkspur)	1.2	Reseda (mignonette)	0.6
Cosmos	1.2	Salvia	1.2
Dianthus (pinks)	1.0	Tagetes (marigold)	0.8
Gypsophila (baby's breath)	2.0	Tropaeolum (nasturtium)	2.4
Helichrysum (strawflower)	0.3	Viola (violet)	1.2
Iberis (candytuft)	1.4	Viola (pansy)	0.3
Lathyrus (sweet pea)	2.0	Zinnia	0.8

Specific methods for collecting fruits and seeds can be found in the individual plant entries in part 2. When you're collecting seeds for fun or experimentation, you can collect as many as you feel like (just remember that wild-collected seeds often have low germination rates). But if you're saving seed of a vegetable or herb so that you can grow enough of that crop next year to feed your family, it's helpful to estimate how much seed you need to save. In general, it's less than you might think. See How Much to Collect? and Flower Seed Yields (at left) for some data on how much seed you can collect from a 2-foot by 5-foot vegetable, herb, or flower patch.

Extracting and Cleaning Seeds

Horticulturists divide seeds into two classes based upon how much they can be dried before storage and still maintain their viability: orthodox and recalcitrant. *Orthodox* seeds can be dried to low moisture levels (below 10 percent of fresh weight) without losing viability. This enables us to store them for long periods of time. *Recalcitrant* seeds cannot be dried below 25 to 50 percent moisture, depending upon the species, or they may die. Among these are the seeds of oaks, chestnuts, buckeyes, and some maples. Because of their relatively high moisture content, these seeds won't store as long as orthodox seeds and are more subject to rots and attack by insects and rodents. You need to know which type of seeds your plants produce so you won't damage them in postharvest treatments. Fortunately for home seed savers, most seeds are orthodox. We present information on seed type in the individual plant entries in part 2.

Extracting Seeds from Fruits

Orthodox seeds can be divided into two groups: those from plants with fleshy fruits, such as tomatoes, squash, dogwoods, cherries, apples, and junipers, and those from plants with nonfleshy fruits, such as beets and peas; most flowers; most herbs; and ashes, elms, lupines, and pines. In most cases, you'll need to extract the seeds from the fruits before you dry the seeds for storage. For example, if you don't extract seeds from fleshy fruits, the pulp may harden around the seeds and promote rots. Excellent ventilation is essential because their high moisture content may promote uncontrolled fermentation, which could heat and damage the seeds.

With nonfleshy fruits, minimal drying is necessary prior to post-harvest storage if seeds are collected dry to begin with. Extract the seeds and spread them on screens to dry and prevent rots and loss of viability. If you're drying cones that haven't opened, allow sufficient space for expansion of cones upon drying.

Whatever type of seed you're working with, handle it as gently as possible. Rough handling that damages seeds often leads to reduced viability in storage, especially for orthodox seeds. Impact damage can bruise an embryo and crack the seed coat, which allows pathogens to enter. Recalcitrant seeds, with their high moisture content, are especially susceptible to damage during handling.

If the seeds are very moist, dry them within the fruits for a few days before extraction to prevent damage. Seeds that are moist can be bruised and damaged during extraction.

Here are some general guidelines about extracting seeds from fruits:

- Fleshy fruits such as cherry, cucumber, and tomato are usually depulped prior to seed extraction.
- Indehiscent (or otherwise unopened) dry fruits, which include the fruit of many flower species as well as black locust (*Robinia pseudoacacia*), are hulled to remove the seeds.
- Dehiscent dry fruits such as those of beans, cabbage, and spruce can be tumbled or shaken to separate the seeds from the fruits.
- Fruits borne in clusters (for example, those of ash and maple) should be separated, or *singularized*, for easier handling.
- Some species have fruits that irritate the skin: for example, eggplant, potatoes, and parsnips. Wear gloves when working with these fruits, and use a knife to cut open a fruit so you can extract the seeds.

WORKING WITH FLESHY FRUITS

There are a variety of methods for removing seeds from fleshy fruits. For large fruits such as pears and peaches, simply open the fruits and pick out the seeds. For many types of fleshy fruits, such as magnolia, tomatillo, and apple, you can use a food blender. Cautious folks cover the blades with plastic tubing to prevent the seeds from being cut, but this is not necessary in most cases. Place about 1 cup of fruit into the blender, add a little water, then blend with about a dozen short (5-second) bursts of power. When the fruits appear to be fully macerated, rinse with water and float away the pulp, unfilled seeds (*floaters*), and germination-inhibiting substances. Repeat this rinsing process with several changes of clean water. Most of the viable seeds will remain at the bottom of the blender.

HARD FRUITS. If the fruits are too hard to macerate in a blender (hawthorn, for example), soften the pulp until it can be smashed between thumb and fingers. Placing the fruits in a jar full of water, which is changed daily until the fruits are softened, works very well for this purpose.

VERY HARD FRUITS. Some fruits, such as crab apple, are quite hard. Drive over them with your car to crush the fruits, or freeze then thaw them, and store them for a short period to allow the fruit pulp to deteriorate. Either method works well for hard fruits.

PULPY FRUITS. For cucumber, squash, melon, and tomato, slice the fruits lengthwise and scoop out the seeds and jellylike pulp into a container. Add an equal amount of water and allow the container to sit for 3 to 6 days at 60 to 70°F (15 to 21°C) out of direct sunlight, stirring twice daily to prevent fungal growth on the surface. Viable seeds will sink, and the placental material will disintegrate by fermentation. If the soluble solids in the water are high, then the specific gravity is also high, which may cause some viable seeds to float. To remedy this, change the water several times. Once the water remains clear, you can be confident that most of the floating seeds are nonviable, and you may pour them off. Keep in mind that just because seeds sink does not mean they will all germinate, but they do have a much better chance of viability than do floaters. After discarding floating seeds and plant material, rinse the good seeds.

An alternative is simply to remove the seeds from the fruits and place them in a strainer. Rinse off all the pulp you can. The germination percentage of seed that you clean this way will be a little lower because this technique doesn't separate out nonviable seed.

Whichever method you use to clean the seeds, dry them afterward on a screen, cookie sheet, or paper towel in a warm, well-ventilated area. They'll stick to paper, but don't worry about it. Stir occasionally so they dry evenly. (For more details on determining when seeds are dry enough to store, see chapter 4.) When all are dry, fold up the paper with the seeds and put it into an envelope in a cool, dry place. The small bits of paper that cling to the seeds will make no difference in germination.

WORKING WITH DRY FRUITS

Nonfleshy fruits or cones with orthodox seeds, such as those of cabbage, spinach, dill, and pines, usually require drying as a first step before you can extract the seeds. The easiest way to do this is to heat ambient air to a temperature that results in the relative humidity dropping to about 30 percent, such as that in a very low

oven, set no higher than 120°F (49°C). Some locations, such as Montana, have air that is already pretty dry, and drying fruit in these areas requires little or no supplemental heat.

Spread the fruits or cones loosely on screens so that air can easily pass over and among the materials. Often, you can tell when the fruits have dried enough by shaking them: you'll hear the seeds rattle inside. If you're dealing with small plants that have nonfleshy fruits (beans and peas, for instance), let the fruits dry on the vine as much as possible in the garden, then pull an entire plant and hang it upside down in a shed or garage to dry for another week or so.

After the fruits have dried, there are various methods for separating the seed from the fruits, such as tumbling, hulling, flailing, threshing, and heat.

TUMBLING OR SHAKING. Place dry dehiscent fruits or cones in a cloth bag and shake it. This technique works well for beans, chestnuts, and parsnips. To extract seeds from dry seed heads or inflorescences, shake them in a bucket, rapping them against the sides.

HULLING. With dry, indehiscent fruits, such as those of peas, walnuts, yucca, hazelnuts, and hickories, you can crush or cut away the pods or husks, and extract the seeds.

FLAILING. In the mood for some exercise? You can work out your frustrations and lower your blood pressure while you extract the seeds of small dehiscent fruits like those

MAKE YOUR OWN THRESHING TOOLS

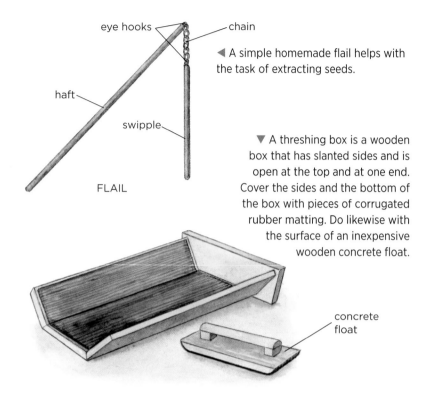

eye hooks — chain

◄ A simple homemade flail helps with the task of extracting seeds.

haft

swipple

FLAIL

▼ A threshing box is a wooden box that has slanted sides and is open at the top and at one end. Cover the sides and the bottom of the box with pieces of corrugated rubber matting. Do likewise with the surface of an inexpensive wooden concrete float.

concrete float

Another variation on threshing is to place dry fruits in a pillowcase and walk on it — gently. Wear sneakers so as not to bruise the seeds.

HEAT TREATMENT. *Serotinous* cones require a brief period of very high temperature to melt resin seals that prevent them from opening. In nature, forest fires provide the burst of heat that melts the seals. Black spruce (*Picea mariana*) trees and some types of pine have serotinous cones. Black spruce cones require a high initial heat, but pinecones generally need only a hot-water dip. Do this right after harvest, and treat for up to 30 seconds with boiling water. As the seals break, you'll hear the cones crackle. When the crackling stops, the cones are adequately treated and ready for seed extraction.

from the cabbage and carrot families. Simply place the dried fruits in a pillowcase or other small cloth bag and hit the bag with a broomstick or length of closet pole. Or you can make a flail especially designed for the purpose. It's fun to use!

Traditional flails had a handle (haft), usually about 40 inches long, and a swipple (also called a swingle), about 24 inches long. The two pieces were hinged together with leather or eel skin. Flails with longer handles gave greater leverage and were generally used for seed removal from grains. The plants or dry fruits with their seeds were spread on a barn floor or rolled up in a piece of canvas or cloth and beaten with the swipple to separate seed from stalk or fruit. To make your own flail, you'll need two poles about 1¼ inches in diameter,

one about 3 feet in length for the handle, the other about 18 inches in length for the swipple. Connect the poles end to end with about 6 inches of chain so that each hinges upon the other. Fasten the ends of the chain to the handle and the swipple with a through bolt or with eye hooks screwed into the top of each wooden piece.

THRESHING. Another way to extract seeds from dry fruits is to place the fruits on a piece of plywood and rub them with a concrete float or a piece of scrap wood. For smaller types of dry fruit, such as capsules and some pods, try using a small threshing box. These can be purchased, but it's easy to make your own, as shown above. Place the dry fruits or the seed heads in the box and rub the float over them, gently separating the seeds.

Cleaning Seeds

Once you've extracted the seeds you want to save, you might think your work is done, but not necessarily. Chances are that there's still some debris mixed with the seeds, and that debris can contain pathogens and insects. There may also be damaged seeds in the mix. Cleaning seeds immediately after extraction removes chaff, soil, and other debris, removing potentially harmful pathogens and making planting easier.

CLEANING SEEDS FROM FLESHY FRUITS

After you extract the seeds from the fruits, remove any juice or flesh that is still adhering to the seeds using clean, cold water and a small brush if needed. If mucilage surrounds the seeds and cannot be removed by washing in clean water, gently rub the wet seeds with coarse sand and then wash off the sand and mucilage in a sieve.

It is also possible to dry the seeds, then rub off the dry mucilage, but the seeds must be well separated to prevent their sticking together during drying, on a surface to which they will also not stick. Dry the seeds in a thin layer, preferably one seed thick, on a fine screen.

CLEANING SEEDS FROM DRY FRUITS

Before you clean seeds from dry fruits, check that the seeds are dry enough to be cleaned without damage; moist seeds (above 16 percent moisture content) can be damaged during cleaning (see page 50 for information on checking moisture content of seeds). Cleaning seeds is

▲ Sieving seeds through a graduated series of screens enables large debris (chaff) to collect in the coarser upper meshes while small debris sifts through to the fine lower meshes. Here, seeds settle on the third screen from the top.

best done by hand using wind, blowers, or graduated sieves to remove the dry debris.

REMOVING TRASH WITH SCREENS. Whereas air is used to clean seeds by weight, screens separate seeds from trash according to width and thickness. You'll need several screens, each one with a different-size mesh. Stack the screens on top of each other, with the largest-mesh size on top and each successive screen the next size smaller. Then pour the seeds onto the top screen and shake the whole stack. The seeds and debris will fall through the screens, and somewhere in the sequence there will be a screen that catches more seeds than chaff.

▶ Swirl seeds and chaff in a plastic cup. An electrostatic charge will trap small debris and dust on the cup's inside surface.

REMOVING TRASH WITH AIR. A gust of air can be used to remove particles that are lighter than the seeds. Dust, small pieces of pine needles, leaves, and wings are examples of lighter materials (collectively known as chaff) that can easily be cleaned out of seeds by blowing air across them, a process known as *winnowing.* Absent a windy day, use a fan or hair dryer set to low temperature to accomplish this, but be sure to adjust the wind force so that the seed doesn't blow away with the chaff. Put the seed in a basket, bowl, or trash-can lid, then toss the seeds and debris gently upward. The seeds, being heavier, will fall back onto the surface of the container; the air will blow away the lighter material.

CLEANING VERY SMALL SEEDS. To create an electrostatic cleaner for very small seeds, rub the inside of a plastic cup or drinking glass with a nylon stocking or other synthetic cloth. Pour uncleaned seeds into the treated cup and rotate the cup to roll them against the sides. Trash will cling to the sides as you remove the seeds. Pour seeds from the cup or glass into a storage container.

CHAPTER 4

SEED-STORAGE KNOW-HOW

THE FIRST BOOK on saving seeds was written in the 1830s, and we have been improving on those old-time techniques ever since. In the simplest form of storage, mature seeds are held for a short period until weather or other factors permit planting. Longer-term seed storage is not difficult as long as you pay attention to a few details. Moisture, heat, and fluctuating temperatures all reduce seed viability. During storage, you must slow metabolism of the seeds as much as possible without damaging the embryos. We usually do this by providing dry, cool conditions. If you cannot control both relative humidity and temperature, controlling just one will go a long way in prolonging seed life. It's also important to prevent microorganisms and insects from attacking the seeds.

How Long Do Seeds Last?

It's ironic (from a gardening perspective) that seeds of wild species seem to live longest, whereas those of most economically important plants have far shorter storage lives. Seeds of many weed species remain viable in soil for a half century. But for most crop plants, average seed life ranges from 3 to 10 years, depending upon storage conditions. Among common garden plants, seeds of the legumes, okra, asparagus, hollyhock, sweet pea, bellflower (*Centaurea* spp.), and some other plants with very hard seed coats store the longest.

Several factors influence how long seeds remain viable, including the seed's basic physiology, the type of seed coat it has, whether the endosperm is oily or starchy, and whether the seed is fully mature at the time it's extracted from the fruit.

Understanding Seed Physiology

How long a seed remains viable in storage depends on not only physiology but also how it was handled prior to storage. As we explained in chapter 3, there are two main classes of seeds, orthodox and recalcitrant. By far most seeds are orthodox in their storage behavior. Orthodox seeds carefully dried to a moisture content of 10 percent or less can be successfully stored at temperatures near or below freezing for long periods of time.

Storing recalcitrant seeds is a bit more complicated. Premature germination or rotting can occur if these are stored in high moisture, but they cannot be dried to the same degree as orthodox seeds. They should not be stored below freezing and cannot be stored for long periods, so place these seeds into cool storage or, better still, plant them right after harvest, as nature does.

Orthodox and recalcitrant seeds each have two subclasses. There are true orthodox seeds and suborthodox seeds, and there are temperate-recalcitrant seeds and tropical-recalcitrant seeds. Let's look at the characteristics of each subclass.

TRUE ORTHODOX SEEDS. Almost all vegetables, herbs, and flowers produce true orthodox seeds. These seeds can be stored for relatively long periods at subfreezing temperatures if they're dried to roughly 5 to 8 percent moisture. Most woody species that are native to northern temperate

regions also have true orthodox seeds, as do many tropical and subtropical woody plants.

SUBORTHODOX SEEDS. The main difference between true orthodox seeds and suborthodox seeds is length of storage. Suborthodox seeds can be stored under the same conditions as true orthodox seeds but for shorter periods. Some common landscape plants, including hickory, beech, poplar, and certain species of walnut, pine, maple, and willow, have suborthodox seeds.

TEMPERATE-RECALCITRANT SEEDS. These seeds must be stored at relatively high moisture content and cannot be stored at temperatures much below freezing. This group includes the buckeyes, chestnuts, oaks, hazelnut, wild rice (*Zizania*), and some species of willow.

TROPICAL-RECALCITRANT SEEDS. Most of us don't grow many plants that have tropical-recalcitrant seeds. Coffee and coconut are two examples that fall into this category. These seeds have the same sensitivity to drying as temperate-recalcitrant seeds and are even more sensitive to low temperatures. Even short periods of exposure to cool temperatures can cause great loss of viability. The minimum safe storage temperature is 50 to 59°F (10 to 15°C), depending on species.

Knowing about these four subclasses of seeds is useful; there is variation among and within the species in some genera, however, so don't put, say, all your maple seeds into one (storage) basket. Seeds of sugar maple can be dried to 5 percent moisture and remain viable for a long time, but seeds of silver maple die at moistures below 30 percent.

As scientists study seeds, they sometimes find that a species once thought to have orthodox seeds really has recalcitrant seeds or vice versa. Beech seeds, for example, were once classed as recalcitrant but are now considered suborthodox. With carefully controlled drying, beech seeds can be kept in extended storage below freezing with a low moisture content.

The Role of the Seed Coat

Hard seed coats, such as those of legumes, okra, dogwood, and hickory, are one factor that helps preserve seed viability. The seed coats keep out moisture and oxygen, which in turn keeps metabolism at a low rate inside the seeds. The hard seed coats also protect the embryos from mechanical damage during collection and conditioning. The thinner or softer a seed coat, as we find in many of our crop seeds, the more likely the seed is to have a relatively shorter storage life.

Oily vs. Starchy

Oily seeds, such as those of hickory, sassafras, and sunflower, do not store as well as starchy seeds, like those of hackberry, ash, and peas. But there are exceptions. Oily seeds of sweet gum and many conifers store very well. Acorns of the black oaks, which are somewhat oily with low carbohydrate content, store longer than acorns of the white oaks, which are full of carbohydrates and very little oil. Even among the black oaks, species with the highest oil content store better than those with less oil. So again, matters concerning seeds are not all black and white.

Seed Maturity

Orthodox seeds that are immature when extracted from their fruits do not store well. Scientists have not figured out why, but it's likely that these immature seeds can't withstand storage conditions because they have not been able to complete their full morphological development. However, species such as European ash (*Fraxinus excelsior*) that naturally disperse seeds that are still physiologically immature often store very well. For some conifers, storage of immature cones for several weeks before seed extraction improves seed maturity and retention of viability.

Getting the Moisture Right

Moisture is inarguably the worst enemy of stored seeds and the most difficult for the home seed saver to gauge accurately in storage. All your work may go for naught if you don't sufficiently dry the seeds before putting them in storage. Plan on drying large seeds longer than small ones, and give those that appear to have dried on the plant additional time after harvest to dry sufficiently.

Moisture and Metabolism

The water content of developing seeds is often 70 to 80 percent. This drops dramatically as the seeds ripen and shed, and seed moisture is the most important factor in maintaining viability during storage. Dry seeds have a low metabolic rate that helps to preserve foods stored in the endosperm and other seed tissues and limits the embryo's demand for water and oxygen. Enzyme activity increases dramatically at a moisture content above 10 to 15 percent and, along with respiration, depletes the seed's food reserves. If moisture content remains above 30 percent, some seeds will germinate even in cold storage. And if storage temperatures are well below freezing, ice crystals can form in seeds that have such a high moisture content, which may kill the embryos.

Moisture content also relates to pest problems of stored seeds. At 10 to 18 percent moisture content, fungi remain active and can rot the seeds. Insects can attack seeds that have a moisture content above about 8 percent. Mites can be a problem at even a somewhat lower moisture percentage. On the other hand, if seed moisture falls below about 5 percent, the seeds will dry out and the embryos will die. Thus, for most orthodox seeds, as mentioned above, the target seed-moisture content is 5 to 8 percent.

The normal practice is to dry all orthodox seeds to these levels and store them in moisture-proof containers such as freezer bags and sealable glass jars. (For more suggestions about storage containers, see page 54.) The moisture-proof container is essential to prevent rehydration, because seed moisture will

RECALCITRANT SEEDS NEED MORE OXYGEN

LOWER-THAN-NORMAL oxygen levels can increase seed longevity, but you won't be able to provide a controlled low-oxygen environment in a home situation. Keep in mind, though, that recalcitrant seeds have a fairly active metabolism and will die if stored in airtight containers. It's best to store them in partly opened plastic bags (leave one corner unsealed) to afford free access to the air while maintaining reasonably high moisture. To be a bit more certain of adequate aeration, some folks even place a small tube or piece of small-diameter pipe through the unsealed corner to prevent suffocation of the seeds. Just don't let them dry out!

equilibrate with the moisture level of the air in the storage atmosphere. In other words, if you put your dried seeds in a container with moist air, the seeds will soak up moisture from the air until the moisture content of the seeds and of the air are equal. Seeds high in proteins soak up the most moisture, followed by those high in carbohydrates and, finally, those high in lipids.

Recalcitrant seeds need different treatment. Do not dry them to lower than 8 percent moisture. This is also true for soybeans and some other seeds. Seeds of orange are damaged when dried to a moisture content below 25 percent, and those of grapefruit may be killed at a moisture content below 51 percent. In the plant entries in part 2, we specify those plants whose seeds need to be stored at higher moisture levels.

Seed-Drying Methods

Drying seeds to the desired moisture level for storage may take a few days to several weeks, depending upon the species, the atmospheric humidity, and the equipment you use. The faster you dry the seeds, the less likely they will be to succumb to pathogens. The lower the humidity of the air in which the seeds are placed, the faster the seeds will dry. Seeds dry quickly at first, then more slowly as their moisture content nears that of the air around them.

PACKAGING SEEDS FOR DRYING

Small coin envelopes work very well for holding seeds while they dry, better than standard paper letter envelopes, which often have small holes at their corners through which tiny seeds can spill out. Another option is to wrap small seeds tightly in pieces of paper towel and secure the seams with tape. Whether you use envelopes or paper towels, package only a few seeds together. If you have a lot of seeds, you'll have to use a lot of envelopes. Be sure you carefully label each envelope or paper towel with the date and the crop or species name (and the variety name if there is one). Note that although plastic bags and glass jars work well for storing dried seeds, they are not good container choices for seeds during the drying process.

DEHUMIDIFIER DRYING

If the air in your home is damp, as is often the case in coastal states and in the South, set up your seeds in a small room with a dehumidifier. You can use an inexpensive hygrometer and a thermometer to estimate how dry seeds are by tracking the relative humidity and maximum and minimum temperatures in the drying room. Measure the relative humidity and temperature daily.

Leave the seeds in the drying area for at least a week to be certain their moisture has come into equilibrium with that in the air. Ideally, follow this rule of thumb: The sum of the relative humidity and the storage

STAY OUT OF THE SUN

AVOID PLACING seeds in direct sunlight while they are drying. The strong ultraviolet light may damage seed embryos, affecting long-term seed viability.

temperature in degrees Fahrenheit must not exceed 100, as long as the temperature is less than 50°F (10°C). For example, if the relative humidity in your drying room is 40 percent and the temperature is 40°F (4°C), then the sum is 80 (40 + 40), which is less than 100, so your seeds should dry just fine.

Most orthodox seeds will dry to an ideal 6 to 8 percent moisture if they're held at 38 to 40°F (3.5 to 4.5°C) and 30 to 35 percent relative humidity. Seeds of recalcitrant species like those of the oaks will dry to their appropriate moisture content if held at 38 to 40°F (3.5 to 4°C) and about 40 percent relative humidity. As a homeowner, you don't have the wherewithal to purchase expensive moisture meters to be absolutely sure of having perfectly dried seeds, but these schemes will get you into the ballpark. The graph on page 52

TO DRY OR NOT TO DRY?

SHOULD YOU DRY THE SEEDS that you've collected from your garden? In most cases, yes. If you plan to store the seeds for a while before planting, begin drying them as soon as you can after you clean them. However, if you intend to plant the seeds soon after cleaning, take care *not* to dry them, because this may induce dormancy that does not exist in some fresh seeds. (We talk more about seed dormancy in chapter 5.)

illustrates the relationship between the relative humidity of the drying room and the approximate moisture content of the seeds. You can easily see that keeping the relative humidity below about 40 percent will reduce the seed moisture to below about 8 percent, although more exact percentages depend upon species and seed lot. The table below gives some specific examples of the expected moisture content of vegetable-crop seeds when dried at 40°F (4.5°C) at 45 percent relative humidity.

OVEN DRYING

If you live in a fairly humid area and have no dehumidifier, you can try drying seeds in your oven instead. This is easy, works well, and is safer for the seeds than is drying them in direct sunlight. Note that this method is primarily for larger seeds

▲ This simple graph shows the relationship between the relative humidity of the drying room and the approximate moisture content of the seeds. Source: Justice and Bass, 1978

DRYING VEGETABLE SEEDS

Vegetable seeds will dry to these approximate moisture contents if held at 45 percent relative humidity and 40°F (4.5°C). Note that all are between 5 and 10 percent, which is considered a good range for storage.

CROP	PERCENT SEED MOISTURE
Cabbage, lettuce, turnip	6
Beet, Swiss chard	6.5
Cucumber	7
Carrot, pepper, water-melon	7.5
Tomato	8
Beans, celery, corn, onion, peas	9
Okra, spinach	10

Adapted from Hawthorn and Pollard, 1954

such as those of squash and beans. Tiny seeds such as those of cabbage and carrot seeds should dry sufficiently without heating in an oven.

To prepare seeds for oven drying, do not put the seeds into envelopes or folded paper towels. Instead, spread the seeds in a thin layer on a cookie sheet and place the cookie sheet in the oven for 24 hours at 100°F (38°C). Stir the seeds once or twice during this time to be sure all sides are exposed to drying air, and that should do it. Once the seeds are dry, you can package them in containers for storage. Some pinecones may require a higher temperature, 130°F (55°C) or more; we note exceptions like this in the plant entries in part 2.

USING SILICA GEL

Following a dry-air treatment with a silica-gel treatment provides extra insurance that your seeds will be dried well enough. This is advisable if you live in the mid-Atlantic or southern states, or in other areas with very high relative humidity, or in general if you want to be absolutely sure your seeds are ready for storage. Silica gel is widely available at photo-supply stores or by mail order from some garden seed companies and is very easy to use. The best bead size for drying seeds is 1/16- to 1/8-inch diameter.

To set up seeds to dry with silica gel, first measure out an amount of the deep blue silica gel equal to the weight of the seeds. Place either the silica gel or the seeds in a porous bag, then place both in an enclosed space such as a glass jar with an airtight seal. Make sure the silica gel is not touching the seeds. Place the containers in a room held at 59°F (15°C) for drying. You'll need to replace the used silica gel with fresh, either daily or when the color turns from deep blue to pale blue or pink (which indicates that the gel has absorbed its fill of moisture).

Allow the seeds to sit, changing the silica gel as needed, for a few weeks to be sure they have dried properly. Larger seed lots and bigger seeds require longer drying times.

Some folks use powdered milk as an alternative drying agent to silica gel, but it is perhaps one-tenth as effective. Use it as you would the gel. Powdered milk can't be redried very well, so dispose of it.

Getting the Temperature Right

Once your seeds are properly dry, you'll need to find a place that is the right temperature for storing them. A cool temperature is good for seeds because it lowers metabolic rates. Moisture and temperature are linked, though, because moisture content determines just how low storage temperatures can be without damaging the seeds. The drier the seeds, the lower the safe storage temperature is. Further, cold air holds less moisture, so cold air is drier air.

Orthodox Seeds

Orthodox seeds with 20 percent moisture can be stored fairly well at temperatures of 32 to 5°F (0 to ⁻15°C); orthodox seeds with 15 percent moisture can be stored below 5°F; and seeds with less than 13 percent moisture can be stored in liquid nitrogen at ⁻321°F (⁻196°C), but that's a bit colder than most home freezers. With this in mind, true orthodox seeds dried to moisture levels in the range of 5 to 8 percent can be safely stored even at room temperature.

Suborthodox seeds can be stored for a few years at temperatures as low as ⁻4°F (⁻20°C). It is important, though, to dry all types of seeds to the moisture contents stated above, or less, when storing seeds well below freezing, as excess moisture will form ice crystals that can kill an embryo.

Whenever you move seeds from cold storage to room temperature, take care to spread them out, because the change in temperature can cause moisture to condense out of the air and onto the cold seed surface, which increases the risk of mold formation.

SAFE MOISTURE % FOR STORING SEEDS AT THREE TEMPERATURES

The desired moisture content of seeds depends on the types of seeds and on how you plan to store them: at room temperature, in a cool storage area, or in a freezer. This chart lists the maximum safe moisture content (percent) for vegetable seeds stored at three different temperatures.

CROP	40–50°F* (4.5–10°C)	70°F (21°C)	80°F (27°C)
Beans	15	11	8
Beet	14	11	9
Cabbage	9	7	5
Carrot	13	9	7
Celery	13	9	7
Corn	14	10	8
Cucumber	11	9	8
Lettuce	10	7	5
Okra	14	12	10
Onion	11	8	6
Peas	15	13	9
Peanut	6	5	3
Pepper	10	9	7
Spinach	13	11	9
Tomato	13	11	9
Turnip	10	8	6
Watermelon	10	8	7

*Surface-dry seeds rapidly when moving from cool storage into warm air to avoid or remove condensation on the seed coat.
Adapted from Hawthorn and Pollard, 1954

REUSING SILICA GEL

YOU CAN REUSE SILICA GEL over and over again as long as you dry it out between each use. To do so, heat used silica gel in a warm oven to drive off the moisture. Set the oven at a temperature above 212°F (100°C) but below 275°F (135°C). Place the gel in a thick-walled Pyrex dish in a layer no more than an inch deep. Stir the gel a few times during the drying process; 1 quart (1.9 pounds) of gel should take about 2 hours to dry. You'll know it's dry when it has turned deep blue again. Alternatively, heat the gel in a microwave oven at a medium or medium-high setting for 3 to 5 minutes. If it is still not dry at the end of the cycle, stir it and heat it again for the same period. It may take about 10 minutes to dry a pound. Store the dried gel in an airtight container for future use.

Recalcitrant Seeds

Store recalcitrant seeds of temperate-zone species at 32 to 27°F (0 to ⁻3°C); lower temperatures for only a few months will kill them. Tropical-recalcitrant seeds such as those of *Araucaria* have a much higher lethal minimum temperature, and chilling damage and death will occur below temperatures of 54 to 68°F (12 to 20°C), depending upon the species. If you must store recalcitrant seeds, then store them near freezing in a refrigerator if they are from temperate species or at about 50°F (10°C) if they are tropical species, in moisture-proof plastic freezer bags to maintain relatively high humidity. Do not seal the bags entirely; leave a small part of the end unsealed to allow for some air circulation.

Good Places to Store Seeds

For most of us, there is no need to store seeds for years. Mainly, you'll want to store your seeds from one growing season to the next. If that's the case and you have a cool, dry, dark room in your house, then you can store most seeds there in paper envelopes. They will maintain their viability nicely.

If your aim is to store about a quart (or less) of seeds for maximum longevity, then your household refrigerator or freezer should work just fine. Keep in mind, though, that dehumidification is a factor when seeds are stored in household refrigerators. Most newly manufactured refrigerators are frost-free, which means that the inside air is dehumidified. In such units, recalcitrant seeds quickly become desiccated if you don't take care to maintain their moisture content by storing in plastic freezer bags packed with moist peat moss.

Many orthodox and suborthodox seeds are best stored in the freezer at ⁻0.4 to ⁻4°F (⁻18 to ⁻20°C). Seeds of temperate-recalcitrant species must be stored at temperatures no lower than 27°F (⁻3°C) and those of tropical-recalcitrant species no lower than about 50°F (10°C). A small refrigerator dedicated to seed storage and set to the appropriate temperature works very well.

Containers

After you've finished drying orthodox seeds, transfer them from paper envelopes or toweling to labeled sealable containers. Cans with sealable lids are by far the best for long-term storage. If you have none, then sealable glass jars work very well, too, but they can break. Mason jars with

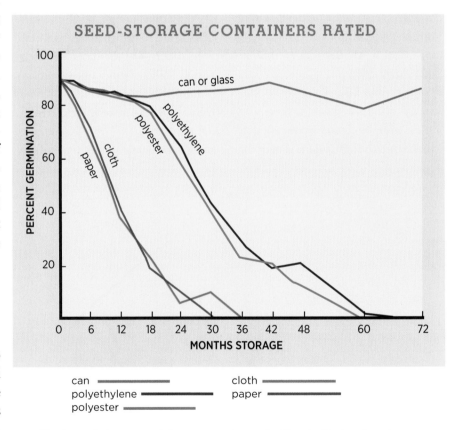

SEED-STORAGE CONTAINERS RATED

The type of storage container you use has a significant effect on how long seeds will last in storage.

Source: McCormack, 2004

good seals are excellent. Some folks use baby-food jars. They work well as long as the seals are in good condition. Be sure to store seed-filled glass jars in the dark, because exposure to light could cause a greenhouse effect and raise the temperature inside the jars to dangerous levels.

Sealable polyethylene bags work, too, but not all plastic is the same. Use polyethylene bags with a minimum wall thickness of 0.075 to 0.1 millimeter (3 to 4 mils). Common household freezer bags work very well. Sandwich bags are too thin and permeable to moisture vapor. For recalcitrant seeds, the maximum bag-wall thickness is 0.25 millimeter (10 mils), as thicker plastics limit gas exchange. There is no maximum thickness for orthodox seeds. Seeds with sharp points such as those of hickory and ash can pierce the bag walls, so double-bag these or store them in another suitable thick-walled container.

Summing Up Storage Recommendations

It's good to know all the ins and outs of how to store seeds, but don't let it get too complicated. We commonly store our dried vegetable seeds in paper envelopes in a dark room at about 60°F (16°C). They store quite well there, but this is short-term storage, and we live in Montana, a place not known for its high humidity. For best storage we recommend that you keep all orthodox seeds dried to a moisture content of 5 to 8 percent in moisture-proof, sealed containers. If the period of storage will be 3 years or less for true orthodox species or 2 years or less for suborthodox species, storage temperatures of 32 to 41° F (0 to 5°C) are fine. For longer periods of storage for both types, use a freezer set at ‾0.4 to ‾4°F (‾18 to ‾20°C).

Store temperate-recalcitrant seeds with a moisture content at least as high as that present when the mature seeds were shed. This moisture level must be maintained throughout storage, which may require occasional wetting of the seeds. Temperature should range from 32 to 41°F (0 to 5°C). Containers should be impermeable to moisture but allow some gas exchange. Polyethylene freezer bags with a wall thickness of 0.075 to 1.0 millimeter (3 to 7 mils) are suitable (remember to leave a corner of the bags unsealed). Do not use dehumidifiers in storage of recalcitrant seeds because of the probability of desiccation damage.

Tropical-recalcitrant seeds are stored the same way as seeds of temperate-recalcitrant species, except that storage temperatures are higher. In fact, the lower limits are generally 54 to 68°F (12 to 20°C). For best results, store these seeds no longer than a year before planting.

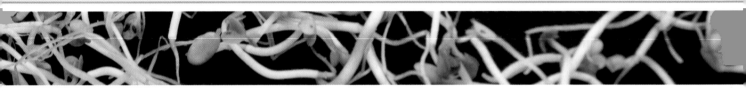

ALL
ABOUT
GERMINATION

$\mathbf{T}$HIS IS WHAT we've been leading up to all along, folks. You saved seeds to plant in your garden or share with gardening friends. If you can't get those seeds to germinate, you have invested much time for nothing! Fortunately, seeds germinate easily if you give them the right conditions and take steps to overcome whatever

dormancies exist. In this chapter, we focus on seed biology again, looking at all the factors that influence germination and the special treatments that some seeds need to allow germination to take place. Then in chapter 6, we get into the how-to details of sowing seeds and raising healthy seedlings.

Nature's Exquisite Logic

Nature has planned it all very carefully. A plant species has the greatest chance of survival if its seeds germinate only at a time and place that's favorable for the seedlings to survive. Nature helps ensure this by sending many types of seeds into dormancy before or shortly after they're dispersed. Dormant seeds won't germinate, even in a favorable environment. For example, when some woody plants shed seeds in late summer, those seeds don't germinate right away, even though conditions are favorable, because the seeds are dormant. This is smart on the plant's part, because if the seeds did germinate immediately, the plants would likely suffer winter damage (unless a gardener covered them with a cold frame or other winter protection).

▲ Viable seeds sprout and grow vigorously when dormancies are overcome.

Usually, subsequent favorable environmental conditions trigger internal physiological processes in a seed that overcome its dormancy and enables it to germinate. Sometimes when we save seeds we want to speed up these processes so we can quickly get our seeds to germinate and grow.

Each species of plant has its own requirements for germination. With most annual vegetables, herbs, and flowers, overcoming dormancy is a simple process. In fact, you won't have to worry about dormancy much at all in most vegetable seeds and many flower seeds. One exception is the seeds of peppers and tomatoes, which will show better germination if they are "cured": that is, stored dry for at least 2 months at about 77°F (25°C) before germination. In some perennial flowers and herbs and most woody species, dormancy is more complex and requires special treatment, such as exposing the seeds to a period of cold temperature or scratching the seed coats. Some species require a combination of treatments in order to prepare the seeds for germination.

In addition to having dormancy requirements satisfied, seeds need sufficient moisture and oxygen and the right temperature to germinate. Seeds of some species may also need light to germinate, whereas for others, light prevents germination. (Fortunately, the presence or absence of light makes no difference to seeds of most species.) Nature takes care of all this in the wild, but you'll have to do it yourself if you want to germinate seeds in a controlled environment. This is another reason it's so important that you correctly identify the species of plant that you're saving seeds from. If you misidentify a plant, the germination treatments that you follow may be ineffective, which means no seedlings!

Types of Seed Dormancy

Nature has come up with a couple of different mechanisms for keeping seeds dormant, and botanists classify them as *primary dormancy* and *secondary dormancy*. Primary-dormancy mechanisms can be external to the embryo (in the seed coat) or internal to the embryo.

Seed-Coat Dormancy

Seed coats of many types of plants are impermeable to moisture and oxygen; this characteristic is called *hardseededness*. Species in the legume family, along with hollies and some other species, are often hardseeded. In other seeds, such as cotoneaster, some layers of the fruit may dry and become additional "seed coats." Under high moisture conditions, spinach seeds develop a layer of mucilage that prevents oxygen from getting to the embryo.

Another type of seed-coat dormancy is caused by the seed coat's mechanically resisting the swelling of the endosperm, delaying the uptake of water (*imbibition*) and interfering with the emergence of the radicle. This type is present in pecans, peaches, and other stone fruits and some oaks.

A third type of seed-coat dormancy involves the presence of chemicals in some seeds that inhibit germination. These chemicals must be leached out before germination can occur, and sometimes tough "seed coats," such as the endocarp in peaches, prevent this leaching. The juice of some fleshy fruits, such as cucurbits, tomatoes, apples, citrus, and grapes, also prevents seeds from germinating, while the hulls of mustard and sweet pea prevent germination in those species. So if we can scratch, puncture, or otherwise breach the seed coat to allow water and oxygen to get to the embryo, then we can overcome seed-coat dormancy. This is called *scarification*. (See page 58 for more details on how to scarify seeds.)

Embryo Dormancy

Embryo dormancy may be caused by the presence of hormones or hormonelike substances in seed embryos that inhibit germination. These inhibitors must be broken down before germination can take place. This type of dormancy is present in various kinds of tree seeds, such as some maples, white ash (*Fraxinus*

> ## COMBINED DORMANCY
>
> WHEN TWO OR MORE primary dormancies, such as seed-coat dormancy and embryo dormancy, are present, the seeds are said to have *combined dormancy*. You must treat each dormancy separately for germination to occur. Many conifers and most other trees and shrubs in temperate and colder zones display this type of dormancy. For specifics on overcoming combined dormancy, refer to the plant entries for trees and shrubs in part 2.

americana), apple, and northern red oak (*Quercus rubra*). Embryo dormancy may be shallow (weak), intermediate, or deep (strong).

In some plants, such as lilies and peonies, the dormancy is specific to the epicotyl. The radicle can grow but the epicotyl cannot. *Radicle dormancy*, wherein the radicle cannot grow but the epicotyl can, exists in *Trillium* and some other herbaceous perennials.

Morphological dormancy refers to a condition in which the embryo is not mature at the time fruit is shed and needs an afterripening period to complete growth. This is common in poppy, American holly (*Ilex opaca*), cherries, and several types of cold-hardy pines.

Secondary Dormancy

Secondary dormancy results from some treatment to seeds after they are shed. In other words, the initial dormancy was nonexistent or broken but the seeds entered a second dormancy. Sometimes secondary dormancy is induced by the way you handle your seeds after collecting them. Pine seeds can enter secondary dormancy if exposed to high temperatures (*thermodormancy*) and moisture after they are shed. Trying to germinate seeds of some species at too-high temperatures can induce a thermodormancy. Those seeds, such as lettuce, celery, and pansy, will not germinate even when temperatures return to optimum. The best way to handle this is to avoid trying to germinate these seeds at too high a temperature. If you have goofed and induced this type of dormancy, it's best to toss out the seeds and start over.

Overcoming Seed Dormancy

Nature has evolved elegant mechanisms for overcoming seed dormancy over millions of years, and the treatments we gardeners have devised to overcome seed dormancy mimic what happens in nature. We scarify seeds with sandpaper to imitate the scratching and scraping of seeds that occur inside a bird's crop. Or we use an acid to scarify seeds chemically, to mirror what happens to a seed as it passes through an animal's digestive tract. *Stratification* is the term for a prechilling treatment that simulates what seeds naturally undergo as they rest on or just under the soil surface during winter. *Warm stratification* is a re-creation of the conditions that a seed is naturally subjected to when it's shed with a few weeks of warm weather left in summer. Nature relies on rain to help wash away seed-coat inhibitors; we use water soaks instead.

Treatments for Seed-Coat Dormancy

All treatments to overcome seed-coat dormancy are designed to breach, or *scarify*, the seed coat. Scarification treatments in some cases may also be combined with other treatments such as stratification for best results. Not all scarification treatments work for all seeds. Remember to consult the plant entries in part 2 to learn about the proper treatment for overcoming the dormancy of the seeds you've saved.

COLD-WATER SOAK. Soaking seeds of some hardseeded species in water at room temperature for 1 to 2 days may be all that's needed to overcome dormancy. This is quite effective for peas and some other legumes. Plant them immediately after soaking or set them out to stratify, according to the species.

HOT-WATER SOAK. To treat seeds with hot water, bring enough water to cover the seeds to a boil in a saucepan, then remove the pan from the heat. Put the seeds into the water and leave them there as the water cools. The hot water softens the seed coats or causes them to crack, and imbibition occurs as the water cools. Hot-water treatment also works for seeds that contain some germination inhibitors: the water leaches out the inhibitors. Seeds of many legumes respond well to hot-water treatment. Those of honeylocust (*Gleditsia*

> ## BREAKING THE SEED-COAT CODE
>
> IF YOU HAVE SAVED SEEDS of a species that's not covered in part 2 of this book, you can experiment to try to figure out the right treatment to break seed-coat dormancy. For some plants, several types of treatment may be effective. Start with the cold-water soak. If that doesn't work, try the hot-water soak on a second batch of seeds. If you still have no success, try another treatment on yet another batch of seeds. And if you're eventually successful, we'd love to hear about it!

triacanthos), black locust (*Robinia pseudoacacia*), and some other species will swell and turn a lighter color when they're ready to be taken out of the water. Put still-dark seeds back into the water to soak longer, or re-treat them with boiling water to start the process over again. Keep in mind that once you've soaked seeds in hot water, you need to plant them immediately. Otherwise, as the seeds dry, dormancy could be reinduced.

HOT WIRE OR BORING. Use a heated needle, electric wood-burning tool, or small drill bit to burn or bore small holes through the seed coats. Seeds treated this way can be returned to storage for a short time after treatment. For example, the hardseeded sweet pea (*Lathyrus*) and moonflower (*Ipomoea alba*) can be burned or bored to admit water. Just be sure you don't damage the embryo, and take care you don't leave seeds treated this way in storage too long or they will dry out.

MECHANICAL TREATMENTS. Mechanical scarification is simple, safe, and used quite often. You can scarify seed coats with a knife or a triangular file or by rolling seeds between two pieces of sandpaper. Do keep your enthusiasm in check: Don't cut your finger or puncture the embryo with that knife! Remember that your goal is to cut through the seed coat only, to avoid damaging the embryo. Some seeds are best treated in this manner by cutting through the seed coat at the radicle end of the seed, near the hilum. Just don't cut the radicle!

Treatments for Internal Dormancy

Treatments to overcome internal dormancy usually involve a moist chilling period. This technique gets its name, stratification, from the old-time method of alternating, or stratifying, layers of seeds and sand or peat in flats set outside over the winter. Some seeds require just one period of cold treatment to overcome dormancy, but other seeds require a warm period (warm stratification) followed by a cold period (stratification) or a cold period followed by a warm period followed by another cold period. These treatments give the seeds a period of afterripening

ACID TREATMENT

PROFESSIONAL SEED PRODUCERS sometimes soak seeds in acid to scarify them. This procedure is very dangerous and can result in serious injury if the acid spills or splatters. We don't recommend that anyone other than a professional try this procedure, which involves soaking seeds in concentrated (95 percent) sulfuric acid for several minutes to several hours, depending upon the species. Professionals always observe the following precautions when undertaking this procedure:

• This procedure is best done outdoors. Indoors, it should be done only in a well-ventilated area.
• Protective clothing, rubber gloves, eye protection, and a respirator are required.
• Water is never poured into acid because this causes the acid to splatter dangerously.
• Seeds must be surface-dry. When poured into acids, wet seeds cause splattering, and the resulting dangerous reaction can cause much heat, which damages both seeds and people.
• Dry seeds are placed in a basket made of copper screening, then the entire basket is lowered into the acid bath.
• Enough acid is used to cover the seeds to at least twice their depth, and the seeds are stirred a few times while they are in the acid bath to reduce overheating.
• No more than 20 pounds of seeds are processed at one time; otherwise, the acid is likely to overheat, which can result in a dangerous situation that causes injury.

Immediately after the treatment, the seed basket is removed from the acid and placed under running water for at least 20 minutes to thoroughly rinse away the acid. The rinsed seeds are then spread onto a screen to surface-dry before planting.

Vinegar is a safer (but far less effective) form of acid that can be used to scarify seeds that do not have an extremely hard seed coat; the technique is the same as with sulfuric acid, but the length of time needed for soaking is far greater.

so that embryos and/or endosperm can mature. It sounds complicated, but before you throw in the towel, we'll point out that most plants are not this fussy. For those challenging plants that require multiple treatments, you can find detailed instructions in the relevant plant entries in part 2.

STRATIFYING SEEDS

All stratifications must be done with imbibed seeds under moist conditions. Moisture is the key. You can't stratify seeds if they aren't moist.

The usual interval for stratification is 1 to 6 months, to simulate the natural winter conditions of temperate seeds lying exposed outdoors. Usually, the longer the winter a species would naturally endure, the longer the stratification period. The length depends also on the type of plant. Most vegetable-crop seeds, for example, don't require any stratification period. Some perennial flower seeds require only a few weeks of cold treatment; those of several woody species may require up to 2 years of stratification.

The easiest way to stratify seeds is simply to plant them outdoors in nursery beds or in the garden in autumn to undergo winter conditions, just as nature does. (Refer to page 78 for how-to instructions for creating a nursery bed and sowing seeds.) Provide them with a winter mulch to reduce heaving. Several inches of leaves or straw will do the trick.

You can also stratify seeds by mixing imbibed seeds with moist sawdust or peat moss and placing the mix in plastic bags, then storing the bags at the proper temperature for the proper length of time according to the species. Or you can simply place imbibed seeds without sawdust or peat moss in a plastic bag and place that in cold storage. This is sometimes called *naked stratification*. In general, you should refrigerate fully imbibed seeds at 33 to 41°F (0.5 to 5°C) for the length of time specified for the species you're working with. The most important things to remember are that during the stratification period the seeds must remain moist and cool and must not freeze. Don't break those rules. Ever.

Small seeds are difficult to handle when wet, so you can sow them in peat moss or sand in flats or pots first, water them, then place the entire containers in the refrigerator or outdoors for stratification. This works particularly well for flower seeds. If you want to do it the old-fashioned way, sow the seeds in flats of sand. Then dig a hole about a foot deep outdoors in your garden or yard and set the flats in the hole. Cover the flats with straw, then refill the hole with soil. Leave the flats there over the winter to remain cold and moist during the

---

DORMANCY DIVERSITY

HERE ARE JUST A FEW EXAMPLES of the range of dormancy requirements among garden and landscape plants.

Silver maple (*Acer saccharinum*). This tree's seeds will germinate right after they're shed, given the right temperature and moisture conditions — they have no dormancy.

Apple (*Malus* spp.). Johnny Appleseed may not have known it, but apple seeds need a long period of stratification, and then, when given the right combination of moisture and temperature, they'll germinate.

Tree peony (*Paeonia suffruticosa*). These seeds begin to form a rudimentary root system, but then they need a cold period to remove epicotyl dormancy before their shoots grow. However, stratifying the seeds without allowing the root initiation first does not satisfy the dormancy requirement.

Trillium (*Trillium grandiflorum*). In this case, seeds must have a stratification period to remove radicle dormancy, then a warm period in which the roots can grow, followed by a second stratification period to remove epicotyl dormancy, followed by a second warm period so the shoots can grow.

Lettuce (*Lactuca sativa*). These seeds germinate readily in cool conditions — about 57°F (14°C) — there is no dormancy in them. However, if the seeds are sown when it's hot — 84°F (29°C) or warmer — they will not germinate. If exposed to 84°F for even a few days, then returned to 57°F, they still fail to germinate. The high-temperature exposure induces a dormancy. However, that dormancy can be overcome by exposure of the seeds to red light.

stratification period. This technique is especially successful with hard, bony seeds. This includes recalcitrant seeds such as are found in the oaks; the seeds of such legumes as honey locust; and those of some shrubs, small fruits, and roses. Hazelnuts, chestnuts, acorns, and hickory nuts can be mixed with sand and stored in a cool cellar until spring planting to help keep the mice away.

Remember those seeds you stratified and intended to plant but didn't? Don't put them back into storage. If you redry stratified seeds below 10 percent moisture, they will often enter a secondary dormancy. It's like opening a can of tuna and not eating it all. It's not wise to put the leftover tuna back into the can and store it for any length of time. If you do, then eat it, you might enter a permanent dormancy!

If the seeds are not imbibed, stratification cannot begin, and if they freeze or dry out, stratification then stops, and the seeds may enter a secondary dormancy, compounding your problems. Additionally, if the temperature drops too low or rises too high during stratification, the effect of the process is diminished. Both the low and high temperatures vary with species but are often around 25°F (¯4°C) and 60°F (16°C).

The optimum length of the stratification period varies greatly among species. Think about it: plants that adapted to southern conditions also adapted to shorter, milder winters. Thus, aim to reproduce those conditions in your treatments. Length of stratification can even vary depending on whether a plant of a given species is growing in a warm climate or a cool climate. Seeds collected from

plants grown in the North, which are adapted to long winters, often need longer stratification periods than do seeds from specimens of the same species that have been growing in southern areas, where winters are short and mild.

WARM STRATIFICATION

Many species of plants, particularly woody plants, that exhibit complex embryo dormancy or morphological dormancy germinate more quickly if given a warm-stratification treatment prior to standard stratification. Seeds that require warm stratification can be treated in a greenhouse, a hotbed, or a warm basement by keeping them covered with moist burlap or paper. Seeds must be kept moist during the treatment. Temperatures for warm stratification vary by species, and you'll find detailed recommendations in part 2.

LIGHT TREATMENT

Some seeds that have shallow physiological internal photodormancy will germinate if, after imbibition, they are exposed to red light at wavelengths of

660 to 760 nanometers. The impulse to germinate will then be reversed if the seeds are exposed to darkness or to far-red light (760 to 800 nanometers). Lettuce seed is a prime example of this type of reaction. (See Phytochrome Physiology, page 64.)

Imbibed photodormant seeds buried deeply in the soil are exposed to proportionately more far-red light than red light, since the red light penetrates soil less deeply. The seeds remain dormant. That's why you should not bury seeds too deep and, in the case of photosensitive seeds, not bury them at all. It's also a reason that it's risky to expose lettuce seed to incandescent lights, which are high in both the red and far-red wavelengths. If you think about it, it would be like coming to a traffic intersection and having both a green light and a red light at the same time. Do you go or not go?

GERMINATION-ENHANCEMENT TREATMENTS

Seeds of peppers and some other species typically have lower than average germination rates. Microwaving has been found to increase germination percentage in such seeds. Place the seeds in a thin layer in a suitable container and microwave them for 10 to 15 seconds. You will have to experiment a bit to find the right length of time for your seed lot. Researchers speculate that microwave radiation may alter permeability of the seed coat and cause other changes in the seed that speed germination. The precise effect depends upon seed moisture content and the species treated. If you want to try this, do so on small lots only until you find what's right for yours. In part 2 we've specified those plants that might benefit from this conditioning treatment.

Another recent finding is that some newly harvested vegetable seeds, especially those of tomato, pepper, and some cabbage family plants, may germinate very slowly or not at all. Fortunately, this dormancy or sluggishness is easy to overcome by storing the seeds for at least 2 months at about 77°F (25°C). Simply holding the seeds at room temperature over the winter should do the trick.

Promoting Optimum Germination

Seed germination is a process with several stages. First, a seed absorbs (imbibes) water, which triggers several phenomena. Water softens the seed coat and kicks up metabolism. Enzymes and hormones become activated, and, in turn, they transform food stored in the endosperm or cotyledons, or both, into energy and "raw materials" for growth. Sometimes the starches and proteins and other storage materials themselves swell, causing the seed coat to crack. This lets even more moisture and more oxygen into the seed. Cell growth in the embryo resumes, tissues develop further and, like the beak of a chick that punctures an eggshell at hatching, the radicle breaks through the seed coat, and the seedling is off and running. The most important environmental factors that influence these happenings are moisture, temperature, aeration, and, sometimes, the absence or presence of light.

Moisture

Moisture is the element that starts germination down the road. Water can enter through the seed coat, through a scarified seed coat, or through the micropyle or hilum.

Constant, adequate moisture during germination is far better than alternating periods of dryness and saturation. Too much moisture can saturate the soil, decreasing aeration, retarding plant growth, and increasing the chances of damping-off fungi ruining your seedlings. In saturated conditions, seed coats split but the embryos might suffocate or rot for lack of oxygen. At the other extreme, too little moisture may result in the medium's drying out and crusting over, which can retard or stop germination. If germination has already taken place, seedling emergence may be hampered by the crusted soil or potting mix, and the seedlings may shrivel and die. Too little or too much moisture can also cause nutrient deficiencies that interfere with plant growth and photosynthesis. For example, improper moisture hampers root growth, and if the roots cannot grow, they cannot absorb sufficient nutrients. A lack of nitrogen, iron, magnesium, and a few other essential nutrients interferes with chlorophyll synthesis; the plants become chlorotic and unable to photosynthesize adequately.

For all these reasons, you should make it a priority to keep the medium around seeds evenly and moderately moist. When you're starting seeds indoors, the best way to accomplish this is by bottom-watering. Set the entire flat or the pots in which you've sown seeds into a shallow pan of water and let the medium absorb water from the bottom. When the medium at the top of the flat is moist, remove the flat from the water.

If you can't bottom-water your seed-starting containers, then the alternative is to cover the medium surface after sowing and before watering. A single sheet of newspaper laid directly on the surface of the seed-starting mix works well for this purpose. The paper breaks the impact of the water from the sprinkling can, and it also serves as a mulch to moderate surface drying. Remove the paper when the seedlings emerge, then use a spray bottle set to

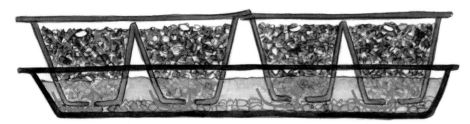

▲ Side view of a flat with sown seeds set into a shallow pan of water; arrows indicate movement of water by capillary action through the drainage holes in the flat up into the soil medium.

fine mist to syringe the young plants daily. If your seeds require light to germinate, then cover the surface of the flat with plastic wrap, loosely applied. The old-timers placed a pane of glass cut to fit the size of the flat directly onto the surface of a soilless mix. Some professionals still do.

Temperature

Seeds of temperate-zone plants germinate over a wide range of temperatures, but the optimum range for germination of seeds of most species is 68 to 86°F (20 to 30°C). The cool-temperature-tolerant group of plants (mostly temperate-zone plants, such as broccoli, cabbage, carrots, and *Alyssum*) germinate from a minimum of about 34°F (1°C) to a maximum of about 104°F (40°C), depending upon the species. The cool-temperature-requiring species are a different group. They are mostly adapted to the Mediterranean area and require temperatures no higher than 77°F (25°C). Examples are celery, lettuce, onion, *Cyclamen*, primrose, and delphinium. The chart below gives approximate temperature ranges for many popular vegetables; different references list slightly different optimum temperatures. We've included temperature recommendations for germination of individual species in part 2 as well.

Seeds of some firs as well as northern red oak (*Quercus rubra*) can germinate in snowbanks at 33°F (0.5°C). Seeds of many species, such as apple and pear, can germinate during stratification at 35 to 40°F (1.5 to 4.5°C) if left for long periods of time. Apple seeds sometimes germinate inside old fruit stored in a refrigerator.

In general, germination is best at the optimum temperature cited for a particular species rather than at either end of the optimum range. However, there are times when achieving optimum germination is not the goal to strive for. The optimum soil temperature for germination of watermelon seeds is 95°F (35°C), for example. But if you intend to direct-sow watermelon seeds in your garden, waiting for the soil temperature to reach 95°F may leave you too short a growing season to enable the melon fruits to ripen. Your germination rate may be terrific, but your harvest could be nil. Likewise, planting lettuce seeds when soil temperature has reached the optimum for germination (77°F [25°C]) will mean planting when air temperatures are too warm for highest-quality leaf production. Thus, the optimum temperature for seed germination is not always the best temperature at which to plant.

TEMPERATURE AND LIGHT RECOMMENDATIONS FOR VEGETABLE-SEED GERMINATION

65–70°F (18–21°C)	75–80°F (24–27°C)	85–90°F (29–32°C)	95°F (35°C)
Broccoli, Brussels sprouts, cabbage, cauliflower, celeriac*, celery†, Chinese cabbage*, collards, escarole, garden cress*, globe artichoke, kale, kohlrabi, leek, lettuce*, mustard*, New Zealand spinach, onion, parsnip, peas, potato, rhubarb, salsify, shallot, spinach, witloof chicory*	Asparagus, beans, carrot, endive, parsley, Swiss chard	Beet, eggplant, gourds, lima beans, muskmelon, okra, pepper, pumpkin, radish, rutabaga, tomato, turnip	Corn, cucumber, squash, watermelon

*Requires light for germination
†Requires that temperature fluctuates nightly to 60°F (16°C) or lower

Adapted from Maynard and Hochmuth, 1997; Gough and Moore-Gough, 2009

Seeds of many perennials and most temperate woody plants germinate most rapidly at alternating temperatures of about 68°F (20°C) at night and 86°F (30°C) during the day. A fluctuation of 10 to 18°F (6 to 10°C) between day and night temperatures may actually improve germination and be more important than the cardinal temperatures, within reason. This is precisely what happens in nature, where days are usually warmer than nights. You may not be equipped to supply precise alternating temperatures, but try to provide a fluctuating temperature to simulate what happens in nature.

Air and Light

Insufficient oxygen is not usually a major problem for germinating seeds, except when seeds are buried too deep, planted in waterlogged or heavily compacted soils, or submerged in water for too long. In any other circumstances, if there was an oxygen deficiency, you'd be in bigger trouble than your seeds.

The role of light in seed germination is complex and in part depends upon how the seed was handled before germination. Most seeds don't need light to germinate, and in some species, such as florist's cyclamen (*Cyclamen persicum*), light actually inhibits germination. But some species do require light for germination; often these are swamp or woodland species, where light is ordinarily a limiting factor. Most grasses, celery, celeriac, rhubarb, and sorrel all require light for germination. Red light does not penetrate the plant canopy as well as far-red

light. Birch seeds won't germinate on the forest floor where red light is limited but will germinate in sunny areas. The far-red light reaching the forest floor prevents germination in this species.

Much depends not only on the species but also on whether seeds were sown fresh or on how they were otherwise handled before sowing. Our rule of thumb is to give very small seeds, such as those of lettuce, petunia, and foxglove, adequate light by very shallow planting or simply sprinkling them on top of the soil and pressing them in with a hand or board. If you plant them so deep that they receive no light, they won't germinate anyway. There is also some evidence that seeds of some of the southern pines germinate better if sown shallow rather than sown deep or sown in dim light. Seeds of these pines covered with only ⅛ inch of medium germinate better than those covered with ¼ inch.

In those seeds that require light to germinate, light quality as well as intensity can be very important. Incandescent lights are rich in red

---

PHYTOCHROME PHYSIOLOGY

IN 1959, A GROUP of U.S. Department of Agriculture scientists extracted a soluble protein pigment from seeds and seedlings. Further study of the pigment, which is called *phytochrome*, revealed that it is a photo-receptor that controls many fundamental processes in plants, including germination.

Phytochrome exists in two forms. One form, called P_{660} or P_r, absorbs red light (660-nanometer wavelength), while the other form, called P_{730} or P_{fr}, absorbs far-red light (730-nanometer wavelength). When the pigment absorbs light in one form, it is converted to the other form. It's sort of like a toggle switch. For example, if P_{660} absorbs red light, it is converted to P_{730}. If P_{730} absorbs far-red light, it's converted back into P_{660}. That conversion proceeds in the dark as well, as P_{660} is formed by dark decay of P_{730}. The ratio of P_{660} to P_{730} in a plant is what triggers a response, such as germination. Red light produces a preponderance of P_{730}, which in turn promotes germination in lettuce seeds. Far-red light, by producing a preponderance of P_{660}, inhibits germination.

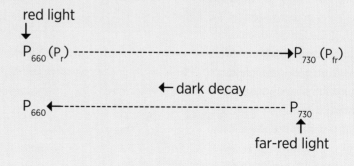

and far-red parts of the spectrum and give off much heat, which itself can become a problem. The high amounts of red light can actually promote germination of some seeds. Cool-white fluorescent lights, which most of us use, are higher in the blue part of the spectrum and lower in the red part. Fluorescent light can retard the germination of seeds of some ericaceous species. But the seeds of most species that require light for germination show no preference for a certain part of the spectrum and do perfectly well under any white light.

You'll find specific instructions for plants whose seeds require light for germination in the plant entries in part 2. Usually, light intensity needed for germination is from 50 to 150 foot-candles, far lower than that required for good seedling growth. When necessary, light is usually supplied for 8 hours per day. If your germination protocol calls for a fluctuating temperature, then expose the seeds to light during the high-temperature portion of the protocol.

The Waiting Game

How long it takes for your seeds to germinate depends both upon environmental conditions and upon species. Radish is renowned for germinating in only a couple of days. Most vegetable and flower species germinate within a week or two. Among the vegetables, parsnip and salsify may take up to 3 weeks to germinate. But that's nothing, considering that some woody species require a year or more. In the wild, some acorns take 2 years to germinate!

Overall, we advise you to give your seeds time. What's the rush? If a few weeks have gone by and you are beginning to wonder whether they are ever going to germinate, take a pencil or a small plant label and use the tip to tease a few seeds out of the medium. If you see the white radicles emerging, tuck the seeds back in and wait a few more days. If no radicle has emerged, squeeze a seed or two. If they're mushy, the seeds have rotted. Correct the problem and replant. If they're not mushy and you see no signs of fungal growth, review your notes to be sure you overcame any dormancy the seeds may have had. If you did, wait a few more days or weeks and see what happens. If nothing happens, try again next year with new seeds.

Germination Highs and Lows

The good news is that the vegetable and flower seeds you save from your garden may show a high percentage of germination. Indeed, any seeds from plants that have been propagated from seeds for a long period will generally give you good results. Don't expect the same level of success from seeds you collect from the wild. No matter how careful you are, more than half of such seeds may not germinate. In some cases, perhaps only 5 percent will germinate. Of those that do, perhaps half of them will not produce usable seedlings. But that's okay. You'll still have more than enough for your needs.

We know that the longer seeds are stored, the greater the decrease in viability, as expressed by lower germination. We can measure that by doing what's called the rag-doll test. This test is easy to do and is commonly used on seeds of vegetable and flower species. Simply count out a number of seeds on which you want to test the germination (10 seeds is a convenient number for working

SEED-STARTING SUCCESS WITH WOODY PLANTS

This list offers a brief sample of germination percentages to show you what you might expect in terms of usable seedlings that result from your seed-starting efforts with woody plants. These numbers reflect how many seedlings you could expect to grow from an initial sowing of 100 seeds.

SPECIES	NUMBER OF USABLE SEEDLINGS
European white birches (*Betula* spp.)	15
Siberian pea tree (*Caragana arborescens*)	33
Hackberries (*Celtis* spp.)	30
Siberian crab (*Malus baccata*)	15
Spruces (*Picea* spp.)	50
Torrey pine (*Pinus torreyana*)	60
Black locust (*Robinia pseudoacacia*)	25
Silver buffaloberry (*Shepherdia argentea*)	12
Elms (*Ulmus* spp.)	12
Nannyberry (*Viburnum lentago*)	25
Lilac chaste tree (*Vitex agnus-castus*)	16

Source: Anon, 1948

out percentage of germination). Fold them up in a paper towel, moisten it, and put the batch into a zipper-closing plastic bag or a jar. Leave the bag or jar (with the "rag-doll" inside) on the kitchen counter or window-sill for a week, then count how many seeds have germinated and figure the percentage of germination. You then know the viability of your seeds.

STORED SEEDS LOSE LUSTER

During storage, seeds can lose their ability to produce vigorous seedlings. This is called *loss of vigor*. Loss of germinability (viability) and loss of vigor are not the same thing. For example, old squash seeds may show 70 percent germination in a rag-doll test, but when you plant some of those seeds out in the garden or in peat pots, they may take a longer than normal amount of time to germinate. And many of the seedlings that do come up will be deformed and too weak to grow into healthy plants. Seed-packet labels indicate the percentage of germination, but they do not indicate the degree of vigor of the seedlings. Seed vigor declines faster than seed viability.

You may find huge differences among species and seed lots in the relationship between viability and vigor due to the way the seeds have been harvested and stored. Nevertheless, the graph below left shows the overall relationship between vigor and viability. At high germination percentages (above about 80 percent), viability and vigor are at their peak, but as the seeds age, vigor declines more rapidly than does viability. When germination percentage has declined to about 20 percent, vigor is almost completely lost, and it does little good to keep the seeds. Our rule of thumb is that when germination of vegetable and some flower seeds has fallen below about 70 percent, there is a very good chance that seedlings will perform poorly. You can try to overcome this by increasing planting density, or you can simply discard the seeds and make plans to save a new batch.

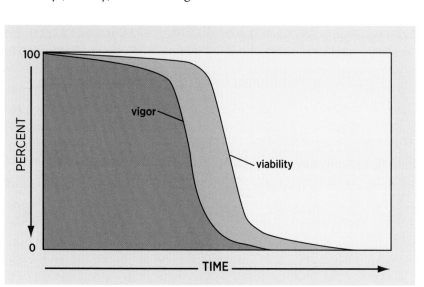

▲ As seeds age, they lose both viability and vigor, but vigor declines more quickly than does the ability to germinate.
Source: Harrington, 1977

SOWING SEEDS AND RAISING TRANSPLANTS

We've discussed seed collecting, seed storing, and overcoming dormancy; now let's look at sowing your seeds and growing transplants for setting outdoors. That includes how to rid your seeds of pathogens easily and quickly, what special soils and soilless potting mixes to use, presprouting seeds, transplanting containers, and the best light intensities for growing transplants. We also discuss some common problems you may encounter with your transplants and how to correct them. We also describe how to set up nursery beds for your woody plants. It's almost time to get those plants into the garden. Let's go!

Heat-Treating Seeds

Diseases such as leaf spot, seed rot, black leg, and some types of blight can be transmitted from generation to generation in infected seeds. If you have not noticed disease problems on your plants, then chances are these pathogens haven't infected your saved seeds. But to be sure, you may want to treat the seeds you've saved to remove any pathogens that are present. Commercial companies treat seeds with hot-water soaks or fungicide applications, or both. For home gardeners, we recommend only hot-water soaks for control of seedborne pathogens. If properly done just prior to planting, this process will kill most external and internal pathogens without damaging seed embryos. Water temperature and precise timing of the soak are critical. You will need a high-quality thermometer that is accurate to within a degree or two, because sometimes water only two or three degrees (1 to 1.5°C) too hot will kill the seeds, and water two or three degrees too cold will not kill the pathogens. Be sure the seeds you are treating are vigorous and have a high germination percentage (see page 65 for a test for germination percentage). Here's the technique.

1. Fill a pot halfway with water and place it on the stove top to heat to the recommended temperature for treating your seeds (see page 68 for temperature recommendations or refer to the plant entries in part 2.)
2. Put the seeds loosely in a cloth bag or a piece of panty hose.
3. Prepare a second pot with water that has been warmed to about 110°F (43°C).
4. When the water in the first pot has reached precisely the right temperature, put the bag of seeds into the second pot (from step 3) for a minute or two.
5. Move the bag to the pot of hot water on the stove for the prescribed number of minutes, stirring the water slowly and

constantly for the entire treatment duration. Use a large strainer to suspend the bag and prevent it from touching the bottom or sides of the pot. Water should be able to circulate all around the seeds.

6. After treating, gradually cool the seeds in tepid water, then in cool water; then spread them out to surface-dry before planting.

When the seeds are dry, do not put them back into storage. Sow them right away, either outdoors in the garden or indoors in a flat or small pots.

HEAT TREATMENT TO CONTROL PATHOGENS

Preheat seeds
in 110°F H$_2$O

↓

Move seeds to H$_2$O at ___°F
(___°C) for ___ minutes*

↓

Move seeds to tepid H$_2$O

↓

Move seeds to cool H$_2$O

↓

Spread seeds to dry

*See Hot-Water-Treatment Requirements, at right

HOT-WATER-TREATMENT REQUIREMENTS

This chart lists the time and temperature requirements for soaking seeds to kill seedborne pathogens. Be sure to monitor water temperature and timing when treating seeds. Unlike other flower seeds, delphinium and larkspur (*Consolida* spp.) seeds can be treated in water ranging from 122 to 129°F (50 to 54°C). Soak these seeds for only 10 minutes.

SPECIES	118°F (48°C)	120°F (49°C)	122°F (50°C)	125°F (52°C)	127°F (53°C)	131°F (55°C)
VEGETABLES						
Broccoli			20 min			
Brussels sprouts			25 min			
Cabbage			25 min			
Carrot			15–20 min			
Cauliflower			20 min			
Celery	30 min					
Celeriac	30 min					
Chinese cabbage			20 min			
Collard			20 min			
Coriander					30 min	
Cress			15 min			
Cucumber			20 min			
Eggplant			25 min			
Kale			20 min			
Kohlrabi			20 min			
Lettuce	30 min					
Mustard			15 min			
Radish			15 min			
New Zealand spinach		60–120 min				
Pepper				30 min		
Rape			20 min			
Rutabaga			20 min			
Spinach			25 min			
Tomato			25 min			
Turnip			20 min			
FLOWERS						
California poppy				30 min		
Foxglove						15 min
Nasturtium				30 min		
Stock						10 min
Zinnia				30 min		

Adapted from Shurtleff, 1966

Direct Seeding

Direct seeding in garden beds is the most economical and efficient method for starting new plants. It can save you the extra effort and cost of artificial stratification treatments and of growing transplants in containers. But there are good reasons to sow seeds indoors, too, especially those of annual flowers and vegetables, and we describe how to do that later in this chapter.

Many of the seeds that you've collected from perennials can be sown while still fresh in fall directly outdoors in your garden or in a cold frame or nursery bed. *Reminder*: Whenever you sow any seeds anywhere, take good care to label them and even make a map of the planting for extra insurance.

When you direct-sow fresh seeds of perennials that have no dormancy mechanisms, they sprout as they would naturally after exposure to warm fall and/or cold winter temperatures. If seeds do require stratification, natural winter conditions outdoors will take care of that for you. Seeds of rose campion (*Lychnis coronaria*) and many other perennials and those of most woody plants will make it just fine sown in this manner, provided you live in a temperate area and you time your planting correctly.

There is a downside to sowing in the fall, especially if you are sowing seeds that need to stratify over the winter. You may lose many of the seeds to depredation by birds and animals and to frost heaving. If you have limited quantities of such seeds, it's safer to stratify them indoors, then sow them outdoors in spring.

Seed Sowing How-To

Unless you're a brand-new gardener, you probably have some experience with sowing seeds in the garden and in flats or pots indoors. In general, the same sowing techniques you use with seed you've purchased from a garden center or seed company works just fine for seed you've saved from your yard and garden. Annual flower and vegetable seeds are usually planted in shallow rows or in hills and lightly covered with soil. Sow seeds of most trees and shrubs in rows, and press them into the seedbed with a roller, a hoe, your thumb, or a board.

Read on for a few of our best tips for success with sowing seeds.

SOWING TINY SEEDS

You may be very good at doing detail work, but when your hands get as old as ours, the fingers are a little stiff. We find we have trouble sowing very fine seeds evenly in a flat. One alternative is simply to broadcast the seeds rather than try to sow them in rows. To do this, first mix the tiny seeds with sharp sand or gelatin. Place the mix onto a piece of stiff paper that you've creased down the middle, then sprinkle half the mix in the flat along one dimension and the other half along the other dimension; that is, you cross-seed the seed/carrier mix. Cross-seeding results in a more even distribution of seeds, and the sand and gelatin are light colored so you can see where you have already sown dark seeds. The gelatin is high in protein, which contains nitrogen, so it can also give the germinating seeds a small, extra boost of nutrients as well.

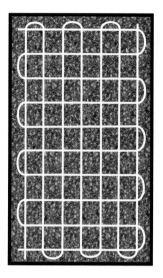

▲ For even coverage, sow tiny seeds in a looping pattern across a flat from left to right, then from top to bottom.

DEPTH OF COVER

The depth of seed covering is critical. Too deep and the seeds won't germinate or the seedlings will die before making it to the soil surface; too shallow and the seeds will dry out. Our rule of thumb is this: Don't cover seeds to more than twice their largest diameter. For example, if a squash seed is ¾ inch long, we would plant it no deeper than 1½ inches. If you're sowing seeds in containers using garden soil, which is heavier than a soilless medium, then sow seeds slightly less deep than you usually would.

The final stage in the sowing process consists of covering the mix in the containers with some type of mulch, such as perlite or coarse vermiculite. Light-colored mulches are best because they reflect sunlight and keep the soil cool. Mulching is not necessary for seeds of vegetables, herbs, and flowers but is particularly important for seeds of woody species that occupy containers or beds for long periods.

CARE AFTER SOWING

Keep newly sown seedbeds moist until germination is complete, usually within 1 to 3 weeks of sowing for vegetables and flowers and 3 to 4 weeks of sowing for trees and shrubs. Covering the soil surface with wet newsprint or moist burlap can accomplish this.

After 3 to 4 weeks, or when germination is complete, thin seedlings to their proper spacing and remove weeds. You can pull out small seedlings, but for larger seedlings, use scissors to clip them off at soil level because pulling them may uproot neighboring plants.

Presprouting Seeds

Presprouting seeds before you sow them may save you a month to even several months, especially with seeds of woody species. The timing of sowing and seed placement is critical and must take into account the proper time for planting outdoors. For example, presprouting viburnums, along with the other pregermination treatments that viburnums require, may take months. Presprouting vegetable seeds, though, takes only days.

You can presprout seeds in pure quartz sand, on cotton or rolled paper towels, or in peat moss. The seeds of some species germinate best in a specific germination medium (see plant entries in part 2). Once you've set up seeds to presprout, put them in the proper temperature and light conditions for germination. Check seeds daily for signs of sprouting (described below), and sow them as soon as they've sprouted.

PLANTING PRESPROUTED SEEDS

You can sow presprouted seeds directly into containers. This is especially useful for seeds requiring long or variable stratification and seeds from large-seeded recalcitrant species like the buckeyes. Stratify or scarify seeds first, following directions in part 2 for the specific plant you're working with. As soon as the seed coats have split and the radicles have begun to emerge, gently prick the seeds out of the container in which they presprouted and sow them in standard seed-starting mix in a pot or larger flat, then cover them with about ½ inch of perlite or vermiculite to prevent drying. Keep the seeds moist, but not so wet as to allow mold to develop. Take care to always transplant presprouted seeds before the radicle becomes too long and starts to curve, a stage at which it is easily damaged. Position the radicle to point downward. The easiest way is to use a dibble or pencil to make a small hole in the medium and gently position the seed in it.

Raising Transplants

Raising seedlings indoors for transplanting into the garden requires some advance preparation of containers, growing mix, and a well-lighted area for seedlings, but it's worth the work because of the benefits it offers.

EXTENDING THE GROWING SEASON. Starting seeds indoors essentially extends the growing season by as

---

SOIL QUALITY COUNTS

IT SEEMS ALMOST TOO OBVIOUS to mention, but your carefully saved seeds aren't going to thrive if you plant them in poor soil. Always try to improve the soil *before* you sow seeds or set out transplants. Vegetables and flowers grown as annuals do well on sandy loams and silt loams. Biennials and herbaceous perennials do well on deep silt loams. The best soil type for tree and shrub crops is sand to sandy loam at least 18 inches deep.

If you can, improve the soil by planting a green-manure crop the year before. Green manures supply organic matter and serve as "catch crops" to capture mineral nutrients such as phosphorus and iron. Till or dig the green-manure crop under in late summer to allow time for the organic matter to decompose before spring planting. If you have some straw or leaf litter, work that into the soil, too, and add any other organic amendments you may have, such as composted manure, pine litter, or peat moss. Sawdust makes a good amendment for tree and shrub soils, as long as it's rotted and you apply extra nitrogen fertilizer to compensate for that used by soil microorganisms to break down the woody material. Check soil pH, too, and adjust to the range that's ideal for the types of plants you want to grow. Incorporate a balanced fertilizer into the soil at this time as well. In spring, continue to work the bed gently to get the soil into good tilth.

much as 2 months, enabling northern gardeners with short frost-free growing seasons to successfully produce a crop of heat-loving crops such as peppers and eggplants. Keep in mind, though, that transplanting sets back plants a bit so that, for example, transplanted tomato plants mature later than field-sown tomatoes. Thus, be sure to start early when you're sowing seeds indoors; otherwise you may lose the timing advantage you hope to gain. Starting seeds indoors can also help ensure success of cool-season crops such as spring broccoli. Often direct-sown broccoli plants do not reach the head-forming stage until the heat of summer, when they are more apt to flower too fast and to develop strong flavors. Seedlings started early indoors and then transplanted to the garden reach maturity sooner and yield a better-quality crop.

BETTER GERMINATION OF TINY SEEDS. Crops such as petunia and foxglove have tiny seeds that are difficult to sow directly into garden beds at the proper depth. Also, they are very prone to being covered too deeply when pelting rain or irrigation water washes them farther into the soil. Starting these indoors in pots allows you to handle them with greater care.

GREATER FLEXIBILITY. Sometimes seeds that you've set up to stratify reach the end of their treatment period before it's possible to plant them outdoors. For example, you may end up with pine seeds ready for sowing in February, when the soil is frozen in your yard. But if you don't sow the seeds, they'll rot, dry out, or enter a secondary dormancy. If that's

the case, you can sow the seeds in containers indoors instead and set out the seedlings when the soil thaws in spring.

The best vegetable and flower transplants should be about 6 inches tall, stocky, a deep green or other color appropriate to the variety, and at the proper stage of development when set out. It's natural that once a plant is growing well in the house, you'll want to keep it there for longer than needed, just because it's doing so well. You hate to move it! But planting a big plant is not as good as planting a smaller plant. Larger plants lose a disproportionate amount of their root system during transplanting and are set back in their development more by transplant shock. Initially, this slows their growth greatly, and those transplanted when smaller usually catch up to and surpass the bigger ones in vigor. Vegetable plants such as tomatoes and peppers that are old enough to have flowered and even set fruit before transplanting never fully recover and are relatively poor croppers. Bigger is not always better! Remember, not all plants should be or need to be transplanted.

Containers for Starting Seeds

Some containers serve better for starting seeds than others. You can use purchased peat or fiber pots, peat or fiber pellets, cell packs, or plastic flats for vegetables and flowers and larger containers for trees and shrubs. Or if you want to save a little money, you can recycle household items, such as disposable coffee cups and egg cartons, or reuse plastic or clay flowerpots. You can even make small pots out of newspaper. It's easy

to build your own wooden flats out of redwood or cedar that will last you a short lifetime. Any container will do as long as it's clean. If you use a seed-starting container more than once, wash it in soapy water, rinse with clear water, dunk in a solution of one part bleach to nine parts water, then rinse thoroughly in clear water.

A relatively new practice used by commercial growers for producing transplants of trees and shrubs is to line containers with copper compounds, such as copper hydroxide, that "chemically prune" the root system to reduce formation of circling roots. Such circling roots could later girdle and kill the plant.

Growing Mixes

The growing medium must be well drained but have adequate water-holding capacity. It must be free of pests and weed seeds, inexpensive, and readily available. If you intend to both sow seeds and grow transplants in a single medium, use a commercial soilless mix. One cubic foot of medium will fill about 275 peat pots that are 2.25 inches square; 60 round peat pots with a 4-inch diameter; or 20 cell packs measuring about 5 by 8 by 2.75 inches.

You can also make your own soilless mix for growing seedlings. Combine the following ingredients to make 1 cubic yard of mix:

> 0.5 cubic yard sphagnum moss
> 0.5 cubic yard horticultural vermiculite (not building grade)
> 5 pounds ground or dolomitic limestone
> 4 pounds of bone meal
> *or*
> 2 pounds superphosphate
> 2 pounds dried blood
> *or*
> 1 pound calcium nitrate

This mix will work well for most herbaceous seedlings. It may also work well for many woody plants, but some woody species are pretty fussy about what they're planted in. Always try to mimic the soil where the seeds of your plant would naturally germinate in the wild. For example, pine seeds grow well in pine-needle litter mixed with a basic soilless mix. Red maple (*Acer rubrum*) seeds and those of the ericaceous (acid-loving) plants such as rhododendrons grow best in an acid soil at a pH about 4.5. Some other species germinate best in sand and then are transplanted to other medium. For specific recommendations, refer to plant entries in part 2.

Garden soil may also be used, but to prevent contamination of the seeds by soilborne fungi and other pathogenic organisms, be sure to sterilize it first by heating in an oven set at 130°F (55°C) for 30 minutes.

Sowing Seeds in Containers

A seed-starting medium is typically hydrophobic, meaning it resists wetting. It's a good idea to water the medium thoroughly the night before seeding so the water has time to soak into the mix. Fill flats or other containers with moist mix to within ½ inch of the lip just prior to sowing.

If you are using flats, sow enough seeds to obtain about four plants per inch of row (for herbaceous species) and perhaps one plant per inch or two for woody species. Most vegetable and flower seeds have very good germination, about 80 to 90 percent; seeds of most woody species have highly variable germination percentages (see page 65). If you are using peat pots, sow two to three seeds per pot and thin excess plants later.

Sow most seeds no deeper than twice their larger diameter. This might be a couple of inches for very large seeds. Sprinkle very small seeds such as those of petunia on the soil surface and press in with your thumb or a board. You'll also find that many small seeds require light for germination, another good reason to simply sprinkle them on the surface. You should have moistened the medium before you plant, but if you find it is still dry after seeding, bottom-water it. Do not soak it, however, as a waterlogged medium will delay or inhibit seed germination. Moist medium is sufficient.

Getting Seeds to Germinate

As discussed in chapter 5, temperature and light conditions influence germination. Keep in mind that a medium is always cooler than the air above it because of evaporative cooling (the moisture in the medium "robs" heat from the medium as it evaporates). You can counteract this by using a heating mat, which provides bottom heat. Without a heating mat, if the room temperature in your home is 68°F (20°C), then the medium temperature may be as low as 65°F (18°C), which in turn may be too low for germination of seeds of some species. If this is the case in your home, then bottom heat will be useful. On the other hand, seeds of some species, such as geranium and delphinium, germinate in temperatures cooler than most rooms. Try putting flats of such species in a particularly cool area only until the seeds germinate.

Many tiny seeds, such as those of mountain laurel and petunia, require light to germinate. For other species, though, light retards germination. Be sure you know what is required for the seeds you've sown. Full-spectrum, cool-white fluorescent light works well for most species that require light to germinate. When light is needed for germination, the intensity need be only from 50 to 200 foot-candles. Incandescent light may inhibit germination in some species and promote it in others, whereas fluorescent light may have the opposite effect in those same species.

Some species are sensitive to certain wavelengths of light more than they are to others. For example, fluorescent lights, which have relatively high amounts of blue wavelengths and low amounts of red and far-red wavelengths, inhibit germination of seeds of some blueberry species, whereas incandescent lights, which have relatively high amounts of red and far-red wavelengths and low amounts of blue wavelengths, promote germination of strawberry seeds. Fortunately, cool-white fluorescent lights work well for the majority of seeds.

Setting Up Your Indoor Growing Area

Once seedlings emerge, they require light (or more light if some was needed for germination) and a slightly lower temperature than that used during germination. In general, good daytime growing temperatures are often from 10 to 15°F (5.5 to 8°C) cooler than optimum germination temperatures. Night temperatures should be from 10 to 15°F (5.5 to 8°C) cooler than daytime growing temperatures. Too high a temperature causes transplants to become leggy, especially with insufficient light; too low a temperature retards growth and induces bolting in some vegetables. Some gardeners place special heating mats beneath the flats or pots in their home to increase the heat to the plants' roots, thereby increasing growth. This may be useful but is often not necessary.

PROVIDING SUFFICIENT LIGHT

Most seedlings need more light than you might imagine. On a clear mid-June day at noon in the eastern and midwestern states, the intensity of sunlight is about 10,000 foot-candles. Intensity in some desert and tropical areas can be as much as 12,000 foot-candles. You need to increase the light intensity that your plants receive as they grow to prepare them for outside light. For example, you may start your seeds in darkness. After seedlings emerge, you should increase the light intensity to perhaps 1,000 to 1,500 foot-candles. As seedlings grow, increase the intensity to 2,000 to 2,500 foot-candles by the third or fourth week of growth; up to 3,000 to 3,500 foot-candles by the sixth to eighth week; and up to 5,000 foot-candles for a week or so before setting out. As a general rule of thumb, give transplants 14 to 16 hours of light daily. You would be wise to invest in a light meter; it is inexpensive and a very handy tool to have. (See the chart on page 74 for foot-candle equivalents.)

Some entries in part 2 include specific light-intensity recommendations for seedlings. If information is not available for the species you're raising, follow the general rules above. In most indoor situations, ambient light cannot provide these intensities, so you'll need to set up an area with supplemental artificial light.

Light intensity decreases as the inverse square of the distance from the source grows. That's a technical way of saying that if you double the distance of an object from the light source, the light intensity reaching that object will be only one-quarter what it was at the original distance. Let's say that your lights are 1 inch above the top of your plants and they provide about 200 foot-candles at that distance. If you move them 2 inches from the plants (double the distance), the plants will receive only one-quarter the amount of light (¼ of 200 = 50 foot-candles). If you move them 4 inches from the top of the plants, those plants would receive only ¹⁄₁₆ the amount of light they received at 1 inch from the source (¹⁄₁₆ of 200 = 12.5 foot-candles). You can readily see, then, why it's important to keep your plants as close to the light source as possible to maintain the correct light intensity.

To accommodate most plants' needs, situate the lights only an inch or so above the top of the seedlings when using cool-white fluorescent lights. You may have to keep them 2 to 4 inches from the plants if you are using incandescent lights, to prevent overheating of the foliage. A simple pulley system can be used to adjust light height. Seedlings to be transplanted outdoors usually can use 2,500 to 5,000 foot-candles of light intensity for good growth, depending upon the species.

There are exceptions to the rules above for light intensity. Seedlings of species that are natural understory plants, such as mountain laurel (*Kalmia* spp.) and rhododendrons, need less-intense light than seedlings of sun-loving plants. Plants grown under light intensities too high for their species may "bleach out," turning yellow or white, and may also show marginal leaf scorch. Symptoms are usually more prominent on the upper foliage.

If, on the other hand, your seedlings start to become spindly or leggy, increase the light intensity by using more or higher-output lights or by moving the plants closer to the light source.

Pricking Off and Stepping Up

Thin seedlings when they are about an inch tall and still in the cotyledon stage: that is, before the first true leaves emerge (see the illustration on facing page). Moisten the medium before thinning, and if the plants removed are themselves to be transplanted, fill new pots and moisten that medium before the "spotting out" or "pricking off" operations begin. When seedlings have reached the primary leaf stage, transplant them to larger containers. Transplanting seedlings is necessary to provide the young plants with additional room to grow. It also breaks some of the root tips, stimulating lateral branching of the remaining roots and increasing root-surface area, making for a more robust plant.

Typically, tree and shrub seedlings are grown in 2- to 4-cubic-inch containers for 4 to 6 months, then transplanted to beds. Vegetable and flower seedlings are grown for several weeks following the pricking-off and initial stepping-up processes until they have attained the right size for setting out and the weather has become cooperative. We give directions for individual species in part 2.

Transplanting is particularly useful for woody plants whose seeds have complex dormancies or are

LIGHT SOURCES AND REQUIREMENTS

This chart gives you an idea of relative light intensities. Note that the illuminance we humans need for reading indoors is only 1 percent of what plants need for good growth. The natural lighting in your home may seem adequate for flowering plant growth, but it isn't.

LIGHT SOURCES AND REQUIREMENTS	ILLUMINANCE (FOOT-CANDLES)
Noon sun in mid-June	12,000
Adequate for most plants	5,000
Cloudy, midday, midsummer	5,000 (may vary 25–75% according to cloud density)
Noon sun in midwinter	4,000
Shady side of two-story house in summer on a sunny day	2,000
Adequate for individual leaf	1,200
Adequate for shade plants	1,000
Cloudy, midday, midwinter	1,000
Shady side of two-story house in winter on a sunny day	500
Fluorescent deluxe cool white light (150 W at 5')	100
Adequate for indoor foliage plants	50 to 500
Incandescent light (100 W at 5') with a 10"-diameter reflector	22
Adequate for reading indoors	20 to 30
Threshold for photoperiodic induction	0.3 to 15
Moonlight	0.02

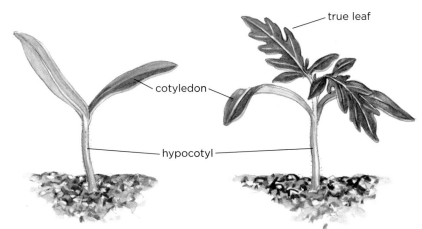

true leaf

cotyledon

hypocotyl

▲ Allow seedlings to develop their first true leaves before "spotting them out" into larger containers.

very small, both conditions that usually result in very low germination. If you sow seeds directly into a nursery bed, you will end up with a lot of wasted space. Therefore, planting into pots first allows you to choose the

healthiest seedlings to be transplanted to the nursery bed and thus allows for maximum use of bed space.

Firm the growing medium around the transplant to ensure good root contact and allow seedlings to grow

▲ A spacious transplant hole is critical for young woody plants. Jamming the roots into a hole that's too small forces the root into a J position, which hampers root growth.

into larger plants before setting out into the garden or field. All plants initially sown in pots or flats may be transplanted (stepped up) more than once, each time into successively larger containers, before being set out. Take special care when transplanting to avoid creating a "J-rooted" position of the seedlings. This happens when a long root system is jammed into a hole too shallow to accommodate it and the bottom part of the root system is bent into the shape of the letter J. J-rooted plants grow poorly and often never overcome the poor handling they received early in life. Clip the bottoms of very long roots or dig deeper holes to make planting easier.

Fertilizing and Watering Seedlings

WATERING. To determine when to water your transplants, take a bit of medium from the top ½ inch and squeeze it between your thumb and forefinger. If water runs out of it freely, the medium is moist enough. If not, use bottom-watering, especially when seedlings are very small. Always use water kept at room temperature, as cold water from the tap can shock roots of some plants and temporarily impede their growth.

Excess water can cause watery swellings on the leaves. These later break open, and the areas become rough and rust colored. This is called *edema*. It's a nonpathogenic problem and won't spread, but it does mean that you are overwatering or that the relative humidity of the air is too high. Increase air circulation and water less frequently.

SEEDLING PROBLEMS AND SOLUTIONS

Whether you're raising vegetables and flowers or woody species, transplants often suffer from various problems while still in the flat or cold frame. The following table will help you identify the causes of these problems and suggest ways to remedy the problems.

DESCRIPTION	POSSIBLE CAUSE	CORRECTION
Leggy, spindly plants	Not enough light	Use full-spectrum lights or a combination of fluorescent and incandescent lighting. Move plants closer to light source.
	Excessive watering	Do not allow transplants to sit in water. Maintain moist but not wet medium
	Excessive fertilizer	Apply according to label directions. Reduce concentration. Reduce nitrogen.
	Plants are too close together	Provide each seedling with enough room to allow for stocky growth.
	Temperature too high	Lower temperature.
	Old seeds	Plant fresh seeds next time.
Stunted plants	Nutrient deficiency (see below)	Apply fertilizers more often but in low concentrations.
. . . with discolored leaves: Stems and undersides of leaves may be reddish. Leaves may be small. Roots may be stunted.	Phosphorus deficiency	Apply a high-phosphorus starter solution according to label directions.
. . . with lack of dark green color in leaves and stems. Slow growth. Leaves chlorotic.	Nitrogen deficiency	Apply nitrogen fertilizer solution according to label directions.
. . . with discolored leaves and discolored roots. Plants may wilt in bright sunlight. Lower leaves turn yellow and drop.	Too much fertilizer	Leach excess fertilizer by running clean water through the medium.
. . . with discolored roots and leaves	Too long/too hot sterilization of medium	Next time after sterilization of the medium, soak or leach the mix before planting.
. . . with no root discoloration: Growth is slowed. Foliage may have purple veins.	Temperatures too low	Maintain proper air and soil temperatures.
Rotting and collapse of the stems near soil surface	Damping-off (caused by fungus in medium)	Next time, do not use garden soil that has not been sterilized. Wash all used pots in soap and water. Rinse with bleach solution, then water.
Slow root growth	Poor drainage	Improve drainage.
	Low fertility	Increase supply of nutrients.
	High fertility	Reduce supply of nutrients.
	Low temperature	Raise temperature.
	Herbicide residue in soil mix	Repot into new mix.
Algal and moss growth on medium surface	High moisture content of medium	Reduce watering.
	Poor medium mix	Add coarse sand to loosen the medium.
	Poor aeration	Increase air movement around plants.
"Bleached" leaves, especially upper leaves, with or without marginal scorch	Light source is too intense	Reduce light intensity by moving plants farther from the light source or changing to lower-output bulbs.
Seed coat adheres to the cotyledons; seedlings may be weak and deformed	Old seeds with low viability	Use fresh seeds next time.

Adapted from Gough and Moore-Gough, 2009

FERTILIZING. Many of the nutrients added to a commercial soilless medium at the time of manufacture will be used by plants during their early stages of growth, so you'll have to add more fertilizer as the plants grow. The plants will let you know when to do this, mainly when the foliage just begins to turn yellow. Don't let it turn completely yellow or allow it to show any amount of yellow for any length of time — get right on the problem. When it's time, dissolve 2 level tablespoons of 20-20-20 or another complete, nitrogen-equivalent fertilizer in a gallon of water and apply this at weekly or biweekly intervals during regular watering operations. Ericaceous plants and some other species have little need for fertilizer, so don't fertilize them on a regular basis, only as needed.

Hardening Off

Plants that have spent their entire life in a protected environment where they wanted for nothing are ill-adapted to the outdoors, so before setting them out, you need to acclimate, or "harden" them. Hardening off involves a slowing of growth and a toughening of the plant tissues. It also results in a net accumulation of carbohydrates, some of which the plants turn into red pigments called *anthocyanins*, causing the stems and leaf veins in some species to turn pink or purple.

The hardening-off process takes about 2 weeks for herbaceous plants and up to several weeks for woody plants. Move the plants outdoors during warmer days and bring them in at night, gradually increasing the length of time they spend outdoors until they remain outdoors day and night for a week before transplanting. Tree and shrub seedlings in greenhouses are often moved to a shade house outdoors to harden. During the hardening period, decrease watering to provide only enough to prevent incipient wilting. This reduction in water forces a plant to thicken the waxy cuticle layer of its leaves and stems, which gives it some protection from drought and wind. If plants were grown in flats, cut or "block" the medium with a knife to divide it into individual soil blocks, so that each plant is a single block. As with transplanting, blocking causes a proliferation in root branching, which ultimately increases the root surface area. Don't block roots of cucurbits, okra, and certain other plants that regenerate their root system very slowly.

When you begin the hardening-off process, protect plants from the intense direct sun of midday by placing them in the shade. But gradually work up to providing them with full sunlight over the course of a week or more, depending on the type of plant. Fertilize minimally or not at all during hardening. This will force the plants to decrease vegetative growth, making the stems woodier and the plants stockier.

FIGURING OUT THE TIMING

IT CAN BE TRICKY to calculate when to sow seeds indoors so you'll have transplants of the right size, properly hardened off, at the right time to plant outdoors in your garden. A common mistake is to start seeds too early in the season. This results in plants that are too large for easy transplanting. They'll suffer undue transplant shock and yield poorly, if they survive the operation at all. Ideally, you should transplant flowers, tomatoes, peppers, and eggplants to the garden just before they're ready to bloom. If fruit have already set, remove them at transplanting time or subsequent fruit development will be severely retarded. It's best to transplant cucurbits to the garden as soon as their leaves are about 2 inches across.

Some woody species will be transplanted a few times before they're set in their permanent spot, and the entire process may take several years. Time your pretreatments, if any, so that their completion, added to the time for germination plus the time needed to grow transplants, coincides with the advent of warmer spring weather, allowing for outplanting to a nursery bed. We provide specific guidelines for some types of plants in part 2.

Setting Out Transplants

When you unpot a transplant, you may find the roots tightly twisted into a firm root-ball. If so, gently break the ball apart or make several light scoring cuts along the surface of the ball before planting. This breaks encircling roots mechanically and promotes formation of lateral roots and rapid development of greater surface area. Exceptions to this are cucurbits and other species with poor root regeneration. Disturb their roots as little as possible when transplanting.

Set plants outdoors at the same depth as they were planted in the pot. It's best to transplant in the afternoon or on a cloudy day to decrease stress on the transplants. After transplanting, firm the soil and water well to rid the soil of air pockets.

Care after Planting

Your new garden plants will need special care for the first few weeks to ensure that their root systems become well established. This may include shading, watering, frost protection, and fertilizing.

WATERING. Estimate the amount of water to apply by checking soil-moisture measurements or setting a pan of water outside and measuring how much evaporates. If the weather turns hot, use sprinklers to cool the soil and plant surfaces while new transplants are still succulent.

FROST PROTECTION. If frost is predicted after you've transplanted tender young plants, set up sprinklers in the bed and leave them running until all danger of frost has passed and the temperature has risen out of the danger zone. You can also use row-cover material to hold in what little soil heat there is in spring.

FERTILIZING. Unless soil tests show specific nutrient deficiencies, nitrogen, phosphorus, and potassium are the only fertilizers we typically apply. Of these, nitrogen is the most important, because it's the nutrient that's most often lacking. Suspected nutrient deficiencies, as indicated by symptoms such as chlorosis (yellowing), should be confirmed by soil or foliage tests.

Creating a Nursery Bed

Growing woody species in nursery beds for a season (or longer) before transplanting them to the garden makes them far easier to handle later on and increases their chances for survival. If you intend to grow large quantities of woody plants from seed, you may want to build your own nursery to grow on the plants for a year or more after the seedling stage.

A nursery-bed area consists of rectangular beds and also usually includes a water source, a compost heap, and a small shed to hold potting equipment, soil amendments, and other materials. It's a good idea to fence the area to keep out roaming animals.

Bed Setup and Orientation

Divide the nursery area into thirds. In one location establish your "seedbeds," into which you will plant seeds directly. In another location construct your "transplant beds," in which seedlings from the seed beds are "stepped up" into more fertile soil and set at greater spacings. In the third area will be your "pot beds," where potted seedlings are set before transplanting to the field.

Situate the beds so that one narrow end faces into the prevailing wind, usually west. Be sure the soil beneath the bed is well drained. Locating the beds on a gentle 1 to 2 percent south-facing slope will provide good drainage and some additional warmth. A thin planting of shrubs around the nursery area will break the force of the wind but still allow some air circulation, very important for reduction of seedling diseases such as damping-off.

Nursery-Bed Construction

Essentially, each nursery bed is a simple raised bed. These are especially important in areas of high rainfall and where soil is poorly drained. If you live in a dry, hot, windy area, construct a sunken bed to conserve moisture. Make each bed 3 to 4 feet wide. You should be able to reach comfortably to the center of each bed, because part of the purpose of creating raised beds is to avoid the need to step on the soil in the bed. A length of about 3 yards seems long enough for most folks. Partition the beds according to species so you can meet the different requirements of each.

Bed sides made of rot-resistant 2×10s of redwood, cypress, white oak, or another suitable species work well. If you can't procure lumber of those species, use regular-dimension lumber (e.g., fir, hemlock) treated with a wood preservative according to label directions. If this is a pot bed, take extra care to make the bottom level so pots will be less likely to tip over. Spreading a few inches of coarse sand over the soil will help create a level surface for pots.

Medium

You'll want to fill beds for seeds and seedlings with a medium that is reasonably fertile, deep, well drained, and rich. Most seeds germinate a bit better in a slightly acidic medium (pH 6.0 to 6.5), so don't incorporate limestone unless it's really needed. Make the upper layers of the medium very fine textured and fertile. A 1:1 mixture of finely screened compost-enriched soil and coarse sand works very well. Some growers also work in rotted sawdust with nitrogen added. Lightly tamp the bed soil so that it won't settle after seeding. The combined depth of good soil should be about 18 inches. Thus, if you construct the bed sides of 2×10s, you will also have to amend the soil an additional 9 to 10 inches below ground level. Incorporate a good fertilizer according to label instructions just before planting.

Shade

Provide for removable partial shade to protect very young seedlings from bright sunlight. This is especially important for understory plants such as rhododendrons that evolved in filtered sunlight. Be sure the shading setup allows rain and some sunlight through and can be easily removed at the right time. There are a couple of easy ways to provide the right amount of shade. First, construct a simple frame of 2×4s over the bed, then place the shading material over the frame.

SHADE CLOTH. This plastic-weave material is available in different grades to provide different percentages of shade, such as 25 percent, 33 percent, and 50 percent. Light shade would be about 25 percent and moderate shade about 50 percent. Fasten the cloth over the frame when shade is needed and remove it for storage when it's not.

LATH. A frame made of wooden lath looks great. Two-inch-wide lath spaced 2 inches apart provides 50 percent shade; spaced 4 inches apart, it supplies 33 percent shade; spaced 6 inches apart, it provides 25 percent shade.

SNOW FENCE. Snow fence, especially the old wooden type, can be rolled out along the frame and then rolled up again for storage. This works about the same way as a lath frame but can look a bit sloppy.

Planting a Nursery Bed

When you're ready to sow seeds in a nursery bed, first be sure the soil surface is level, then drag a rake over it to gently loosen the upper inch or so. This removes some weeds and allows for better water penetration.

Broadcast very small seeds by sprinkling them over the soil surface and pressing them in lightly with your hand or a board. Row seeding (planting in rows) makes for a neater bed and greater ease in weeding operations. Board seeding is useful in spacing various seeds the proper distance apart within and between rows. To make a planting board, drill holes through a suitably sized board or piece of plywood to create a grid pattern, with the holes 2 inches apart. Place the planting board on the soil surface as a template, and push a small dibble through each hole in turn. Then drop seeds into the holes.

A similar device, called a *spotting board* or *dibble board*, has pegs fastened to the board at desired intervals so that when the device is pressed into the soil and then removed, it will leave premade holes. Both devices are rapid ways to make evenly spaced planting holes.

Pregerminated seeds planted into the planting beds as soon as they are sprouted are called *germinants*. Using germinants is especially useful when a species has a long stratification or germination time. Take care not to damage the emerging radicle during

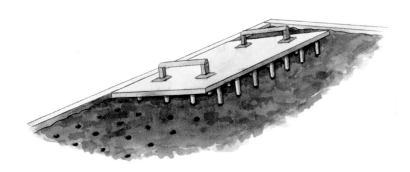

▲ Using a dibble board is a real time-saver when you want to sow lots of seeds uniformly across a bed.

planting. Both seeds and germinants need to be covered with ¼ to ½ inch of clean sand to reduce drying out and damping-off.

After sowing, cover the seeds with soil, sand, sawdust, or some combination, depending upon the recommendations for each plant in part 2. Sand and sawdust act as mulches of sorts, the former to keep the soil surface dry and reduce damping-off, the latter to keep it moist for germination. Cover seeds with mulch when indicated.

You can also transplant small seedlings that you've raised to the first-leaf stage in a greenhouse or hotbed into the outdoor beds.

Care of Nursery-Bed Plants

Keep the beds weed-free and water regularly. If plants look especially weak, apply a complete liquid fertilizer according to directions on the label. Adjust the watering schedule so the soil several inches deep remains moist. The soil surface should dry out between waterings.

A few weeks before transplanting nursery-bed plants out into the garden, block the plants by severing the roots all around a plant several inches from the stem. This promotes hardening off as well as root branching. At the same time, rogue the beds, removing weak, diseased, and damaged seedlings so only healthy, strong, woody plants 10 to 16 inches tall remain.

Transplanting

Slowly decrease watering for a few weeks before you want to transplant, then water the plants deeply a day before transplanting. This lets plants imbibe the water before they are moved and also softens the soil, making it easier to take plenty of roots with the plants. Transplant on cool, cloudy, damp, or rainy days, preferably in the late afternoon. Select plants that have at least six leaves, and move them with as much of a root-ball as possible to their new location. If taproots are long and spindly, snip them to reduce the chance of J-root planting. When you've cleared out a seedling bed, it's a good idea to replace the soil mix in that bed to reduce the incidence of soilborne diseases.

Most woody seedlings are large enough to move after 3 or 4 months, but let size, not date, be your guide. Transplant into pots filled with rich garden soil or to a well-prepared planting bed. Plant spacing should be equal to the expected height of the plants when they reach 2 years of age. Water the plants well after transplanting to settle the soil around the roots. You will probably be able to move these plants into their permanent locations in the landscape after 2 to 3 years.

Nurserymen use a two-digit code to indicate the age of bare-root seedlings. It indicates the number of years a plant has been in the seedbed (the first number) and the number of years it has been in the transplant bed (the second number). For example, a 1+0 plant has been in a seedbed for 1 year and has spent no time in a transplant bed. A 2+2 plant has spent 2 years in a seedbed and then 2 years more in a transplant bed (or in a pot in a pot bed).

The sum of the numbers equals the total age of the plant. For example, a 2+2 plant is 4 years old.

CHAPTER 7

BREEDING YOUR OWN VARIETIES

M ANY GARDENERS will be satisfied with saving seeds of their favorite varieties to replant year after year. But some gardeners enjoy stepping into the role of amateur plant breeder, experimenting with home hybridizing of garden plants to see whether they can develop a gardenworthy new variety of a crop or flower. If you find yourself asking, "Why bother to go through all the work of breeding your own plants when there are so many good selections available on the market?" — then plant breeding is not for you, and that's fine. But if you're intrigued by the notion of developing a new variety and seeing the wonderful array of flower, fruit, and leaf colors you can unlock, then go for it!

Many gardening books written from the midnineteenth to the midtwentieth century had chapters on amateur plant breeding. Back then,

many folks saved their own seeds routinely and found it a great hobby to breed their own varieties. (To learn more about this era in our gardening history, turn to page 279.) But after World War II, as hybrid seed became widely and cheaply available, interest in plant breeding at the home-garden scale died out. There are some good reasons to take up the hobby of amateur plant breeding again. It's a way to avoid reliance on the whims of huge corporate seed companies. It also may save you the expense of buying hybrid seed year after year, and it's a much cheaper way to amuse yourself and enjoy the natural world than going on a vacation away from home.

We all crawl before we walk, so begin your breeding venture with plants that are easy to work with. For starters, try some annual flowers, such as snapdragon, morning-glory,

cosmos, lobelia, marigold, and pansy, and, in the vegetable garden, squash and cucumber. In your home breeding program, consciously or not, you will be selecting for plants with traits that are most appropriate for your garden. In other words, while you may manipulate genotype, you actually make your selections based upon phenotype: that is, the interaction of the genetic makeup of your plants with your local environment. You are developing unique strains of plants that do best in your local conditions. These are called *landraces*. These plants may not grow nearly as well for your relatives who live a hundred miles away or for friends in other states because they garden in different environments. But one or more of those varieties may become your own personal "heirloom" crop that you can share with your gardening neighbors and pass along to your children and grandchildren. By convention, an heirloom variety is at least 50 years old, is open-pollinated, and often (but not always) has an interesting history attached to it.

Plant-Breeding Basics

It's pretty simple and very exciting to breed your own varieties as long as you've learned how to identify the sexual parts of flowers (if you need a review of flower structure, see pages 13 and 14). Identifying those structures is crucial because you'll need to control the pollination process very closely. You'll be identifying particular plants to serve as the seed source (the "mother," or female plant) and the pollen source (the "father," or male plant). And for plants that have perfect flowers, you'll need to remove the anthers from flowers on your seed plants to ensure that they are pollinated only by the desired pollen-donor plant.

You will also need to check on whether the plant you want to breed is naturally cross-pollinated or naturally self-pollinated. Naturally self-pollinated plants, such as beans and peas, have a limited hereditary makeup. Other types of plants are typically self-pollinated but can be cross-pollinated as well. And among woody plants, self-pollination is not commonly found in nature but may occur in landscape specimens where only a single plant of the species is present in an area. A self-pollinated variety of a plant is referred to as a *line*.

Some plants are self-sterile and therefore must be cross-pollinated. Self-sterility in its broadest sense may be the result of a biochemical self-incompatibility that prevents self-fertilization or it may be a physical incompatibility. For example, the pollen may ripen before or after the stigma is receptive, and hence self-fertilization cannot take place. This is called *dichogamy*. The type of dichogamy wherein the pollen ripens before stigma receptivity is called *protandry*, and when the stigma is receptive before the pollen ripens, *protogeny*. Dichogamous plants are perhaps the easiest to work with in breeding projects, because you don't have to worry about selfing.

One of the first steps is to select the flower to be used as the female parent, or seed parent, and the one to be used as the male parent.

Guarding Against Self-Pollination

If a plant is dioecious, with male and female flowers on separate plants, then simply bag the pistillate flowers or cage the entire female plant to prevent pollination by a source other than the designated male parent. On monoecious plants that bear separate male and female flowers, bag the pistillate flowers. When monoecious plants have perfect flowers, then things become a little more involved (and for many gardeners, this is the fun part).

PREPARING THE FEMALE PARENT FLOWER

When a plant has perfect flowers, you'll need to emasculate the female parent flowers before they open to guard against self-pollination. This is done by removing the stamens. Of course, this is not necessary with self-sterile flowers, but it's good insurance nevertheless, and it's pretty simple. Ultimately, you render the stamens nonreproductive. Here are the basic methods for emasculating flowers:

- If the stamens are *exserted* (the anthers protrude beyond the corolla), then simply pinch them off using your fingers or a pair of tweezers.
- If the anthers are set down (recessed) into the corolla, try first to pinch them off with sharpened tweezers. If you're like us, you may have to use a magnifying glass to better see

the stamens as you do this. To be sure you don't inadvertently transfer unwanted pollen between flowers, dip the tweezers into alcohol each time you pinch out a stamen or an anther.

- If you cannot remove the anthers from an intact flower successfully, carefully remove the corolla and then try again to pinch off the stamens. Sometimes instead of removing the entire corolla, it is sufficient to make a longitudinal cut in the corolla and fold it back. Then, after removing the anthers, you can fold the corolla back

▲ Sharp-pointed tweezers work well for removing stamens from a perfect flower to prevent self-pollination.

into place and fasten it over the stigma with tape to prevent cross-pollination.

- In the flowers of some plant families, the stamens are attached deep inside the corolla and not easily reachable. In this case, you will have to remove the entire corolla, along with the anthers, then bag the remaining part of the flower to prevent cross-pollination.

Making the Cross

Once you have emasculated the flowers of the seed parent, enclose those of both the seed and pollen parents in paper or cloth bags to guard against contamination by foreign pollen. Remove the bags daily and examine the stigmas. Once they have become receptive — that is, moist and sticky — go ahead and hand-pollinate by dusting the stigmas with pollen from your selected plants. With a fine camel-hair brush, daub some of the pollen from your selected male sources onto the stigmas of the emasculated female flowers. *Note*: Don't use a brush with synthetic bristles because pollen won't adhere well to them. You can use inexpensive watercolor brushes, as long as they are made of natural bristles.

Instead of using a camel-hair brush, some folks like to smear pollen on their hand, then rub their hand over the inflorescences of the female flowers. This must be done carefully so as not to damage the tiny pistils. This works quite well with members of the carrot family. Always wash your hands well, and swab them in alcohol to remove the pollen from one source before you continue to pollinate with another source. Or you can pinch off the stamens of the male flowers and shake them onto the stigmas of the female flowers, or even shake the entire flower or inflorescence that is your pollen source onto the female flowers. Do what works best for your situation.

Whatever hand-pollinating technique you've used, be sure to rebag or otherwise isolate the female flowers to continue protecting them against contamination. Make a note right on the bags or in your journal of when you pollinated the flowers. Once the fruits have set (that is, once you see noticeable swelling of the ovary and the abscission of the stigma), you can remove the bags. When the fruits have ripened, collect them and extract and clean the seeds, and sow them for the next generation.

Plant-Breeding Hints by Family

Following are some special hints for breeding plants in different vegetable and flower families. We have not included instructions for many of the woody-plant families, since it is unlikely you will be interested in breeding, for instance, beech and coconut trees. Nevertheless, the basic techniques remain the same and can be applied to many families. Corn is a special case, and we have covered corn breeding both in chapter 2 and in the Corn entry in part 2. The general descriptions that follow may not pertain to all species or genera within a family.

APIACEAE. The flowers are borne in simple or compound umbels or, rarely, in a head and are perfect or, rarely, imperfect. The five stamens are typically well separated from the style and are easy to remove.

ASTERACEAE. In this large family, what we gardeners call the "flower" is actually a group of small flowers, or florets, packed together into a head. Some of the florets are perfect, some are pistillate. This family includes zinnia, aster, cosmos, marigold, and dahlia.

There are at least four types of zinnia flowers. In the most primitive, "medicine-hat" type, the head has a single row of pistillate ray flowers, while the disk flowers are all perfect. The disk flowers form the central portion of the flower head, and the ray flowers are arranged around the periphery. In extremely double zinnias, all the flowers are pistillate and hence cannot be a source of pollen. A third type has several rows of pistillate flowers and a few perfect flowers, whereas the fourth type has far fewer perfect flowers and closely resembles the extreme double.

Aster flowers have both pistillate ray flowers and perfect disk flowers as well. Snip out all the tiny perfect flowers to effect emasculation. The marigold flower is similar to that of the aster.

Cosmos has many flower types as well, but all florets are perfect, which means emasculation becomes paramount. You must remove the corolla tube very carefully. You can make a simple instrument to do this by hammering flat the head of a pin and then inserting the pin into a wooden handle. Then, carefully use the head to "cut away" the corolla tube and stamens.

Dahlia also has a number of flower types, from ordinary single (with heads similar to the medicine-hat zinnia) to completely double, which bears only pistillate flowers. However, many garden dahlias are self-sterile, which means you don't have to emasculate the flowers. That makes things a bit simpler.

BRASSICACEAE. The flowers are generally small and perfect, with the anthers easy to see against the four petals. There are usually six stamens, two of which are shorter and inserted lower than the other four. Be careful not to miss these during emasculation.

CAPRIFOLIACEAE. Flowers in this family are usually perfect flowers. Each blossom typically has five stamens that are exserted above the style. The flowers are easily emasculated.

CARYOPHYLLACEAE. This family has a large range of flower types. Flowers usually have 8 to 10 exserted stamens. Emasculation is fairly simple and straightforward in the single-flowered types.

CUCURBITACEAE. Plants in this family are monoecious with imperfect flowers, for the most part. The female flower can easily be identified by the ovoid ovary at its base. The flowers are often large and easy to work with. In fact, the corollas are so large they can be taped together to prevent unwanted pollen from landing on the stigma. The male flowers usually have three stamens but may have as few as one or as many as five.

FABACEAE. These usually perfect flowers are difficult to emasculate without causing damage to the style and stigma. The anthers, usually 10 but sometimes more or fewer, are either all separate or nine are fused at their base, with the uppermost remaining separate. The fused anthers form a closed column around most of the ovary; they branch out above it to surround the stigma. Removal of this column of anthers frequently damages the tender and brittle female parts of the flower. If you want to breed plants in this family, you will have to either take special care in reaching into the corolla with tweezers or else slice gently through the corolla, fold it back, remove the anthers, pollinate the stigma, and tape the corolla back into place.

GERANIACEAE. Flowers in this family are perfect. The 5, 10, or 15 stamens, some sometimes lacking anthers, are often separated from the style and are easily accessed.

IRIDACEAE. This family of monocots, which includes irises, has perfect flowers with three or six large stamens that are easily removed without damage to the style.

LAMIACEAE. Flowers of species in this family, which includes salvias and mints, often have an unusual arrangement in that the four (rarely two) stamens are held out from the flower wall by a sort of brace and the stigma protrudes beyond them. Emasculation should be easy.

LILIACEAE. Large anthers and strongly exserted, or plumelike, styles are common in flowers in this family, making emasculation easy. There are usually six stamens, sometimes just three, arranged in the perfect flowers.

MALVACEAE. Flowers in this family are usually perfect, with many stamens often inserted below the stigmas and around the style. Be careful you do not damage the style during emasculation. To reduce the risk of injuring the style, slice open the corolla before emasculating. Then tape up the corolla or remove it completely.

POACEAE. Most flowers in this family are perfect; corn is the exception. Corn plants are monoecious with imperfect flowers, the pistillate flowers being the silk. Each strand of silk is actually a style connected to an ovule, which matures into a kernel. If that strand of silk is not pollinated effectively, no kernel develops. The tassel is composed of the staminate flowers, which shed a profuse amount of pollen. See page 30 for detailed instructions on hand-pollinating corn.

POLEMONIACEAE. Flowers in this family, which includes phlox, have anthers fastened to the inside of the corolla tube. To emasculate them, you must remove the corolla tube. Do this soon after color has begun to show in the petals but before the small buds open. Give the corolla a slight tug, and it should separate from the ovary, leaving the style intact.

RANUNCULACEAE. Numerous anthers fit closely around the stigmas of the compound ovary in these perfect flowers and can be removed fairly easily; larkspur is an example. Just don't miss any!

ROSACEAE. Flowers in this family vary tremendously from single to double forms. To emasculate the more complex forms, clip away as many of the petals as necessary to expose the anthers, which can then be easily removed.

SCROPHULARIACEAE. The anthers of the perfect flowers in this family do not shed pollen until a flower is fairly large. The corolla can be easily removed before pollen starts to shed without damaging the style or ovary. Snapdragon, spurred snapdragon, and monkeyflower are a few of the plants in this large family that you might have in your garden.

SOLANACEAE. Most flowers are perfect in this family, similar to those in the Convolvulaceae, and so are easy to work with. However, some are a bit different. The five stamens are greatly reduced or absent in the extreme double types of flowers, as in some petunias, making those flowers useful only as seed parents.

VERBENACEAE. The flowers in this family are usually perfect and arranged in modified cymes, racemes, or spikes. The normally four stamens (rarely two or five) are easily excised in some genera, such as chaste tree, but are fused to the inside of the corolla tube in others, such as *Verbena bipinnatifida* and common lantana (*Lantana camara*). To emasculate a flower, you can simply reach down into the tube with tweezers and pinch out the anthers. Or you can split the tube, remove the anthers, then tape the tube back up.

Evaluating Your Results

If your emasculation and hand-pollination work is successful, the plants will set fruit. However, the results of crossing usually do not show up in the fruit of the parental generation because the fruits are composed of maternal material.

SELF-FRUITFUL OR NOT?

IF YOU DECIDE TO VENTURE into the world of breeding woody plants, you'll need to become familiar with the varieties you're growing and whether or not they're self-fruitful.

A variety is said to be self-fruitful if it will set fruit by its own pollen. This is not exactly the same thing as being self-pollinating. Flowers of a particular variety may be able to be pollinated by their own pollen, but no fruit may result. Hence, it is self-pollinating but not self-fruitful.

'Earliblue' is a highbush blueberry variety that can set fruit with its own pollen. It is a self-fruitful variety. 'Delicious' apple does not set fruit by its own pollen and is said to be self-unfruitful or self-incompatible. You can use pollen from a 'McIntosh' apple tree to pollinate a 'Delicious' apple tree and successfully have fruit set. This makes 'Delicious' and 'McIntosh' apples cross-fruitful or cross-compatible. Almost all commercial sweet cherry varieties are self-unfruitful, and only certain varieties fertilize other varieties, leading to a great deal of cross-incompatibilities. Checking on compatibility of cultivars is important even if you aren't planning on doing any home plant breeding, because if you plant two varieties that are incompatible, you won't end up with anything to harvest for eating!

The seeds in those fruits, however, will reflect the results of the cross, because the seeds contain genetic material from both parents. If you cross a 'Sugar' pumpkin with a 'Zucchini Elite' squash, for example, the fruits of the squash and the pumpkins will look entirely normal, since they are ripened ovaries and are, genetically, maternal tissues. The results of the cross are in the seeds and will appear the following year, when those seeds are planted.

Corn provides a slightly different example. Let's say you pollinate the silk of a white sweet corn variety with pollen from a yellow variety. Yellow color is the dominant trait. The kernels on the white ears that were pollinated by pollen from yellow corn will be yellow, and those pollinated by pollen from a white corn will be white. Usually, the indiscriminate mixing of pollen in the garden from white and yellow corn results in an ear with bicolor kernels. The color is noticeable in the year of pollination. If you were to pollinate the white variety with pollen from another white-kerneled variety only, you would end up with all white kernels. So in corn, you can actually see in the same year color changes in the kernels affected by the cross.

Other genetic changes that cannot be seen, such as changes in sugar manufacture and tenderness, also occur, but you will have to plant and grow out the seeds from the cross the following year to express some of those changes.

In most cases, though, you will have to harvest the fruits and extract and plant the seeds to fully evaluate the results of your crosses. The plants that grow from those seeds are called the first filial generation, which is commonly abbreviated F_1.

Continuing the Selection Process

Let's say you sow the seeds from the fruit that you "created" above and grow out a set of seedlings. You'll select the best of those plants for making the next round of crosses. You could continue cross-pollinating, but ultimately, it may be more productive to try selfing the plants or backcrossing instead.

SELFING

Selfing is simply deliberately using the pollen from a plant to pollinate itself; you'd place the pollen on the stigmas of the same plant. Selfing tends to increase *homozygosity* (that is, "sameness") and decrease genetic diversity. Breeders use this technique to enhance certain traits in a population. For example, let's say you are selecting for red flower color in geraniums. The plants may have the genetic makeup to produce red, pink, or white flowers. By rogueing the plants with pink or white flowers in each generation and constantly selfing only the red-flowered plants, you will eventually end up with geranium plants that produce red flowers only. Unfortunately, repeated selfing, while it stabilizes some desirable traits, also tends to decrease vigor in each succeeding generation. Thus, you might end up with puny red-flowered geranium plants.

Sometimes you'll need to self your parent plants for several generations before you make a controlled cross between the two plants. This is true for plants that are naturally cross-pollinating because of their innate genetic diversity. Working through several generations of selfing serves to "purify" the plants genetically, more or less stabilizing the trait you want to pass to the F_1 generation. This is not necessary with self-pollinating plants, but you can do it anyway, just to be sure.

BACKCROSSING

You may want to try the backcrossing technique if most of the plants in your F_1 generation are disappointing. Take flowers from the F_1 or even later generations (F_2, F_3, etc.) and cross them back to the original parent plant with the bright red flowers. When this technique is repeated for several generations, it will serve to enhance the trait for which you are selecting while also increasing genetic diversity (heterozygosity) a bit.

Breeding Hybrids

Many people object to the high cost of hybrid seeds, but they're expensive because of the labor required for the hand-pollination involved in their production. To make this clear, we'll walk you through a simplified example of how a hybrid variety is developed.

Developing an Inbred Parent

Let's say you find a sweet corn plant in your garden that produces very large ears. You decide you would like to select for that trait, with the goal of creating a hybrid variety that consistently produces large ears. You probably remember from biology class that animals and plants contain two alleles for every genetically determined characteristic: one allele contributed by the maternal parent and one contributed by the paternal parent. The genes for ear size can exist in two alleles: one form results in large ears, the other form in small ears. Let's use the letter *E* to represent the gene for ear size. And we'll use a capital *E* to represent the allele for large ears and a small *e* for the allele for small ears. Let's say further that *E* is a dominant trait, so whenever a plant contains the *E* allele, it will produce large ears.

Now let's return to our example of the corn plant in your garden that produced very large ears. Your first task as a breeder is to reduce the chance variation in this corn plant by selfing, which we also call *inbreeding*, for several generations, and selecting each generation for homozygosity. We know that inbreeding for six generations results in about 95 percent sameness in the offspring. In the case of your corn plant, that will mean large ears.

A corn plant can have four possible genetic combinations related to ear size: *EE, Ee, eE,* and *ee. EE* and *ee* are both homozygous. *Ee* and *eE* are heterozygous. An *EE* plant would have large ears, and an *ee* plant would have small ears. Heterozygous plants (*Ee* and *eE*) would have large ears, because large ears (*E*) are the dominant trait.

Your large-eared corn plant cannot be an *ee* plant, because it does not have small ears. Your large-eared plant must be *EE, Ee,* or *eE,* but you don't know which one yet. To find out, you would start selfing this corn plant, and in each generation you would rogue out all the plants that

produce small ears (go ahead and enjoy eating those ears, though!) and self the plants that produce large ears. Fairly soon, you'll be able to determine the genetic makeup of the parent plant; here's why.

EE × EE: If the genetic makeup is *EE*, then all the offspring of the very first selfing would have large ears. If this is the case, continue to inbreed for several generations, selecting only kernels from the *largest* ears for planting

WHY TRAITS REVERT

A BRIGHT RED PEPPER FRUIT on a hybrid pepper plant carries genes not only for red fruit color, but also for yellow fruit color. The yellow fruit gene just isn't being expressed on that plant. This phenomenon is true for all varieties of plants: they contain a large number of genes that produce traits that are not expressed.

If you save the seeds from that bright red pepper and plant them, some of the resulting offspring will be plants that bear red fruit, but others will bear yellow fruit. In fact, in the second generation (F_2) in a large-enough population, one in four plants will bear yellow, or even orange, fruit. And the offspring of the F_2 generation will produce yellow, orange, and red fruit.

By the time you reach a generation somewhere between the sixth (F_6) and the tenth (F_{10}) , you will no longer be able to predict what color fruit the plants will bear. The offspring will produce all the variations that are found in wild types. In other words, heterozygosity (diversity) increases in each successive generation: you will have lost the original set of traits you worked so hard to select. This is sometimes called *reversion of traits*. Because of this tendency to reversion in each generation from which you save seeds, you must be vigilant to select only plants with the desired traits. Do not save seeds from any off-type, no matter how small the variation seems to you!

In addition to genetic segregation, natural crossing and mutations contribute to reversion of traits. In all cases, no matter the cause, when we lose the traits of the original variety, we say that the variety has "run out."

So be vigilant and rogue off-types whenever you find them to keep your selection as genetically pure as possible. This task is easier in naturally self-pollinated crops such as tomato than in naturally cross-pollinated crops such as pepper and corn. If you are losing the desired traits in your variety, then reselect the desired parents and self the parental lines for 6 to 10 generations. That will give you more than 95 percent homozygosity (genetic purity), and then you can make your hybrid cross once again.

each time. If you don't see any variation from year to year in large ear size, then you have a population of corn that is homozygous for large ears and will always produce them.

***EE* OR *ee*:** If your corn is heterozygous for ear size, then one-quarter of the plants that result from the first selfing will produce small ears. Here's how we figure that: The combinations from the unions will be *E* or *e* × *e* or *E* = *EE, Ee, eE,* and *ee.* The plants that have the genetic combinations *EE, Ee,* and *eE* will have large ears, so you will have a ratio of 3:1 — that is, three plants that produce large ears to one that produces small ears. Discard or eat the seeds from the small-eared plants, and plant only the seeds from plants that produce large ears. The seeds with *EE* genotype will continue to produce all large-eared plants through the generations, while seeds with the *eE* or *Ee* genotype will continue to produce three large-eared plants to one

small-eared plant. Continue to rogue out the small-eared plants, and eventually you will get rid of the small-eared (*ee*) types and all, or almost all, of your corn will produce large ears. You will then have an inbred line.

Two Parents Needed!

Now, to make a hybrid, you'll need two inbred parents, not just one. So let's say that at the same time you decided to select for large ears, you also found a corn plant in your garden that produced very sweet kernels but had small ears. Let's assume the genetic makeup for sugar content of that corn was *SS, sS,* or *Ss,* with a large *S* allele giving high-sugar kernels and a small *s* allele giving starchy kernels. You inbreed that corn, selecting in each generation for high sugar (sweetness), and eventually you have corn plants that mostly have high sugar but small ears. Due to the inbreeding, the vigor of each generation decreases. When you are done, you have a population of large-

eared plants and another population of plants with very sweet kernels, but all the plants look pretty rundown. This is called *inbreeding depression,* and it is more common in some species than in others.

Here's the Hybrid

Now that you have your two inbred lines, you can cross-pollinate them. The seeds you collect from that cross will be F$_1$ hybrid seed. When you sow that seed and grow out the plants, they will produce ears that are both large and sweet, and the plants will show hybrid vigor, also called *heterosis;* the plants will be far more robust than those of the inbred parental types. But that hybrid vigor will be there only in the F$_1$ generation. You can't repeat your success by saving seed from the F$_1$ plants. You'd need to go back and create more inbred parents! If this sounds like a lot of work, it is, and that's why hybrid seeds are expensive.

PART 2

The
Handbook:
From Vegetables
to Nuts

From a small seed
a mighty trunk may grow.

— Aeschylus

SAVING VEGETABLE SEEDS

CHAPTER 8

SAVING VEGETABLE SEEDS is a wonderful and satisfying way to put food on your table for a minimum cost. It's not difficult to learn how to save most types of vegetable seeds. To ensure success, you'll need to pay a little extra attention to how you manage the crops from which you want to save seed. Here are some factors to keep in mind.

What Varieties to Grow

Choosing open-pollinated varieties helps to ensure consistent seed-saving results. If you are depending on the seed to feed your family, it's best not to save seed from F_1 hybrids. However, you certainly can experiment. Seeds saved from hybrids do not usually produce plants with the same attributes as the parent, but the plants may yield some very interesting, unusual, and delicious vegetables.

How Much to Plant

Some kinds of vegetables suffer greatly from inbreeding depression, some hardly at all. Check the entries in this chapter for the plants you want to save seed from. If the entry includes the category "Save seed from," your goal will be to save seed from at least that many plants when possible. We list the ideal numbers of plants, but some of these numbers are unattainable for home gardeners. For example, professional seed

producers plan to put in at least 200 corn plants, saving seed from no fewer than 100 to avoid the severe genetic depression that happens quickly in corn. That's a lot more corn than you may have room to grow! If your garden isn't large enough to plant the recommended numbers, do your best, and know that you'll have to watch for undesirable plants in the following generation.

Whether to Adjust Plant Spacing

Some vegetables, particularly biennial crops, require more space during their reproductive phase than they do when they're simply vegetative (i.e., when you're growing them to eat). If you live in an area where biennials will overwinter, thin your crop to the recommended spacing for seed production while you're harvesting for the table and leave the seed plants where they are. Folks who live in colder locations should dig the plants in the fall, store them as described in the plant entries, and replant at seed-production spacing the following spring.

What Kind of Flowers a Crop Bears

Vegetable crops may have perfect flowers or bear separate male and female flowers on the same plant. Some crops bear male flowers and female flowers on different plants. It's important to know the flowering habit of the plants you're growing and to be able to tell which flowers are male and which are female, especially if you plan to hand-pollinate.

Avoiding Unwanted Pollination

Even if you grow an open-pollinated variety, you can still end up with undesirable results if one variety pollinates another. Some species, such as beans and peas, are self-pollinating, and others, though typically self-pollinated, can be cross-pollinated by insects. Self-sterility can occur as well. The seeds we eat, such as beans, peas, and corn, may be inedible if undesired crosses occur. And in the case of squash-family plants, one type of crop can pollinate another. If you allow your butternut squash plants to be pollinated with pumpkin pollen, seeds you save from those butternut squash won't produce butternut the following year.

With crops that aren't usually grown to the seed stage, such as kale and collards, cross-pollination is less of a concern, because it's unlikely that there will be any kale or collards flowering in your neighborhood other than the plants you're raising.

YOU CAN'T SAVE SEEDS FROM THESE

SAVING VEGETABLE-CROP SEED is generally easy and very productive, but there are a few crops that just won't cooperate with seed savers.

"All-male" asparagus. Asparagus produces male flowers and female flowers on separate plants, and varieties such as 'Jersey Giant' have been bred to produce male plants only. That means no seeds! See the Asparagus entry for more details on this.

Garlic. Garlic (*Allium sativum*) and elephant garlic (*A. sativum* var. *ophioscorodon*) are grown by planting individual cloves; seeds are not produced reliably. Pull and cure garlic bulbs out of direct sunlight for 2 to 3 weeks, then select the bulbs that appear to have the largest outside cloves to plant for the next crop. Garlic cloves have distinct polarity; replant them right-side up.

Horseradish. Technically speaking, you can save seed from horseradish (*Armoracia rusticana*), but in many locations the plants will not flower at all. When they do, they produce a 2- to 3-foot flower stalk with small, white blooms. However, seed probably will not form or mature well. Horseradish is usually grown from root cuttings, and that's your best bet.

Jerusalem artichoke. Also known as girasole or sunchoke, Jerusalem artichoke (*Helianthus tuberosus*) is a perennial in the sunflower family (Asteraceae) that flowers but doesn't form any viable seed. No worries about propagating it, though. It's typically grown from tubers, which produce so prolifically that the crop can become a weed.

Multiplier onions, shallots, and Egyptian onions. These crops are related to the common garden onion, but unlike regular onions, they aren't propagated from seeds. Egyptian onions (*Allium cepa* [Proliferum group]), also called topset or walking onions, are propagated by bulbils that form in the inflorescence. Multiplier onions and shallots (*A. cepa* [Aggregatum group]) are usually propagated by shoots of the underground bulb or lateral bulb, because their seeds are often sterile.

Sweet potato. Sweet potatoes (*Ipomoea batatas*) are propagated from shoots or tubers, not from seeds. The plants do produce seed, but growing plants from seed is useful only for breeding purposes, not for producing a harvestable crop. "Yam" is a common name used for the type of sweet potato that is soft, sweet, moist, and darker orange, but true yams are *Dioscorea* spp. and belong to a different plant family from that of sweet potatoes.

In the plant entries, check the Isolation Requirement category to learn how to avoid unwanted pollination. Bagging and caging are often good techniques to use. Isolation can also be accomplished by planting different varieties of a plant at different times of the year or by planting varieties with different days to maturity. We also list recommended isolation distances to prevent cross-pollination. Keep in mind that isolation distances vary by authority and that this information is more important for market gardeners and farmers than for home gardens. Maintaining isolation distances of a half mile or more is unrealistic in neighborhoods with small yards and lots of gardens. See chapter 2 for more information on isolation distance.

Managing Biennials & Perennials

As mentioned above, to harvest seed from biennial crops, you'll need to overwinter them. Consult the plant entries for these overwintering instructions. Also refer to page 34 for a description of the technique for overwintering biennials.

Using the Right Seed-Collection Technique

Some types of seedpods are easy to collect, but other types may burst open and shed their seeds before collection. Seed heads of certain crops, such as carrots, shatter easily. You may need to bag these crops to avoid seed loss. With other crops, such as tomatoes and eggplants, you'll have to separate the seeds from the soft flesh. Each plant entry covers the recommended procedures for collecting seeds.

Once you've successfully harvested and dried vegetable seed, you'll find that most types keep longest in a cool, dry, dark location in a sealed container. When you're ready to sow your saved seed from a particular crop, check the plant entry for that crop to see whether it includes a Seed Treatment category. There you'll find information on whether the seeds may benefit from a dip in water heated to a specific temperature for a given length of time to remove pathogens. You'll also find information on the right conditions for seed germination and raising transplants from your home-saved vegetable seed.

ARTICHOKE, GLOBE

Cynara scolymus

FAMILY: Asteraceae; sunflower family
PLANT TYPE: Perennial, grown as annual in some regions
SEED VIABILITY: 5–7 years
SPACING FOR SEED SAVING: 3–4'

FLOWERING: The edible part of an artichoke plant is composed of short, thick-stemmed bracts that enclose the flower parts. As the purplish blue flower opens, it strongly resembles a large thistle flower. Artichokes are insect-pollinated.

ISOLATION REQUIREMENT: Globe artichokes will cross with cardoon (which is a closely related crop grown for its edible stems rather than flower heads). Prevent crossing by bagging individual flowers. The pollen is shed about 5 days before the stigmas become receptive. Shake or tap bagged heads daily to help the pollen contact the stigmas; otherwise, pollination probably will not take place.

SEED COLLECTION: After the feathery pappus starts to show, bag the head until it's dry, then cut it from the stem. Or remove the head from the stem at that time and dry it in a warm spot until the head is brittle.

SEED CLEANING: Flail individual heads in a bag. Separate the plumes and plant material from the seed, then winnow the seed to remove any remaining plant material.

SEED TREATMENT: Soak seed overnight prior to sowing.

GERMINATION: Optimum soil temperature for germination is 70 to 75°F (21 to 24°C). If seeds have not germinated in 2 weeks, try presoaking additional seeds for an additional 48 hours, then stratify at 35 to 40°F (2 to 4°C) for 4 weeks in moistened sphagnum moss (for aeration) that has not been shredded, then return them to 70°F (21°C) for germination.

A SEED-SAVING ADVENTURE

ARTICHOKES ARE DECIDEDLY delicious, but they're not the most practical choice for seed saving. For one thing, artichoke is a cool-season plant that needs a long growing period. When grown as an annual in the North, artichokes sometimes don't mature to the table-ready stage, let alone ripen seed. One way to overcome this is to dig up the roots in the fall and store them in a root cellar during the winter, then replant them in the spring (for more details on how to do this, see Special Handling for Biennials, page 34).

Also, artichokes grown from seed express tremendous variability. Thus, professional artichoke growers use crown divisions or harvest shoots from desirable plants rather than save seed. If you save seeds from your artichoke plants, view it as an adventure.

TRANSPLANTING: Sow seeds 6 to 8 weeks before setting to the garden. Transplant globe artichoke after all danger of frost has passed and air temperatures are not dropping below 60°F (16°C). Some authorities recommend vernalization of transplants for 2 to 4 weeks at about 40°F (4°C).

ARUGULA

Eruca vesicaria ssp. *sativa*

Arugula, gargeer, roca, rocket, roquette, rucola, rugula

FAMILY: Brassicaceae; mustard family
PLANT TYPE: Annual
SEED VIABILITY: 3 to 4 years

FLOWERING: Arugula bears white or yellow, perfect flowers that are primarily cross-pollinated.

ISOLATION REQUIREMENT: Provide as much isolation as you can since this plant is insect-pollinated. Although there are a few varieties that will cross, most gardeners do not allow their arugula to go to seed, nor does arugula cross with any other species in the Brassicaceae, so it is unlikely there will be much chance crossing in your garden. Recommended isolation distance is ½ mile.

SEED COLLECTION: The fruit is an oblong silique with two rows of seeds in each half.

Harvest the fruits when they are brown and dry but before they open, then dry them further for 1 to 2 weeks in a warm, dry place. If necessary, you can pull entire plants by the roots and hang them upside down to dry further. The pods will split open when dry, so be careful not to let seeds drop during collection.

SEED CLEANING AND STORAGE: After the pods are fully cured, flail them in a pillowcase or burlap bag, or strike the plants against the inside surface of a garbage can. Take care not to break the seed coats, because damaged seeds will produce inferior plants or may not germinate at all. Winnow, then store the clean, dry seeds in paper bags or envelopes in a dark, dry area.

GERMINATION: Optimum temperature range for germination is 60 to 77°F (16 to 25°C).

TRANSPLANTING: Arugula is usually direct-seeded to the garden as a spring or fall crop.

ASPARAGUS

Asparagus officinalis

FAMILY: Liliaceae; lily family
PLANT TYPE: Perennial
SEED VIABILITY: 3 years

FLOWERING: Older, standard asparagus varieties are dioecious. Flowers are bell-shaped, whitish green, and small, with male flowers more conspicuous than female flowers. The female plants produce thicker and fewer spears than do the male plants. Some newer varieties are "all-male" clones that produce only staminate flowers and, thus, no seeds. If you can't find berries on any of your asparagus plants, chances are you're growing an all-male variety. Asparagus is insect-pollinated, and cross-pollination with wild, weedy types of asparagus may occur, so scout your neighborhood and remove any wild plants you see.

ISOLATION REQUIREMENT: Few varieties of asparagus are available to home gardeners, so isolation, other than from weedy plants, is usually not needed. Recommended isolation distance is ¼ mile.

SEED COLLECTION: The fruit is a small red berry appearing on female plants. When they start to dry, cut the ferns with berries and allow them to dry for a few days in the field if the weather permits. Otherwise, dry them indoors. Keep in mind, though, that to maintain a strong, healthy asparagus stand, it's important to allow the ferns to die back naturally each year. Another method of collecting seed is simply to pick individual berries prior to the first frost and leave the fronds standing to die back.

SEED CLEANING: Here are two ways to clean asparagus seeds. 1. Soak berries and remove the "skins." Place the seeds on a paper towel to cure and dry for a week. 2. Crush berries in a paper bag or rub them over a screen, then place them in a bowl to wash. The skin, pulp, and nonviable seeds will float and should

MALE ASPARAGUS FRONDS

be discarded. Dry the seeds as above before storing them.

GERMINATION: Germination is erratic. Optimal soil temperature for germination is between 70 and 80°F (21 and 27°C).

TRANSPLANTING: Sow seeds 6 to 8 weeks prior to setting transplants to the garden.

A BALANCING ACT

IN A SEASON WHEN YOU PLAN TO COLLECT asparagus seed, don't harvest as many spears as you usually would. Early in the season, harvest some spears from each plant, enough to determine which are the superior plants. Then mark those plants, and don't cut any more spears from them. Allow the ferns to grow so they can produce berries.

If you're eager to start harvesting asparagus as soon as possible after planting, you'll probably want to plant 2-year-old crowns rather than seeds. Starting plants from seed in the field means you'll end up waiting an extra year or two for your first harvest.

BEAN

Phaseolus vulgaris

Dry bean, French bean, frijol, green bean, haricot, kidney bean, runner bean, snap bean, string bean, wax bean

FAMILY: Fabaceae; legume family
PLANT TYPE: Annual
SEED VIABILITY: 3 years

FLOWERING: Beans bear perfect flowers that are arranged in racemes. They are primarily self-pollinated. Flowers usually open at night, with the anthers dehiscing just prior to opening. The first ovules are fertilized within 12 hours of pollination.

ISOLATION REQUIREMENT: There is little outcrossing with snap beans, but planting a tall crop between patches of different varieties may provide added protection against cross-pollination.

SEED COLLECTION: The fruit is a dehiscent legume or pod. Plants may be determinate (bush type), indeterminate (runner or pole), or semideterminate (half-runner). Determinate types will flower, fruit,

and mature over a short period. Indeterminate plants will flower and fruit over the life of the plants; thus, flowers and pods will be at various stages of development at any given time.

Pods are ready for harvest when some are dry and the remainder have turned yellow, about 6 weeks after the pods were of fresh-eating quality. Cure the pods for 1 to 2 weeks in a warm, dry place. If necessary, you can pull entire plants by the roots and hang them upside down to dry further. The pods will split open when dry, so be careful not to let seeds drop during collection.

SEED CLEANING AND STORAGE: After the pods are fully cured, shell the bean seeds. Shell small quantities by hand. Flail larger amounts in a pillowcase or burlap bag, or strike plants against the inside surface of a garbage can. Take care not to break the seed coats, because damaged seeds will produce inferior plants or may not germinate at all. Winnow, then store the clean, dry seeds in paper bags or envelopes in a dark, dry area.

GERMINATION: Optimum temperature range for germination is 60 to 85°F (16 to 29°C). Soil temperature should be at least 60°F to produce a good stand; beans planted in warmer soil emerge more quickly.

TRANSPLANTING: Beans do not transplant well and are usually direct-seeded to the garden.

SAVING SCARLET RUNNER BEANS

SCARLET RUNNER BEANS (*Phaseolus coccineus*), also called Dutch case-knife, are delicious both for fresh eating and as dried beans. But if you want to save seeds for replanting, keep in mind that unlike snap beans, scarlet runner beans require bee pollination and must either be caged or isolated from other types of beans (especially other varieties of runner beans) by at least ¼ mile to prevent unwanted cross-pollination. Seed collection, storage, and germination for scarlet runner beans are the same as for other beans.

Botanists call bean blossoms *papilionaceous* flowers: their petals are folded back and enclose the sexual parts of the flower, in the same way that butterfly wings enclose a butterfly's body. This flower structure prevents or highly reduces cross-pollination.

BEAN, FAVA (FABA)

Vicia faba

Fava bean, broad bean, horse bean, English bean, European bean, Windsor bean, field bean, tick bean

FAMILY: Fabaceae; legume family
PLANT TYPE: Annual
SEED VIABILITY: 3 years

Fava beans require similar treatment to that of snap beans for seed production and collection. Refer to the Bean entry for details. See below for some key points to remember when you're saving seeds from fava bean plants.

FLOWERING: These perfect-flowered plants are usually self-pollinating but sometimes are also cross-pollinated by insects.

ISOLATION REQUIREMENT: Cage plants to prevent cross-pollination. Recommended isolation distance is 1 mile.

SEED CLEANING: Fava bean pods are harder than those of snap beans. Hand-shell small quantities or put into a bag and thresh.

GERMINATION: Optimal soil temperature for germination is 59 to 77°F (15 to 25°C). Seeds will not germinate in strongly acid soil.

TRANSPLANTING: Fava bean plants usually do not transplant well and are direct-seeded.

BEAN, LIMA

Phaseolus lanatus

Butter bean, civet bean, Carolina bean, sewee bean, sieva bean

FAMILY: Fabaceae; legume family
PLANT TYPE: Annual
SEED VIABILITY: Unknown

FLOWERING: These perfect-flowered plants are usually self-pollinating but sometimes are also cross-pollinated by insects.

ISOLATION REQUIREMENT: All varieties of lima beans can cross among themselves but not with other bean species. If you're growing more than one variety, or if anyone else in your neighborhood grows lima beans, plan to bag or cage plants for seed saving. Recommended isolation distance is at least 1 mile.

SEED COLLECTION: If possible, allow the seedpods to dry on the vine. Hand-pick carefully, as the pods

may shatter. You can also pick the pods when they're mature but not dry, then dry them on screens in a warm, well-ventilated location.

SEED CLEANING: These seeds are more fragile than those of snap beans. Thresh gently to avoid damaging the seed coats.

SEED TREATMENT: After cleaning, freeze the seeds for 2 days if bean weevil is a problem in your area. This will destroy any weevils that have infested the seeds.

GERMINATION: Lima beans require warm soil temperatures for germination. Optimal soil temperature range is 77 to 86°F (25 to 30°C); minimum soil temperature for germination is about 61°F (16°C).

TRANSPLANTING: Sow seeds in place; bean seedlings do not transplant well.

───────── CROP ALERT ─────────

Lima beans need a long, warm growing season to produce a good crop. In short-season locations, they seldom ripen enough for eating, let alone saving seeds for replanting. If you're a northern gardener, consider this crop a seed-saving challenge.

BEET

Beta vulgaris (Crassa group)

FAMILY: Chenopodiaceae; goosefoot family
PLANT TYPE: Biennial
SEED VIABILITY: 4 years
SPACING FOR SEED SAVING: 18"
SAVE SEED FROM: At least 6 plants

FLOWERING: Beet is a biennial crop that flowers after exposure to a cool period. The flowers are perfect and held on panicles. They are wind-pollinated.

ISOLATION REQUIREMENT: Beet varieties will cross-pollinate one another, and beets can also be fertilized by sugar beet and Swiss chard pollen. Beet flower stalks may be bagged or caged if the plants are

OVERWINTERING BEETS

BEETS USUALLY DON'T FLOWER until they've been exposed to winter cold. Thus, saving seeds from beets requires two growing seasons. Here's a rundown on how to overwinter the roots. Be sure you save enough roots to yield at least 6 plants as parent stock to avoid the problem of inbreeding depression.

YEAR 1

In cold regions. Beet roots are not greatly cold-hardy, so if you live in a colder part of the country, you'll have to overwinter the roots indoors and replant them the following spring. To start the process, plant beets midway through the growing season and carefully pull them when the roots are 1 to 2 inches in diameter. Avoid damaging the roots; wounded roots are susceptible to rot. Select for superior qualities such as shape, zoning, color, and size. Remove the tops to within about 1 inch of the root shoulder. Store the roots in a cool (about 40°F [4°C]), dark, damp place, such as in a box of moist sand in a basement or root cellar.

If you live in a warmer climate, plant beet seeds when you usually would for an eating crop. At harvest time, carefully dig the roots, remove the tops, select for superior qualities, and replant the roots right away where they were previously growing. A winter mulch will provide adequate protection. For insurance, it's a good idea to store some extra roots indoors in case your best roots don't make it through the winter in the garden.

YEAR 2

When replanting stored roots, discard any that are damaged, rotted, or shriveled. Those replanted will then resprout, flower, and set seed.

closely spaced. Be sure to put the enclosure around at least six plants to ensure adequate pollination, because beets often will not self-fertilize. Beet pollen is light enough to travel for miles on the wind. Isolate uncaged varieties by ½ to 2 miles. Separate unprotected beets from flowering sugar beet and Swiss chard plants by at least 5 miles.

SEED COLLECTION: After most of the flowers have turned brown, cut the stalks at the soil line, but keep them bagged because the seed heads shatter easily. Move the entire combination to a cool, dry location to cure for 2 to 3 weeks.

SEED CLEANING: Remove the seed balls (these are actually dry, shriveled fruits) from the stalk by hand, or place the stalk in a bag and flail, then winnow to separate seed balls from plant debris.

SEED TREATMENT: Soak seeds in water overnight before planting to leach the germination inhibitor found in the seed coats.

GERMINATION: Optimum range of soil temperature for germination is 50 to 85°F (10 to 29°C); minimum soil temperature for germination is 40°F (4°C).

TRANSPLANTING: Beets are usually direct-seeded to the garden but can be started indoors and transplanted if desired.

BROCCOLI

Brassica oleracea (Italica group)

FAMILY: Brassicaceae; mustard family
PLANT TYPE: Biennial
SEED VIABILITY: 3 years

Broccoli is closely related to cabbage and requires similar treatment to that of cabbage for seed production and collection. Refer to the Cabbage entry for details, but note the following key differances when saving broccoli seed.

Do not harvest any heads from broccoli plants intended for seed production. After all, broccoli heads are made up of flower-bud clusters. If you eat the flowers, there can be no seeds.

FLOWERING AND SEED PRODUCTION: Although broccoli is biennial, spring-planted early broccoli varieties may produce seeds in the fall of the first year of growth. Most main-crop and late-season broccoli varieties, however, will produce mainly abortive flowers and few seeds in the first year. If you have selected main-crop or late-season varieties and live in a cold locale, start seeds early indoors and transplant seedlings to the garden. Do not harvest the flower heads, and a few may produce some seed in the fall. Dig the plants late in the fall and store them between 32 and 40°F (0 and 4°C) and at about 90 percent relative humidity. Rotting may be excessive. Replant the plants in early spring, and the plants will produce tall seed stalks with abundant seeds in summer.

In areas where winter low temperatures remain above about 28°F (−2°C), direct-sow broccoli seeds in the garden in early fall and do not harvest the flower heads for food. The plants will overwinter, flower in the spring, and produce seeds the following summer.

SEED TREATMENT: Just prior to planting, soak broccoli seeds in water held at 122°F (50°C) for 20 minutes to destroy seedborne diseases if they have been a problem in your garden. This may also reduce germination rate.

Broccoli raab

Brassica rapa (Ruvo group)

FAMILY: Brassicaceae; mustard family
PLANT TYPE: Biennial or annual
SEED VIABILITY: 5 years

Broccoli raab is related to turnips and requires similar treatment to that of turnips for seed production.

Refer to the Turnip entry for details. Early varieties of broccoli raab sometimes set seed in the first year of growth, but midcrop and late varieties of broccoli raab are usually biennial and thus need to be overwintered for seed harvest in the second year of growth. This is similar to the procedure for turnips, but in areas where winters are cold, after you dig up plants in the fall, store them in layers in a barrel kept at about 35°F (2°C). The plants are very succulent, and rots are quite prevalent. Replant viable plants in the garden the following spring.

Brussels sprouts

Brassica oleracea (Gemmifera group)

FAMILY: Brassicaceae; mustard family
PLANT TYPE: Biennial
SEED VIABILITY: Unknown

Brussels sprouts are closely related to cabbage and require similar treatment to that of cabbage for seed production and collection. Refer to the Cabbage entry for details. See below for some key differences between Brussels sprouts and cabbage.

SEED TREATMENT: Just prior to planting, soak Brussels sprouts seeds in water held at 122°F (50°C) for 25 minutes to destroy seedborne diseases if any were present in your garden. This may reduce germination rate.

Brussels sprouts plants that you grow for seed production will still produce sprouts, and it's okay to harvest them lightly in the fall to eat. If you dig plants to store over winter, though, the sprouts will desiccate easily during winter storage. If they do dry out, the plants will be useless for spring planting. To prevent drying, store your Brussels sprouts plants in a cool, damp location such as covered with damp sand in a box.

Cabbage

Brassica oleracea (Capitata group)

Cabbage, Savoy cabbage

FAMILY: Brassicaceae; mustard family
PLANT TYPE: Biennial
SEED VIABILITY: 4 years
SPACING FOR SEED SAVING: 2'
SAVE SEED FROM: At least 6 plants

SAVOY CABBAGE HEAD

FLOWERING: Cabbage requires a cool period before plants will flower. The flowers are perfect but will not self-pollinate (self-incompatible), so be sure to plant several plants to ensure adequate fertilization and genetic diversity. The perfect flowers are insect-pollinated; the pollen is viable only for a short time.

ISOLATION REQUIREMENT: Several common vegetable crops are varietas or "groups" of *Brassica oleracea*, and they can all cross-pollinate each other. These include broccoli, cabbage, cauliflower, collards, kale,

OVERWINTERING CABBAGE

CABBAGES GENERALLY don't flower until after exposure to winter cold. In mild-winter areas, you can overwinter them in the garden, but in short-season areas, you'll need to dig the plants and store them in a protected area.

YEAR 1

In cooler areas, plant cabbages to be saved for seed later in the season than you would for eating but early enough that plants will have passed the juvenile stage before winter or cold treatment. Otherwise, they will not produce flower stalks. Plants with stems larger in diameter than a pencil or with four or five true leaves have passed the juvenile stage. In early to mid-October, dig the seed plants carefully (including the roots) and remove excess large leaves. Store the trimmed heads and stalks buried in damp peat or damp sand in a cool (35 to 45°F [2 to 7°C]), humid area, such as a root cellar, or bury them 2 feet deep in an outdoor pit and cover them with a few inches of straw overlain with several inches of soil and that covered with several inches of leaves or straw (see page 34. Maintain them in these conditions until spring planting, even though the plants need only

a month or so of cold temperatures to initiate seed-stalk development. Watch carefully for rotting, and discard any plants that develop rot.

In warmer climates with cool (40 to 50°F [4 to 10°C]) winters, harvest plants for eating selectively, so that plants left in place to overwinter for subsequent seed production are at least 2 feet apart. Mulch these plants or partially cover them with soil for winter protection.

YEAR 2

In the spring, carefully replant stored heads with the bottom of the heads touching the soil surface. Space plants at least 2 feet in all directions.

If you left plants in place in the garden, remove winter mulch as soon as you detect the first new growth on the plants.

As the plants resume growth, a flower stalk will arise from the core and emerge from the top of a cabbage head. Some folks cut an X in the top to aid the emergence of the stalk. If you try this, avoid cutting too deeply or you may damage the growing point. Staking may be necessary as the seed stalk elongates.

and kohlrabi. If you want to save seed from any of these crops, you'll need to either bag the plants or plant a tall-growing, nonrelated crop between different *Brassica* "cousins." Minimum isolation distance to prevent crossing is 1 mile.

If you use bags, be sure to put several plants in each bag, because cabbage and related crops are self-incompatible. Also, you'll need to introduce pollinators into the bags. Alternate-day bagging is an alternative, but seed production may be reduced.

SEED COLLECTION: The fruits on a cabbage plant are pods called siliques, which ripen from the bottom of the plant upward. Pods at the bottom may shatter before those at the top are ripe. Wait until most of the pods turn light brown before picking. Do not harvest green pods, as they will contain mostly nonviable, immature seeds. Alternatively, pull entire plants when most of the pods are brown and hang them upside down to cure in a warm, dry place. The siliques will split open when fully dry, so place something under the plants to catch the seeds.

SEED CLEANING: Place pods in a pillowcase or other bag and crush by flailing or walking on them. *Brassica* seeds are small enough to fall through a sieve or other screen. Sift them from the chaff, or remove chaff by winnowing.

SEED TREATMENT: Just prior to planting, soak cabbage seeds in water held at 122°F (50°C) for 25 minutes to destroy seedborne diseases if any were present in your garden. This process may reduce the germination rate slightly.

GERMINATION: The optimum soil temperature for germination is 65 to 70°F (18 to 21°C). Seeds germinate in 3 to 4 days.

TRANSPLANTING: Cabbage and its close relatives are all easy to transplant, and in shorter-season areas, transplants are a necessity. Sow seeds indoors 4 to 6 weeks prior to setting the transplants in the garden. Transplant while plants are still juvenile to avoid same-season bolting.

CABBAGE, CHINESE

Brassica rapa (Chinensis group and Pekinensis group)

Celery mustard, Chinese mustard, pak-choi (*B. rapa* [Chinensis group])

Celery cabbage, Chinese cabbage, pe-tsai (*B. rapa* [Pekinensis group])

FAMILY: Brassicaceae; mustard family
PLANT TYPE: Biennial
SEED VIABILITY: 3 years
SPACING FOR SEED SAVING: 2'
SAVE SEED FROM: At least 6 plants

FLOWERING: Chinese cabbage bears perfect flowers that are insect-pollinated. The plants require exposure to a cold period, usually winter, before they will flower. Procedure for overwintering Chinese cabbage is the same as for regular cabbage; see page 100.

ISOLATION REQUIREMENT: Chinese cabbage varieties will cross-pollinate one another, and Chinese cabbage will also cross with other *Brassica rapa* crops: some mustards, broccoli raab, and turnips. If you plan to grow more than one variety or a closely related crop for seed simultaneously, you can cage plants with introduced pollinators or use alternate-day caging. Recommended isolation distance is 1 mile.

SEED COLLECTION AND CLEANING: See the Cabbage entry for instructions.

SEED TREATMENT: Just prior to planting, soak Chinese cabbage seeds in water held at 122°F (50°C) for 20 minutes to destroy seedborne diseases.

GERMINATION: Optimum soil temperature for seed germination is 65 to 70°F (18 to 21°C).

TRANSPLANTING: Chinese cabbage is easy to transplant. Start seeds indoors 6 to 8 weeks prior to setting transplants in the garden.

Carrot

Daucus carota var. *sativus*

FAMILY: Apiaceae; parsley family
PLANT TYPE: Biennial
SEED VIABILITY: 3 years
SPACING FOR SEED SAVING: 24–30"
SAVE SEED FROM: 10 or more plants

FLOWERING: Carrot flowers are perfect and borne in compound umbels made of secondary clusters called umbellets. Carrot plants require cool temperatures to flower. The flowers are insect-pollinated.

ISOLATION REQUIREMENT: Carrot varieties will cross-pollinate, and carrots will also cross with Queen Anne's lace (*Daucus carota*), so be sure to keep your carrot plants for seed saving at a safe distance from other cultivated or wild carrots that are in bloom. Caging with introduced pollinators or alternate-day caging may be used if you plan to save seeds from more than one variety. Recommended isolation distance is 2 miles.

Carrots and other members of the Apiaceae family may be hand-pollinated.

SEED COLLECTION: The "seed" is a fruit called an indehiscent mericarp, which is paired with another to form a schizocarp. Carrot seed is small and very light. Harvest individual umbellets as they start to brown; the top one (primary) will ripen first. Or you can harvest the entire umbel when the primary and secondary umbellets are brown, and hang it upside down in a dry location to allow the rest to dry for 1 to 3 weeks, depending on your location's humidity.

If you wait to harvest until all of the umbellets are brown, the first to ripen will shatter in the garden. If you choose to leave the plant in the garden until all umbellets dry, bag the entire umbel to catch seeds as they shatter. In fact, you might cover hanging plants with bags as well to catch shattered seed. The primary and secondary umbellets usually produce the best seed. Cure from 4 to 5 days in hot, dry locales and for at least 2 weeks in more humid locations.

SEED CLEANING AND STORAGE: When stems become brittle, rub seed heads by hand to remove seeds or place the heads in a bag and flail. Screen or winnow carefully to remove plant material. You may debeard the seeds if desired, but doing it will not affect germination.

SEED TREATMENT: Prior to planting, hold carrot seeds in 122°F (50°C) water for 15 to 20 minutes to control seedborne diseases if they are an issue in your garden.

OVERWINTERING CARROTS

CARROTS REQUIRE exposure to a cold period, usually winter, to induce flowering. Here's how to manage your crop for successful flowering and seed production.

YEAR 1

In northern areas, plant carrots in May and June. Carefully dig the stecklings in mid-October and remove the tops, leaving 1 to 2 inches above the roots, then store in a cool (about 35°F [2°C]), dark, humid location, such as in damp sawdust or sand or in barrels with layers of straw between the roots.

In areas with mild winters, you can plant carrot seeds in midsummer and overwinter in place, covered with mulch. Using this method, however, does not allow you to select for desired characteristics, such as size, leaf development, and zoning. You may choose to carefully dig and select stecklings in the fall, then replant them right away where they were previously growing.

YEAR 2

If stecklings have been stored, replant sound ones in early to mid-spring slightly deeper than they were growing previously, about 2 feet apart in the row. If roots overwintered in the garden, be sure to thin them to this spacing. The plants will resume growth and send up flower stalks.

GERMINATION: Carrot seed in general has a low germination percentage. Optimum soil temperature for germination is 50 to 85°F (10 to 29°C). Seed is slow to germinate, so many home growers mix carrot seeds with radish seeds to mark the rows.

TRANSPLANTING: Carrots are usually direct-seeded to the garden, not transplanted.

CAULIFLOWER

Brassica oleracea (Botrytis group)

FAMILY: Brassicaceae; mustard family
PLANT TYPE: Biennial
SEED VIABILITY: 3 years

Cauliflower is closely related to cabbage and requires similar treatment to that of cabbage for seed production and collection. Refer to the Cabbage entry for details. We should warn you, though, that getting cauliflower to produce seeds can be tricky in cold, northern areas. Read on to learn some key differences between cauliflower and cabbage.

--------- CROP ALERT ---------

Do not harvest curds from cauliflower plants intended for seed production. Those curds are flower-bud clusters, and if you eat them, those plants will never go to seed.

FLOWERING AND SEED PRODUCTION: Like cabbage, cauliflower requires overwintering to stimulate seed-stalk production, but unlike cabbage, cauliflower plants won't keep well in a root cellar over winter. Fall planting in a cool greenhouse is an option; the plants need to be exposed to temperatures below 45°F (7°C) for at least a couple of months. The following spring, transplant outdoors to the garden.

In warmer areas, you can sow cauliflower seeds in summer and follow the same overwintering protocol as you would for cabbage.

SEED TREATMENT: Just prior to planting, soak cauliflower seeds in water held at 122°F (50°C) for 20 minutes to destroy seedborne diseases if they are

CAULIFLOWER, FLOWER-BUD CLUSTER

present in your garden. This may reduce the germination rate.

CELERY & CELERIAC

Apium graveolens

Celery (*A. graveolens* var. *dulce*)

Celeriac (*A. graveolens* var. *rapaceum*)

FAMILY: Apiaceae; parsley family
PLANT TYPE: Biennial
SEED VIABILITY: 3 years
SPACING FOR SEED SAVING: 30"
SAVE SEED FROM: 10 or more plants

FLOWERING: The flowers are perfect, small, white, and borne in compound umbels with secondary clusters called umbellets. Both celery and celeriac flower after a cool period, usually provided over the winter. Plants will bolt the first season if exposed to temperatures below 50°F (10°C) for 10 days or more but such plants will produce inferior seeds or no seeds at all. Celery and celeriac are insect-pollinated.

ISOLATION REQUIREMENT: Celery and celeriac will cross-pollinate one another. If you're growing both for seed, try hand-pollination as described in the Carrot entry for best results. Recommended isolation distance is 1 mile.

SEED COLLECTION: The "seed" produced by celery and celeriac is a mericarp, which comes from a split schizocarp (the fruit). Collect seed as you would from carrots; see the Carrot entry for details.

SEED CLEANING AND STORAGE: When the stems become brittle, rub seed heads by hand to remove seeds or place the heads in a bag and flail. Screen or winnow carefully to remove plant material.

SEED TREATMENT: Just prior to planting, soak celery seeds in water held at 118°F (48°C) for 30 minutes to destroy seedborne diseases if they are a problem in your garden.

OVERWINTERING CELERY

CELERY AND CELERIAC need exposure to cold, usually winter cold, to induce flowering. Here's how to make sure your plants survive winter.

YEAR 1

In cold areas. Overwintering the plants can be problematic in cold-winter locations. In fall, carefully dig the best plants. For celery, cut the tops back (the stems should be good for eating) and dig the roots and soil around them, making a good root-ball. Overwinter the roots and root-ball in damp sand in a cool root cellar at about 35°F (2°C). For celeriac, trim tops to 2 to 3 inches above the root crown, remove all side roots, and store the roots in damp sand at 33 to 40°F (1 to 4°C).

Where winters are mild, mulch plants left in the garden over winter. In warmer locales, start transplants in mid- to late summer and set the plants to the garden in midwinter, giving them just enough cool weather at season's end before the warm spring arrives.

YEAR 2

Remove rotted vegetation from stored roots and replant the roots to the garden, spacing them 1½ to 2 feet apart. The plants will resprout and send up seed stalks.

GERMINATION: Celery-seed germination has been much studied because of its complex patterns of thermodormancy and light requirements. (We encourage the serious celery-seed saver to read the literature for additional information.) The optimum range of soil temperature for germination is 70 to 75°F (21 to 24°C). Don't sow until soil temperature is at least 45°F (7°C); warmer is better. Seeds require light for germination unless grown under temperatures ranging from 50 to 59°F (10 to 15°C). Seeds sown in temperatures of 86°F (30°C) or higher will remain dormant.

TRANSPLANTING: Start celery seeds for transplanting 10 to 12 weeks before you expect to set them in the garden. Celery grown from transplants will have a smaller root system than those directly sown, so take care during transplanting. To increase germination, grow under diffuse light with diurnal temperature fluctuations of 85°F (29°C) during the day and 60°F (16°C) at night. In addition, keep the soil very moist. Do not allow temperatures to fall below 60°F during transplant growth to avoid bolting after setting in the garden, and do not set out the plants before the temperature has warmed sufficiently.

CHARD, SWISS

Beta vulgaris (Cicla group)

FAMILY: Chenopodiaceae; goosefoot family
PLANT TYPE: Biennial
SEED VIABILITY: 4 years
SPACING FOR SEED SAVING: 18"
SAVE SEED FROM: At least 6 plants

FLOWERING: Swiss chard is a biennial crop that flowers after a cool period. The flowers are perfect and held on panicles. Swiss chard is wind-pollinated. Its pollen is light and fine and easily carried on the wind.

ISOLATION REQUIREMENT: Chances are that none of your gardening neighbors, near or far, will have chard in bloom at the same time you do. Thus, as long as you are saving seed from only one variety of chard, you don't need to worry about keeping your chard isolated while it's flowering. If you want to

SWISS CHARD

experiment with saving seed from several varieties of chard, set up a sequence of saving a different variety each year. The seeds should remain viable for several years as long as you store them properly.

If you do decide to bag or cage your chard, put the enclosure around six or more plants to ensure adequate pollination, because chard is sometimes self-incompatible; that is, it won't always accept pollen from its own flowers.

Recommended isolation distances for chard range from ½ to 2 miles.

SEED COLLECTION: After most of the flowers have turned brown, remove the stalk and further cure in a cool, dry location for 2 to 3 weeks.

SEED CLEANING: Remove the seed balls (these are actually dry, shriveled fruits) from the stalk by hand, or place the stalk in a bag and flail. Then winnow to separate the seeds from the plant debris.

SEED TREATMENT: Soak seeds in water overnight before planting to leach the germination inhibitor found in the seed coats.

GERMINATION: Optimum range of soil temperature for germination is 50 to 85°F (10 to 29°C); the minimum temperature is 40°F (4°C). There are usually sev-

eral true seeds within each seed ball, so each seed ball you plant will produce a cluster of seedlings.

TRANSPLANTING: Swiss chard is usually direct-seeded to the garden but may be transplanted if you want.

OVERWINTERING SWISS CHARD

SWISS CHARD PLANTS usually need to survive through the winter and into a second season of growth to produce a seed stalk. The technique for overwintering varies depending on the severity of winter in your area.

YEAR 1

If you live in a cooler part of the country, plant chard midway through the growing season. Then in the fall, carefully dig up the plants, taking care not to damage the roots, as wounded roots are susceptible to rot. Select for superior qualities, such as leaf shape, petiole coloring, and size. Be sure to save more than six plants in case some don't survive the winter. Remove the tops to within an inch or two of the root. Store the roots in a box of moist sand placed in a cool (about 40°F [4°C]), dark, damp spot such as a basement or root cellar.

If you live in a warmer climate, plant chard at the time you usually would for eating. When you last harvest your chard at the end of the growing season, leave 1 to 2 inches of stem attached to the roots. Let the roots overwinter right in the garden, covering them with a good mulch.

YEAR 2

If you've overwintered roots indoors, replant them in the garden in spring. Discard any that are damaged, rotted, or shriveled. If you've overwintered roots in the garden, remove the mulch once you see new growth beginning. The plants will produce some leaves first and then send up a seed stalk. Do not cut any leaves for eating; let them stand to nourish the plant. Chances are they will be tough and bitter anyway.

CHERVIL, SALAD

Anthriscus cerefolium

FAMILY: Apiaceae; parsley family
PLANT TYPE: Annual
SEED VIABILITY: 3 years
SPACING FOR SEED SAVING: 8–12"

FLOWERING: Chervil is a 1- to 2-foot-tall annual plant with small, self-pollinating, white, perfect flowers held in compound umbels.

ISOLATION REQUIREMENT: Insect cross-pollination among varieties of chervil may occur. Use alternate-day caging to prevent this. You may also want to hand-pollinate, as described in the Carrot entry.

SEED COLLECTION: Chervil umbels shatter easily. Either bag individual heads after pollination or pull entire plants as the seeds ripen and hang them to dry over a tarp to collect fallen seeds.

SEED CLEANING AND STORAGE: Winnow excess plant material.

CHERVIL FLOWER UMBEL

GERMINATION: Optimum soil temperature for germination is 50 to 68°F (10 to 20°C). Seeds need light for germination, so surface-sow them only.

TRANSPLANTING: Chervil is typically direct-seeded to the garden.

CHICKPEA

Cicer arietinum

Chickpea, Egyptian pea, garbanzo, gram

FAMILY: Fabaceae; legume family
PLANT TYPE: Annual
SEED VIABILITY: 3 years
SPACING FOR SEED SAVING: 1–2'

FLOWERING: Chickpea is an annual plant with flowers that are perfect and self-pollinating.

ISOLATION REQUIREMENT: Some crossing of different chickpea varieties will occur due to insect activity. If more than one variety will be flowering simultaneously, cage the plants from which you want to save seed. Recommended isolation distance is ½ mile.

SEED COLLECTION: Withhold water after pods have set. Pull entire plants if you live in a damp area and hang them to dry in a warm, dry spot. Each pod will contain one or two seeds.

───────── CROP ALERT ─────────

If you have sensitive skin, wear gloves when handling chickpea plants. In some people, contact with the plant causes a rash.

SEED CLEANING: Flail and winnow as you would any type of bean (see the Bean entry).

GERMINATION: Optimum soil temperature for germination is 70 to 85°F (21 to 29°C).

TRANSPLANTING: Chickpeas are usually direct-seeded to the garden and do not transplant well.

CHICORY, WITLOOF

Cichorium intybus

Witloof chicory, Belgian endive, blue sailors, radicchio

FAMILY: Asteraceae; sunflower family

PLANT TYPE: Perennial treated as biennial

SEED VIABILITY: 4 years

SPACING FOR SEED SAVING: 8–12"

FLOWERING: Chicory has self-incompatible perfect flowers. Cross-pollination is usually accomplished by insects.

ISOLATION REQUIREMENT: Various chicories, including weedy species, will intercross, and chicory can be fertilized by endive. If you're growing more than one variety of chicory or you're concerned about crossing with nearby weedy chicory, then use alternate-day caging, or bag and hand-pollinate flowers as described in chapter 2. Recommended isolation distance is ½ mile.

SEED COLLECTION: Stop watering when most of the flowers have set, and allow plants and achenes to dry in the field. The achenes should be as firm, dry, and brown as possible before harvest.

SEED CLEANING: Place seed heads in a bag, flail, and winnow the extra plant material. Or you can leave the seed heads intact for storage and winnow just prior to planting.

GERMINATION: Optimum temperature for sowing is 68°F (20°C). If seeds do not germinate, stratify them for 4 weeks and return to 68°F.

TRANSPLANTING: Chicory is usually direct-seeded to the garden.

——————— CROP ALERT ———————

Chicory is an invasive weed in some areas.

OVERWINTERING CHICORY

FOR SEED-SAVING PURPOSES, treat your chicory as if it was a biennial.

YEAR 1

Plant as usual in the spring. During the growing season, take only a light harvest of leaves from plants intended for seed saving.

In areas with cold winters, dig plants prior to the first hard freeze. Clip small and secondary roots and trim the top to 2 to 3 inches above the roots. Store over winter in a box with damp sand.

In mild-winter areas, mulch and leave the plants in the ground over winter.

YEAR 2

If you stored plants indoors, select the best roots and replant them the following spring. If you mulched plants in the garden, remove the mulch at the first sign of growth in the spring. Plants will resprout and quickly produce seed stalks.

COLLARDS

Brassica oleracea (Acephala group)

FAMILY: Brassicaceae; mustard family

PLANT TYPE: Biennial

SEED VIABILITY: Unknown

SPACING FOR SEED SAVING: Rows 3' apart, 18–24" within a row

Collards are closely related to cabbage and require similar treatment to that of cabbage for seed production and collection. Refer to the Cabbage entry for details. The one difference in seed-saving protocol between cabbage and collards is in the timing of hot-water treatment of seeds to kill pathogens. Soak collard seeds for just 20 minutes in water held at 122°F (50°C). We also want to mention here that it's okay to take a light harvest of leaves from collard plants intended for seed production.

CORN

Zea mays

Sweet corn (*Z. mays* var. *rugosa* or *Z. mays* var. *saccharata*), popcorn (*Z. mays* var. *praecox* or *Z. mays* var. *everta*), flint corn (*Z. mays* var. *indurata*), dent corn (*Z. mays* var. *indentata*)

FAMILY: Poaceae; grass family
PLANT TYPE: Annual
SEED VIABILITY: 2 years
SAVE SEED FROM: As many plants as possible

FLOWERING: Corn plants are annuals and monoecious. The tassels are the male inflorescence. The silks are actually stigmas; they are covered with tiny hairs and are part of the female inflorescences — what we call the "ears." The silks are receptive to pollen along most of their length. The corncob is a receptacle for the ovaries, which will eventually become kernels. Tasseling occurs about 3 weeks after planting, depending on variety and geographic location.

Corn is wind-pollinated, and pollen must land on each silk to fertilize the one ovule attached to it. Poor tip fill occurs when the silks at the tip of the ear fail to protrude from the husk until after pollination has occurred and when high temperatures kill either the pollen or the silks. If there are off-type plants in your stand, either remove them completely or remove the tassels so they don't contribute pollen to your breeding efforts.

ISOLATION REQUIREMENT: Corn varieties will cross-pollinate, and sweet corn, popcorn, flint corn, and dent corn will all cross-pollinate. If you're planting a mixed stand of corn, select varieties and types that have different ripening seasons. Putting in tall plants such as sunflowers between different varieties may help prevent cross-pollination. Recommended isolation distance is 1 mile. If isolation by time or distance isn't possible, you will have to take matters in hand and assist with pollination, as described in Hand-Pollinating Corn, page 30.

Keep in mind that it's best to avoid letting corn plants self-pollinate, which would increase the inbreeding depression that is a problem with corn. One way to prevent self-pollination is by using corn pollination bags, which are readily available from gardening supply companies and through the Internet. You can use plain brown lunch bags, but plan to replace them if it rains. Water-resistant, not water-proof, bags are best.

Identify alternate rows of plants to be the pollen donors and seed producers. Bag the tassels of the female seed producers and the young ears of the male pollen donors, and allow for normal wind pollination. This method may also be used to force a cross between two varieties. If you prefer, you can leave the ears of the pollen-donor plants unbagged so that they'll be pollinated and develop to harvest for eating. But be sure you tag them to make sure you don't accidentally collect them for seed saving instead.

SEED COLLECTION: A corn seed is an indehiscent fruitlet called a *caryopsis*. Corn seeds sometimes germinate on the plant under wet conditions (vivipary).

Select ears for such desired characteristics as size, early bearing, and tolerance to drought or temperature conditions. The uppermost ear is often the largest. Ears are usually ready to be harvested for seed from 4 to 6 weeks after they have reached eating stage. In short-season locales, pick ears when the husks are brown. If seeds are mature, a light freeze shouldn't hurt them. But do harvest before a hard freeze for best seed quality. In longer-season locations, allow ears to dry on the plant. Then cut the entire stalk and shock it along with others right in the garden for a couple of weeks to be sure the ears are fully dried.

DOMINANT AND RECESSIVE TRAITS

IN CORN, the normal sugary (su), sugary enhanced (se), and supersweet (sh2) types are all recessive. In kernel color, black or blue is dominant, and all colors are dominant over white. Flint corn characteristics are dominant over sweet corn characteristics.

SWEET CORN

After harvesting, pull back the husks to expose the seed. Hang to dry further indoors in a cool location until completely dry, 2 weeks or longer. Seeds that are not completely cured may heat in storage, reducing germination percentage. Husk the ears, and place them on a screen to dry for another couple of weeks. Then, twist the ears or rub two together to release the hard, dried kernels into a bucket. For large quantities, mechanical corn shellers are available.

SEED CLEANING AND STORAGE: Winnow any plant debris. Store seeds in a cool, dark place in paper bags.

SEED TREATMENT: Many commercial producers treat corn seed with synthetic fungicides, especially when they're sowing corn seed in cold soil. If you've had problems with corn seedling emergence in the past, be sure the soil is fully warm before planting.

GERMINATION: Optimum soil temperature for germination ranges from 60 to 95°F (16 to 35°C). The minimum soil temperature for sowing corn is 50° (10°C). Do not sow too early in the season or the seed will rot in the ground.

TRANSPLANTING: Corn is not usually transplanted. If you live where the growing season is short, use peat pots or similar containers for growing the transplants so that they can be placed in the ground without disturbing the roots.

Corn is usually direct-seeded, but if you live in a short-season location and intend to save corn seed, you may have to start your seed early and transplant to the garden. Use peat pots, fiber pellets, or other containers that enable you to plant to the garden without disturbing the roots. Sow your seed 4 weeks prior to setting to the garden. Use bottom heat if possible.

Another challenge with corn is its strong tendency to inbreeding depression, a reduction in quality of the plants over time if seed is not genetically diverse enough. It's recommended that professional seed savers plant at least 200 plants, select the best, and save seed from at least 100 plants. If you save seed from a smaller sample, irreversible inbreeding will show up quickly, resulting in unsatisfactory plants.

CORN SALAD

Valerianella locusta

Corn salad, common; lamb's lettuce; mache

FAMILY: Valerianaceae; valerian family
PLANT TYPE: Annual
SEED VIABILITY: 5 years

FLOWERING: Corn salad bears inconspicuous flowers in headlike cymes that are quick to set seed. This species can self-sow and become weedy in some areas.

ISOLATION REQUIREMENT: Varieties of corn salad will cross by insect pollination among themselves and with some wild species. Use alternate-day caging if more than one variety is grown.

SEED COLLECTION: The fruit are dry and indehiscent. Pull entire plants and place them in a pillowcase. Shake to release the seeds. Let the plants dry further in the bag and repeat. Allow the seeds to further dry in a warm location for a few more days before storage.

SEED CLEANING: Winnow extra plant material.

vegetables

GERMINATION: Optimal soil temperature for germination of corn salad seed is 50 to 70°F (10 to 21°C).

TRANSPLANTING: This species is usually sown in place.

CRESS, GARDEN

Lepidium sativum

Upland cress

FAMILY: Brassicaceae; mustard family
PLANT TYPE: Annual
SEED VIABILITY: 5 years
SPACING FOR SEED SAVING: 2–3'

FLOWERING: The inconspicuous perfect white or greenish flowers are borne in racemes and are cross-pollinated.

ISOLATION REQUIREMENT: Crossing between varieties is possible due to insect visitation, but isolation information is not available.

SEED COLLECTION AND CLEANING: The fruits are short, broad silicles. Collect and clean the seeds as you would turnip seed; see the Turnip entry for details.

GERMINATION: Optimal soil temperature for germination is 50 to 60°F (10 to 16°C).

TRANSPLANTING: Cress is not transplanted.

CUCUMBER

Cucumis sativus

FAMILY: Cucurbitaceae; squash
PLANT TYPE: Annual
SEED VIABILITY: 5 years
SAVE SEED FROM: At least 6 cucumbers from 6 plants

RIPE (GREEN) AND OVERRIPE (YELLOW) CUCUMBERS

FLOWERING: Cucumber plants are usually monoecious with imperfect flowers. Male flowers appear first, and plants usually produce more male flowers than female. Though male flowers appear first in the season, female flowers are more numerous than male flowers at the end of the season. Plants may also be gynoecious (all female), andromonoecious (male and perfect flowers), androecious (all male), and hermaphroditic (all perfect flowers). Cucumber fruits may also be parthenocarpic gynoecious (seedless), and it's not possible to save seeds from such varieties, so be sure you know which sexual type you are planting.

ISOLATION REQUIREMENT: All varieties and types of cucumbers will intercross, but cucumbers will not cross with melons or squash. If you're growing more than one variety, or if your neighbors are growing cucumbers, you'll want to hand-pollinate to ensure seed purity. Follow the hand-pollination instructions on page 29, but note that cucumber vines tend to abort their fruit during drought or high temperatures, so don't attempt hand-pollinating during hot, dry spells. Recommended isolation distance is 1½ miles.

SEED COLLECTION: Select large, overripe fruits for seed saving, and allow them to ripen on or off the vine in a cool, dry place for at least 5 weeks after eating stage, until they have turned yellow, white, orange, or brown in color (depending on cucumber type) and have softened. In frosty areas of the North, let the vines winter-kill to reveal the fruits. Select what you want and store them in a cool, dry place for several more weeks to complete their ripening.

SEED CLEANING: Extract the seeds from the fruit by scooping them out and rinsing them or fermenting them as described under Pulpy Fruits, page 44.

SEED TREATMENT: Prior to planting, hold cucumber seeds in 122°F (50°C) water for 20 minutes to control seedborne diseases if they are an issue in your garden.

GERMINATION: Germination and emergence are limited by low temperatures. Do not plant before the soil has reached at least 60°F (16°F); 95°F (35°C) is the maximum optimal soil temperature for planting cucumbers. Freshly collected cucumber seeds may have a dormancy that must be broken by storage for at least 84 days.

TRANSPLANTING: Cucumbers may be transplanted to the garden for earlier production. Take care not to disturb the roots, as the root tips do not regenerate easily. Use fiber pots, fiber pellets, or other types of pots that will decompose when buried. Allow about 4 weeks to grow good transplants.

EGGPLANT

Solanum melongena var. *esculentum*

Mad apple, melongene, aubergine

FAMILY: Solanaceae; nightshade family
PLANT TYPE: Perennial grown as annual
SEED VIABILITY: 4 years
SAVE SEED FROM: At least 6 plants

FLOWERING: Eggplant flowers are perfect and primarily self-pollinating.

ISOLATION REQUIREMENT: Insect pollination of eggplant may occur, and different varieties will cross-pollinate. If cross-pollination is a concern, cage the plants from which you plan to save seeds. Recommended isolation distance is ¼ mile.

SEED COLLECTION: From a botanical standpoint, an eggplant fruit is a fleshy berry, and the majority of the flesh is placenta, in which the seeds are embedded. Seed saved from eggplants at the eating stage is usually not viable. As eggplant passes the edible stage, the skin dulls and turns brown, yellow, or orange. When fully ripe, the eggplant fruit will be hard and will fall from the plant. Hold the fruit for about 2 additional weeks before proceeding. If you have a long growing season, it's okay to harvest some fruit for eating from the plants before allowing the remainder to overripen on the plants for seed; gardeners in short-season areas will want to save the first fruits for seed to select for early ripening.

SEED CLEANING AND STORAGE: There are a few ways to separate eggplant seeds from the placental material. You can scoop the seeds with a spoon into a bowl with water and rub between your hands to separate the seeds. Or you can use a coarse grater to grate the bottom of the fruit into a bowl with water and rub between your hands to separate the seeds. Still another option is to cut the seeded part of the fruit into cubes and put them into a blender with water. Give it a few short pulses, then allow the good seeds to settle. You may need to repeat with fresh water.

Whichever method you choose, the seeds that sink will have the greatest viability. Pour off the extra plant material. Put the good seeds into a strainer and rinse, then drain and place on paper towels or screens in a warm, ventilated area to dry. Stir them daily so they don't stick together. When completely dry, store in a cool, dry location.

SEED TREATMENT: Prior to planting, hold eggplant seeds in 122°F (50°C) water for 25 minutes for control of seedborne diseases if they are an issue in your garden.

GERMINATION: Optimum range of soil temperature for germination is 80 to 90°F (27 to 32°C). Minimum soil temperature for setting out transplants is 66°F (19°C).

TRANSPLANTING: Eggplants are typically set to the garden as transplants, particularly in the North, as they require a long season. Sow seeds 8 to 10 weeks before the plants are to be planted outdoors.

ENDIVE & ESCAROLE

Cichorium endivia

FAMILY: Asteraceae; sunflower family
PLANT TYPE: Annual, sometimes biennial
SEED VIABILITY: 5 years

Endive and escarole are the same species, but they have a somewhat different appearance from each other. Endive has deeply cut and curled leaves; escarole leaves are broad and somewhat crumpled-looking. Don't confuse endive with Belgian endive, which is a different species (*Cichorium intybus*). For instructions on saving seed of Belgian endive, see the Chicory, Witloof entry.

OVERWINTERING ENDIVE AND ESCAROLE

ENDIVE AND ESCAROLE need exposure to a cold period, usually winter, to stimulate flowering.

YEAR 1
Where winters are severe, sow seeds in spring. During the growing season, it's okay to harvest a few outside leaves from each plant for eating. In the fall, dig up the plants prior to a hard frost and trim off the tops to 2 inches above the roots; also clip off small secondary roots. Store the roots in a dark location in damp sand at 33 to 40°F (1 to 4°C).

Where winters are mild, sow endive seed in fall, and mulch plants to overwinter in place.

YEAR 2
In early spring, replant stored plants in the garden; remove mulch from plants left in the ground.

FLOWERING: Flowers are perfect and mostly self-pollinating. There are 18 to 20 blossoms in each flower head. Plants need exposure to cold to stimulate flowering, and bolting may occur after a period of cold early in the season. Flowering occurs in the morning, with flowers usually closing by midday.

ISOLATION REQUIREMENT: Chicory and endive can cross-pollinate. If grown in the same garden, cage one of them. Recommended isolation distance is ½ mile, or cage with introduced pollinators. Leave cages in place until most seedpods have set.

SEED COLLECTION: The fruit is an achene. Wait for the heads to dry, then pull and dry the plants for at least 10 days.

SEED CLEANING: Flail the dry heads in a pillowcase, then winnow to remove extra plant material.

GERMINATION: Soil temperatures should be between 45 and 95°F (7 and 35°C) for best germination.

TRANSPLANTING: Endive and escarole are usually direct-seeded.

FENNEL

Foeniculum vulgare

Florence fennel, finocchio, sweet fennel

FAMILY: Apiaceae; parsley family
PLANT TYPE: Perennial grown as annual
SEED VIABILITY: 4 years
SPACING FOR SEED SAVING: 1'

FLOWERING: The tiny flowers are held in umbels and are insect-pollinated. This species can become weedy in some areas.

ISOLATION REQUIREMENT: If you'll have more than one variety of fennel in flower in your garden, cage the plants or plan to hand-pollinate as described on page 29. Recommended isolation distance is ½ mile.

SEED COLLECTION: Select seeds from late-bolting plants that have the size bulb you prefer. Use care when collecting seeds because the dry umbels shatter easily.

SEED CLEANING: Clean seeds as you would carrots; refer to the Carrot entry for details.

GERMINATION: Optimum soil temperatures for fennel seed germination are 50 to 75°F (10 to 24°C).

TRANSPLANTING: Fennel is typically direct-seeded to the garden.

KALE

Brassica oleracea (Acephala group)

FAMILY: Brassicaceae; mustard family
PLANT TYPE: Biennial
SEED VIABILITY: 4 years
SPACING FOR SEED SAVING: Rows 3' apart, with plants 18–24" within a row

OVERWINTERING: Kale is closely related to cabbage and requires similar treatment to that of cabbage for seed production and collection. Refer to the Cabbage entry for details. Kale plants to be used for seed

KALE IN FIRST SEASON OF GROWTH

production may also be harvested lightly for food. Read on to learn a couple of differences in method for saving seed from kale plants.

Since kale is the most winter-hardy of the *Brassica oleracea* crops, it will survive over winter in the garden even in cold-winter areas, as long as it has a heavy mulch or consistent snow cover. In the South, kale may be planted in the late summer or fall. Select for cold-hardiness by overwintering. Stake the seed stalk if necessary when it appears in spring.

SEED TREATMENT: Just prior to planting, soak kale seeds in water held at 122°F (50°C) for 20 minutes to destroy seedborne diseases if these are an issue in your garden. This may reduce the germination rate.

KOHLRABI

Brassica oleracea (Gongylodes group)

FAMILY: Brassicaceae; mustard family
PLANT TYPE: Biennial
SEED VIABILITY: 3 years
SPACING FOR SEED SAVING: Set plants 18 to 24 inches apart in the row

Kohlrabi is closely related to cabbage and requires similar treatment to that of cabbage for seed production and collection. Refer to the Cabbage entry for details. As with cabbage, you can't take a harvest for eating from kohlrabi plants intended for seed saving. Read on for some specific differences regarding management of kohlrabi for seed production.

OVERWINTERING: Timing of planting kohlrabi for seed production depends on where you live. In warmer locales, sow seeds in the fall and allow plants to overwinter in the garden. In cooler locations, time the sowing of seeds so the plants will not be fully mature when the first frost hits. Manage the overwintering of the plants as you would for cabbage. Replant in spring.

SEED TREATMENT: Just prior to planting, soak kohlrabi seeds in water held at 122°F (50°C) for 20 minutes to destroy seedborne diseases if they are an issue in your garden.

LEEK

Allium ampeloprasum (Porrum group)

FAMILY: Alliaceae; allium family
PLANT TYPE: Biennial
SEED VIABILITY: 2 years

FLOWERING: Leek flower heads are umbels, each comprising several thousand tiny perfect flowers. Plants flower after exposure to a cool period, and the flowers are insect-pollinated.

ISOLATION REQUIREMENT: Leek varieties will cross, but leeks will not cross with onions or with other *Allium* species. Time plantings to avoid having two different varieties flowering at the same time. Alternate-day caging and caging with introduced pollinators is also successful. Recommended isolation distance is 1 mile.

SEED COLLECTION: Pick umbels in the fall when the seeds begin to show, and dry well in a bag or on sheets in case the umbels shatter. Drying may take several months in humid locations. To avoid loss of quality, dry in a warm, well-ventilated location.

SEED CLEANING AND STORAGE: The seeds are in capsules and will require rubbing or flailing for removal. Clean by winnowing.

GERMINATION: Optimum range of soil temperature for leek germination is 70 to 75°F (21 to 24°C).

TRANSPLANTING: Transplant as you would onions; refer to the Onion entry for details.

OVERWINTERING LEEKS

LEEKS WON'T PRODUCE FLOWERS until after a cold period, usually winter. Sometimes leeks will bolt after 4 to 6 weeks of cold weather (41°F [5°C] or below), but those plants should not be used for seed production, as you will be selecting for the tendency to bolt early in the season. Here's how to help your leeks survive through the winter.

YEAR 1

In the North, dig the plants, remove the roots, and trim the tops by half. Overwinter in temperatures of 32°F (0°C) with 80 to 90 percent humidity, such as in a root cellar.

In warmer locations, plant leeks for seed production in midsummer to fall. Leave the plants in the garden and mulch them for the winter.

YEAR 2

Replant stored plants in the spring and pull back mulch from plants left in the garden. You may notice the plants developing small bulblets around the base if they were overwintered in place. You can remove these and plant them.

LENTIL

Lens culinaris

FAMILY: Fabaceae; legume family
PLANT TYPE: Annual
SEED VIABILITY: 3 years
SPACING FOR SEED SAVING: 6"

FLOWERING: Lentil is an annual plant with indiscrete, self-pollinating flowers.

ISOLATION REQUIREMENT: There is little outcrossing with lentils, but planting a tall crop between patches of different varieties may provide added protection against cross-pollination.

SEED COLLECTION: Seedpods shatter easily, so collect them with care before they dehisce.

SEED CLEANING: Shell lentils and process them as you would beans. See the Bean entry for details.

GERMINATION: Optimum soil temperature for germination ranges from 65 to 85°F (18 to 29°C).

TRANSPLANTING: Lentils are direct-seeded.

LETTUCE

Lactuca sativa

FAMILY: Asteraceae; sunflower family
PLANT TYPE: Annual
SEED VIABILITY: 6 years
SPACING FOR SEED SAVING: 1'

FLOWERING: A lettuce flower stalk may grow from 2 to 5 feet tall and usually appears during long days and/or high temperatures. The flowers are perfect and self-pollinating and form a cymose cluster of heads. The terminal flowers usually appear first, followed by the laterals. For head lettuce varieties, cut an X 1 to 2 inches deep in the head at or just prior to market maturity to facilitate flower-stalk emergence. Do not cut so deeply as to damage a stem's growing point.

ISOLATION REQUIREMENT: There is little outcrossing with lettuce, so isolating your seed-stock plants usually isn't a concern. Plant a tall crop between varieties if desired for extra protection against crossing, or separate flowering varieties by 20 feet. An isolation distance of 200 feet is needed if prickly lettuce (*L. serriola*), a common weed, is growing nearby, because it will cross with garden lettuce.

SEED COLLECTION: Lettuce fruits are achenes. Seeds begin ripening 10 to 24 days after the flowers shed their pollen. But since flowering may occur over a period of as long as 40 days, the seed heads that form first may shatter by the time flowering is complete at the bottom of the stalk.

Collect achenes only from plants that are slow to bolt, or you will inadvertently be selecting for fast-bolting lettuce. Harvest the seed heads as they begin to dry, or shake them into a bag to avoid loss from shattering. Alternatively, pull the entire plant when enough seed heads have formed to satisfy you. Hang plants upside down. Professionals harvest when the seed stalk is in 50 percent feather (the white pappus is the "feather"). Dry indoors for about a week after harvest.

SEED CLEANING: Lettuce seeds are small and light. Rub the heads in your hands or flail them, then carefully screen the plant material. Use a screen of the proper size so the seeds will fall through. This will leave you with a mixture of seeds and seed fuzz. If you want your seeds to be cleaner, select a screen size smaller than the seeds and gently rub the seeds and fuzz against the screen to separate the fuzz.

SEED TREATMENT: Lettuce seeds are dormant immediately after harvest. Hold for 2 months before planting and they will be released from dormancy. Immediately prior to planting, hold lettuce seeds in 118°F (48°C) water for 30 minutes to control seed-borne diseases if they are an issue in your garden.

GERMINATION: Lettuce seeds have thermodormancy, which inhibits germination. If they are exposed to temperatures above 77°F (25°C) for even a day, either in soil or in storage, dormancy will be induced and must be broken with a period of cold temperatures (50 to 59°F [10 to 15°C]). Light is required by some varieties for germination. The interaction of light and temperatures is complex. The range of soil temperatures for germination is 40 to 80°F (4 to 27°C), though 75°F (24°C) is optimum.

TRANSPLANTING: Lettuce is usually direct-seeded but may be raised as transplants. For an extra-early crop, sow seeds for transplants 3 to 4 weeks before you plan to set transplants to the garden.

OVERWINTERING LETTUCE

IN THE NORTH, lettuce is a spring-planted crop and seed saving occurs the same year. But in the warmer South, lettuce for seed production is planted as a fall crop, with the plants going to seed the following spring, so the plants don't need special protection over the winter. In mild regions in between, time planting of crisphead lettuce so plants are about 2 inches tall before it turns cold in the fall. Mulch after the first heavy frost.

MELON

Cucumis melo

Honeydew melon, Crenshaw, casaba
(*C. melo* [Inodorus group])

Muskmelon, Persian melon (*C. melo*
[Reticulatus group])

FAMILY: Cucurbitaceae; squash family
PLANT TYPE: Annual
SEED VIABILITY: 5 years

FLOWERING: Melon plants are mostly andromonoecious or monoecious. Identify male and female flowers by the presence or absence of a small, melonshaped ovary at the base of the pistillate flower. Other sexual types exist: plants with only female flowers, plants with only male flowers, and perfect-flowered plants. In the most common types, the first flowers to form are male, with female flowers following several days later. These first female flowers are the most likely to set fruit. There are more male flowers than female in all types of melons.

Melons are insect-pollinated, so the presence of honeybee pollinators is important. Professional seed producers import hives to ensure adequate pollination.

ISOLATION REQUIREMENT: All *Cucumis melo* plants will cross with each other, but they will not cross with watermelon (*Citrullus lanatus*), cucumbers, or squash. You can hand-pollinate melon flowers to ensure seed purity; see the hand-pollination directions for squash on page 29. Be careful: The procedure is a little trickier for melons than for squash because melon flowers are small and more delicate. If you do hand-pollinate, remove any fruits that form that were not hand-pollinated to encourage continued flowering. Recommended isolation distance is 1½ to 2 miles from other varieties of *C. melo*.

SEED COLLECTION: Fruit appearance depends on the variety, but botanically speaking, all melons are fleshy pepos. Fruits generally ripen 6 to 8 weeks after pollination, depending on the variety, location, soil, and other environmental variables, and seeds are

MELON SEEDS READY FOR CLEANING

ripe when melons are at table-ready stage. Harvest muskmelons at the "full slip" stage, or when a melon cleanly separates from the pedicel when pulled only gently. Other melons must be cut from the vine when mature. Select for early ripening and disease resistance. Cut melons in half, scoop out the seeds, and enjoy eating the melon.

SEED CLEANING AND STORAGE: Separate seeds from the pulp in a bowl or bucket of water and rinse. Repeat if necessary. Rinse seeds in a sieve, then spread them on a cookie sheet, screen, or paper towel to dry thoroughly.

GERMINATION: Melons require warm soil temperatures. Optimum range for germination is 80 to 90°F (27 to 32°C). Minimum soil temperature for germination is 60°F (16°C). Seed should germinate in 4 to 8 days.

TRANSPLANTING: Sow seeds 3 to 4 weeks before setting plants to the garden. Be very careful when transplanting melons, as the root tips do not regrow well when damaged. Sow seeds in containers such as peat pots that can be set intact into the garden. Use bottom heat if possible. Harden transplants slightly, and be sure the soil temperature is at least 65 to 75°F (18 to 24°C) when you plant.

MINER'S LETTUCE

Claytonia perfoliata (syn. Montia perfoliata)

Claytonia

FAMILY: Portulacaceae; portulaca family
PLANT TYPE: Annual
SEED VIABILITY: 5 years
SPACING FOR SEED SAVING: 6"

FLOWERING: Miner's lettuce has small, perfect, self-pollinating flowers.

ISOLATION REQUIREMENT: Several varieties of miner's lettuce are available, and they will cross with one another. Cage your plants if purity is desired. Miner's lettuce will not cross with purslane (*Portulaca oleracea*).

SEED COLLECTION: The fruit is a three-valved capsule with one to three shiny black seeds. Watch the plants carefully, as the seeds will form and drop prior to a plant's fading. Plant in plastic mulch to catch the seeds, or pull entire plants and hang them to dry, or dig and move flowering plants to where the seeds may fall and sow themselves naturally to produce a crop for the following year.

SEED CLEANING: Collected seed should be dried in a warm spot for 3 to 5 days before storing.

GERMINATION: Optimum soil temperature for seed germination is 50°C (10°C). Seeds need light to germinate.

TRANSPLANTING: Miner's lettuce is typically direct-seeded.

MUSTARD

Brassica juncea

Mustard greens, brown mustard, Indian mustard, leaf mustard

FAMILY: Brassicaceae; mustard family
PLANT TYPE: Annual
SEED VIABILITY: 4 years
SPACING FOR SEED SAVING: 18–24"

FLOWERING: Mustard green flowers are perfect and resemble those of other *Brassica* crops. The flowers are insect-pollinated.

ISOLATION REQUIREMENT: Plants are usually self-sterile, but all varieties of *B. juncea* will cross-pollinate, including weedy mustards such as Indian mustard. It is reported that mustard will cross with Chinese cabbage as well. If accidental cross-pollination is a concern, cage the plants. Recommended isolation distance is 1 mile from other mustards or *Brassica* species.

SEED COLLECTION AND CLEANING: The fruits, which are siliques, are smaller than those of other cabbage-family crops, but seed collection and cleaning is the same as for cabbage. See the Cabbage entry for details.

SEED TREATMENT: Prior to planting, hold seeds in 122°F (50°C) water for 20 minutes to control seed-borne diseases if they are an issue in your garden.

GERMINATION: Optimum soil temperature for germination is 77 to 95°F (25 to 35°C). Do not sow seed when the soil temperature is below 50°F (10°C).

TRANSPLANTING: Mustard is usually direct-seeded in the garden.

NEW ZEALAND SPINACH

Tetragonia tetragonioides

FAMILY: Tetragoniaceae; carpetweed family
PLANT TYPE: Annual
SEED VIABILITY: 3 years
SPACING FOR SEED SAVING: 18–20" within the row, with 3–5' between rows

FLOWERING: New Zealand spinach has inconspicuous perfect flowers and is mainly self-pollinated.

N

Flowers slowly mature from the bottom of the plant upward, as do the fruits.

ISOLATION REQUIREMENT: New Zealand spinach can be cross-pollinated by wind or insects, but only one variety is available, so cross-pollination is not a concern.

SEED COLLECTION: The fruit is a capsule containing several seeds. It is the fruit that is normally dried and planted. Fruits are horned, hard, and angular, turning brown when mature, but they may be picked by hand when green or brown, when they have reached mature size and been allowed to dry off the vine. Alternatively, wait until enough seed has set to satisfy you, cut down the plants, and dry on a tarp to catch the seedpods that shatter. Then shake the plants to remove the remaining seed.

SEED CLEANING AND STORAGE: Hand-picked fruits require no further processing.

SEED TREATMENT: Prior to planting, hold New Zealand spinach seeds in 120°F (49°C) water for 1 to 2 hours to control seedborne diseases if they are an issue in your garden. Soak overnight in unheated water just prior to planting to increase germination.

GERMINATION: Wait until the soil temperature reaches at least 50°F (10°C) to sow. Optimum soil temperature for germination is 60 to 70°F (16 to 21°C). Germination is slow and may take 2 weeks or longer.

TRANSPLANTING: New Zealand spinach is usually direct-seeded, but in northern gardens it may be sown indoors and transplanted. This is a slow-bolting, long-season crop.

—————— CROP ALERT ——————

Even though New Zealand spinach is grown as a "spinach substitute," it is in a separate family from garden spinach. Unlike spinach, New Zealand spinach is a warm-season crop, and the techniques for growing it and saving seed from it are quite different from those for regular spinach.

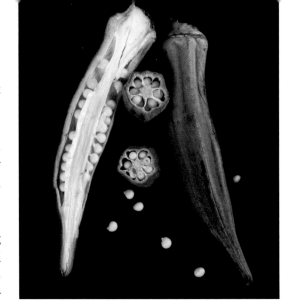

IMMATURE OKRA CAPSULES (SEEDPODS)

OKRA

Abelmoschus esculentus

Gobo, gumbo, gombo, lady's-finger

FAMILY: Malvaceae; cotton or mallow family
PLANT TYPE: Annual
SEED VIABILITY: 2 years
SPACING FOR SEED SAVING: 1½'

FLOWERING: Okra flowers are very ornamental and perfect. The plants are usually self-pollinated.

ISOLATION REQUIREMENT: Although okra is self-pollinating, insects are strongly attracted to the flowers, so cross-pollination can occur among different varieties. To prevent this, cage entire plants or individual flowers to exclude pollinators. To bag individual flowers, watch the flower-opening habit and bag buds that appear to be ready to open the next morning. Secure carefully but tightly to exclude insects. Flowers are pollinated immediately after opening and shortly become unreceptive to additional pollen. Let the bags remain on the plants for 2 days. Mark the bagged flowers so you don't forget which they are after removing the bags. Recommended isolation distance for okra is 1 mile.

SEED COLLECTION: The fruit is a capsule (pod) that is usually ridged, but there are also smooth varieties. When mature, the fruit turn dark brown. Harvest okra pods lightly the year you plan to save seeds to be sure you have enough left over for seed extraction. Wear gloves to avoid skin irritation. Pods will be mature and ready to eat 4 to 6 days after flowering. Allow pods for seed production to remain on the plant until they've "gone by," then pick and allow them to further dry out of direct sun. If there is no threat of frost, leave them on the plants until they are fully dry. The pods will usually split when the seed is ready to be collected, but you'll have to collect seeds just before they're shed. Using gloved hands, split the pods over a bucket to catch the seeds.

SEED CLEANING AND STORAGE: Flail any stubborn pods that refuse to give up their seeds. Allow the seeds to dry another week or so before storing. Winnow to remove excess plant material.

SEED TREATMENT: Soak seeds in 110°F (43°C) water for 1 to 2 hours prior to planting to soften the hard seed coat and improve germination percentage.

GERMINATION: Okra requires warm soil temperatures to germinate. The minimum soil temperature for sowing seed is about 60°F (16°C), but the optimum range is from 80 to 90°F (27 to 32°C).

TRANSPLANTING: In short-season areas, start transplants early. Provide bottom heat if possible, and sow seeds 4 to 5 weeks prior to setting in the garden. Be careful not to disturb roots when transplanting; use peat pots, fiber pellets, or another vessel that allows planting of the entire container to the garden.

Onion

Allium cepa (Cepa group)

Common garden onion

FAMILY: Alliaceae; allium family

PLANT TYPE: Biennial

SEED VIABILITY: 1–2 years

SPACING FOR SEED SAVING: 3–4" within the row, with 3' between rows

SAVE SEED FROM: At least 2 plants

FLOWERING: Onion plants will not flower until after a cool period, usually over the winter. The flower head is an umbel with many tiny perfect flowers, and the flowers are insect-pollinated. Bolting may occur if plants are subjected to temperatures below 50°F (10°C) when they are beyond the five-leaf stage. Don't save seed plants that bolt during their first year of growth.

OVERWINTERING ONIONS

ONIONS REQUIRE EXPOSURE to winter cold to induce flowering. Here's how to bring your onion plants through the winter.

YEAR 1

In cold regions. Plant, harvest, and cure your onions as usual the first year. Cut the tops back to about 6 inches from the bulbs, and store the bulbs in crates or bags at 32 to 40°F (0 to 4°C) in a dry room. Bags are preferred, as they enable better ventilation. Rogueing in the field prior to storage allows for removal of off-types such as those with thick necks or an off-color, and doubles.

In milder climates, the onions may be left in the ground to overwinter. Another option in mild-winter areas is to plant onions for seed production in the fall.

YEAR 2

Sort and select the best onions from your storage stock. This enables you to select for the good keepers as well as disease resistance and size. When replanting, cut off any rotted, dried, or diseased tissue; cut an X in the top of the onion to facilitate the seed stalk's emergence; and cover the entire bulb with about ½ inch of soil. The plants will resprout and grow; rogue out those plants that are early to bolt.

If you planted your onions in the fall and overwintered them in the ground, harvest them in late spring or early summer. Store them for the summer at 32 to 40°F (0 to 4°C) — in a spare refrigerator, for example — and replant them in the fall.

SWEET SPANISH WHITE ONIONS, RED ONIONS,
SWEET SPANISH FLAT YELLOW ONIONS

ISOLATION REQUIREMENT: Onion varieties can cross with one another, but onions will not cross with leeks or chives. Occasional crossing with *Allium fistulosum* (Japanese bunching onion) can occur. If you're concerned about cross-pollination, use alternate-day caging or caging with insect-pollinators. Recommended isolation distance is ½ mile.

SEED COLLECTION: The fruit is a capsule with three cells. Each cell contains one or two black seeds when mature. Carefully cut the umbel from the stalk when the seeds are visible. Onion umbels shatter easily. Either bag the umbels or bend them into a bag when you cut them. Allow the umbels to dry for 2 to 3 weeks.

SEED CLEANING: Umbels readily lose their seeds, but you may want to rub them over a screen, then winnow the plant material. Or put the seeds and crushed plant material in a bucket of water. The seeds will sink and the chaff will float. Do not soak the seeds. Dry well before storing.

GERMINATION: The optimum soil temperature range for onion-seed germination is 50 to 95°F (10 to 35°C).

TRANSPLANTING: Sow seeds indoors 8 to 9 weeks prior to setting plants to the garden, or up to 14 weeks ahead if you have a very short season. Seeds germinate better in darkness.

ORACH

Atriplex hortensis

French spinach, mountain spinach

FAMILY: Chenopodiaceae; goosefoot family
PLANT TYPE: Annual
SEED VIABILITY: 5–6 years
SPACING FOR SEED SAVING: Variable

FLOWERING: Orach is a monoecious plant with flowers borne in clusters like spinach. The flowers are wind-pollinated.

ISOLATION REQUIREMENT: Raising more than one variety of orach is an unlikely scenario, but if you do, you'll need to isolate the varieties by bagging several plants of one variety together during pollination. Recommended isolation distance is ¼ mile.

SEED COLLECTION: Fruits are utricles with seeds contained in bracts. Cut plants prior to maturation and seed drying. Drying of the plants should be as quick and thorough as possible.

SEED CLEANING: No seed cleaning is necessary.

SEED TREATMENT: Orach plants produce seeds of two different colors. Those that are yellowish brown germinate quickly and those that are black have a low germination percentage. Scarification or exposure to light may improve germination of black seeds, but why not use the lighter-colored seeds exclusively? You'll save time and avoid trying your patience.

GERMINATION: Optimal soil temperature for orach germination is 50 to 75°F (10 to 24°C).

TRANSPLANTING: Orach is usually direct-seeded.

GIVE ORACH SPACE TO SPREAD

UNLIKE MANY OTHER PLANTS, orach grows larger in the cool northern areas than in regions where winters are mild and the air is humid. In the North, orach plants will grow to 6 feet or even 10 feet tall, and their spread can be as much as 4 feet. Keep this in mind if you decide to embark on the adventure of saving orach seed!

Parsnip

Pastinaca sativa

FAMILY: Apiaceae; parsley family
PLANT TYPE: Biennial
SEED VIABILITY: 1 year
SAVE SEED FROM: 10 or more plants
SPACING FOR SEED SAVING: 3'

FLOWERING: Parsnip flowers are perfect and held in compound umbels. Plants flower only after exposure to a cool period, usually after winter. The flowers are insect-pollinated.

ISOLATION REQUIREMENT: There are very few varieties of parsnips grown, but varieties will cross if allowed to flower simultaneously. If needed, hand-pollination is a great option. Follow the directions on page 29, under Hand-Pollinating Carrots.

Parsnip will cross with wild parsnip (*Pastinaca sativa* var. *sylvestris*). If this weed is present, isolate your parsnips by 200 feet.

SEED COLLECTION: Follow the instructions in the Carrot entry for collecting seed, but keep in mind that parsnip umbels shatter more easily than do those of carrot.

SEED CLEANING AND STORAGE: See the seed-cleaning instructions in the Carrot entry.

GERMINATION: Parsnip seed germinates best at soil temperatures from 50 to 70°F (10 to 21°C). Don't plant seed when soil temperatures are below 35°F (2°C). Seed germinates slowly over a period of up to 3 weeks.

TRANSPLANTING: Parsnips are usually seeded directly to the garden.

OVERWINTERING PARSNIPS

PARSNIPS USUALLY NEED exposure to winter cold to induce flowering. Here's how to keep your plants alive through the winter.

YEAR 1
In the fall, carefully dig and select those roots with the best characteristics, such as size and minimal zoning. After removal of the tops, replant them right away where they were previously growing, at 3-foot centers, with soil covering the shoulders. Apply winter mulch. Parsnips will usually overwinter even in the North, but in areas of extreme cold, dig them in the fall, clip back the tops to 2 to 3 inches above the roots, and store them in a cool, humid location packed in damp sand or sawdust at 32 to 40°F (0 to 4°C).

YEAR 2
If you stored stecklings out of the garden, replant them in the spring. For plants left in the ground, remove the mulch when growth begins.

Pea

Pisum sativum

Garden pea, English pea, green pea, common pea (*P. sativum*)

Edible-podded pea, sugar pea, snow pea (*P. sativum* var. *macrocarpon*)

FAMILY: Fabaceae; legume family
PLANT TYPE: Annual
SEED VIABILITY: 3 years

FLOWERING: The papilionaceous flowers are perfect and self-pollinated. Flowering begins on the lower nodes of the plants and continues up the stems.

ISOLATION REQUIREMENT: There is little out-crossing in peas. Plant a tall crop between varieties if desired for added protection against cross-pollination. Recommended isolation distance is 50 feet.

PEAS IN ROUND-POD STAGE

SEED COLLECTION: The fruit is a dehiscent legume (pod) that goes through two stages of development. The first is the flat-pod stage, in which the pod elongates and widens; the second is the round-pod stage, in which the seeds fatten and mature. The plants are generally indeterminate, and lower pods become ready for harvest prior to the upper pods. Some varieties flower and fruit near the end of their shoots, and this heavy fruiting slows the stems' development, giving the plants the appearance of being determinate. A color change in the seeds and pods indicates harvest maturity for seed collection. Seeds dry from the outside to the inside. Harvest pods when they are dry and brown and when the peas rattle inside if the pods are shaken. Entire plants may also be pulled and hung upside down in a cool, dry place until the pods turn brown.

SEED CLEANING: Shell peas from the pods by hand after 1 to 2 weeks of storage in a warm, dry place. Flail larger quantities in a burlap bag or pillowcase, or strike the plants on the inside surface of a garbage can. Be careful not to damage the seeds or seed coats.

GERMINATION: The optimum soil temperature for germination ranges from 40 to 75°F (4 to 24°C). Soil should be at least 40°F for best results.

TRANSPLANTING: Peas are usually direct-seeded to the garden.

PEANUT

Arachis hypogaea

Goober, groundnut

FAMILY: Fabaceae; legume family
PLANT TYPE: Annual
SEED VIABILITY: 4 years
SPACING FOR SEED SAVING: 1'

FLOWERING: Flowers are perfect and self-pollinating. The fertilized flower forms a peg containing the ovary, which grows downward 1 to 2 inches into the soil, where the pods form.

ISOLATION REQUIREMENT: Some cross-pollination may occur due to insect activity. If you're growing more than one variety of peanuts, cage the plants to ensure purity. Recommended isolation distance is 1 mile.

SEED COLLECTION: Peanuts are a long-season crop. Fruits are indehiscent legumes that ripen underground. Dig plants (including the roots and peanuts) when the leaves start to turn yellow and the weather is dry. Hold the plants in a well-ventilated, very dry area out of direct sun to cure for 2 to 3 weeks or more. Keeping the vines loose and dry will help prevent postharvest fungi from attacking the fruits. Peanuts need to cure slowly, so don't try to rush the process with heat or direct sunlight or they will become brittle.

SEED CLEANING: Pick or cut the peanuts off the plants.

SEED TREATMENT: Virginia-type peanuts cannot be planted immediately after harvest because they have a 1- to 3-month dormancy, during which they should be kept in a cool, dry, dark place. Spanish and Valencia types have no dormancy.

GERMINATION: Remove the netted husk just prior to planting, taking care not to damage the seed coats of the seeds inside. There will be two to four seeds in each pod, depending on type. The optimum temperature for germination is 77 to 86°F (25 to 30°F).

TRANSPLANTING: Use bottom heat if practical and heat soil to 70 to 80°F (21 to 27°C). Like other legumes, peanuts require special care when transplanting. Use peat pots or similar containers that can be placed intact into the garden. Do not transplant into cold soil in the garden.

PEPPERS

Capsicum spp.

Bell pepper, sweet pepper, chili pepper, red pepper (*C. annuum* var. *annuum*)

Tabasco pepper (*C. frutescens*)

FAMILY: Solanaceae; nightshade family
PLANT TYPE: Perennial treated as annual
SEED VIABILITY: 2 years
SAVE SEED FROM: As many plants as possible

FLOWERING: Flowers are perfect and similar in form to those of tomato, except the anthers do not completely enclose the stigma, allowing for some wind and insect pollination. Flowers are generally self-pollinated.

ISOLATION REQUIREMENT: Different varieties of peppers will cross-pollinate, and some insect and wind pollination may occur. If cross-pollination is a concern, cage the plants from which you want to save seed. Recommended isolation distance is ½ mile.

—————————— CROP ALERT ——————————

Peppers may set parthenocarpic fruit when night temperatures are cool, between 54 and 60°F (12 and 16°C). Such fruit will be small and will not contain seeds. When night temperatures are below 53°F (11.5°C), peppers will not set fruit at all.

SEED COLLECTION: The fruit is a podlike berry. Most open-pollinated peppers and some hybrid peppers turn red at maturity. Pick fully ripe (red) peppers usually about 3 weeks after mature green fruits would have been harvested. Scrape out the seeds and spread them on a screen or paper towel to dry for about 2 weeks.

SEED CLEANING: Pepper seeds do not need any type of cleaning. Once they're dry, they're ready for storage.

SEED TREATMENT: Prior to planting, hold pepper seeds in 125°F (52°C) water for 30 minutes to control seedborne diseases if they are an issue in your garden.

GERMINATION: Optimum soil temperature for seed germination is 70 to 75°F (21 to 24°C) for sweet peppers and 70 to 85°F (21 to 29°C) for hot peppers. Seed dormancy has been studied in several varieties of peppers. Seed germination in fresh seeds normally takes 20 to 50 days at 75°F (24°C), but seeds dried and stored for 2 to 3 weeks germinated in 20 days. Seeds also germinated more quickly if they were extracted from overripe fruit — about 10 days after fruit would be considered physiologically ripe. Storage of tabasco seeds for 3 weeks after extraction increased the percentage of seeds that germinated. The exact details of dormancy mechanisms have not been worked out. Priming has been shown to shorten germination time but not improve emergence percentages.

TRANSPLANTING: Sow seeds about 8 weeks prior to setting transplants into the garden. Peppers transplant with moderate ease.

POTATO, IRISH

Solanum tuberosum

FAMILY: Solanaceae; nightshade family
PLANT TYPE: Annual
SEED VIABILITY: Unknown
SPACING FOR SEED SAVING: 1'

FLOWERING: Potato plants bear perfect and usually self-pollinated flowers. Despite their flowering, though, seed balls form only occasionally.

ISOLATION REQUIREMENT: Since potato seed won't produce plants that are true to type, there's no need to worry about isolation techniques.

SEED COLLECTION: The fruit of the potato is a small, hard, poisonous berry that resembles a tiny green tomato. Wait to hand-pick for a month or two, just when they start to soften. They'll likely fall off a plant into your hand.

SEED CLEANING AND STORAGE: Squeeze out the seeds into a container and add enough water so the bad seeds float and the good ones sink. Pour off the water and floating material and dry the good seeds on a paper towel or cookie sheet. Fermentation may increase germination. See the Tomato entry for a discussion of fermentation-cleaning of seeds.

SEED TREATMENT: Seed is dormant immediately after collection. Treatment with gibberellic acid (available through garden-supply stores and on the Internet) will improve germination if immediate planting is desired.

--------- CROP ALERT ---------

Potatoes are usually propagated vegetatively by planting small tubers called seed potatoes, but potato breeders work with true seed to develop new varieties. Keep in mind that potatoes grown from seed potatoes are genetic clones of the parent potato, but those grown from true seed will not be replicas of the parent variety. In addition, most seed-grown potato plants will not produce a worthwhile crop of tubers. Be sure to plant a row of seed potatoes for backup.

GERMINATION: Optimum range of soil temperatures for germinating potato seeds is 65 to 80°F (18 to 27°C). Germination percentage is unpredictable because of great genetic variability.

TRANSPLANTING: Transplanting is probably less problematic than direct-seeding. Start seeds 6 to 8 weeks prior to setting transplants to the garden. Use bottom heat if practical. The plants will produce small tubers by fall. Dig and cure the tubers; store over winter at about 35°F (2°C) in a dark, humid location; and replant the following spring for your eating crop.

RADISH

Raphanus sativus (Reticulata group and Longipinnatus group)

Spring radishes, summer radishes (*R.* [Reticulata group])

Daikon, winter radishes (*R.* [Longipinnatus group])

FAMILY: Brassicaceae; mustard family
PLANT TYPE: Annual and biennial
SEED VIABILITY: 5 years
SAVE SEED FROM: The more plants the better
SPACING FOR SEED SAVING: 6–9"

FLOWERING: Perfect flowers are borne in racemes and are white, pink, or light purple. Flowers are insect-pollinated, and radishes are self-incompatible. Seed production is reduced in hot, dry weather, and if another crop is flowering simultaneously, bees may prefer those flowers over a radish plant's small ones.

ISOLATION REQUIREMENT: Radish flowers will cross with any other radish variety blooming in the garden at the same time, as well as with wild radishes (*R. raphanistrum*). Radish will not cross with cabbage plants or other *Brassica oleracea* crops. To avoid unwanted crossing, you can use alternate-day caging

PREPARING FOR FLOWERING

SPRING-PLANTED SUMMER RADISHES will produce seeds the same season they are planted. When radishes are table-ready, dig the plants; cut the tops ¾ to 1 inch above the roots; select for color, size, and shape; and replant so soil covers the shoulders of the swollen taproots. Further select for late bolting. In the South, plant in the spring or fall.

Winter radishes, which are biennial, should be dug in late fall and replanted the following spring, as you would turnips. See the Turnip entry, page 133, for details.

or caging with introduced pollinators. Recommended isolation distance is 1 mile.

SEED COLLECTION: Radish seedpods are indehiscent siliques, 1 to 3 inches long. No special care is needed in the field prior to harvest. Collect pods individually when brown or pull the entire plant and allow to dry in a well-ventilated location.

SEED CLEANING AND STORAGE: Rub dry pods in your hands, or gently crush or flail. The seedpods are tougher to open than those of cabbage. Winnow to separate seeds from any extra plant material.

SEED TREATMENT: Prior to planting, hold radish seeds in 122°F (50°C) water for 15 minutes to control seedborne diseases if they are an issue in your garden.

GERMINATION: Radish seeds readily germinate in soil temperatures of 45 to 90°F (7 to 32°C). Minimum temperature for germination is 40°F (4°F).

TRANSPLANTING: Radishes are usually direct-seeded.

RHUBARB

Rheum rhabarbarum

Rhubarb, garden; pie plant; wine plant

FAMILY: Polygonaceae; buckwheat family
PLANT TYPE: Perennial
SEED VIABILITY: Unknown
SPACING FOR SEED SAVING: 2–3' within the row, with 4–6' between rows

FLOWERING: Rhubarb's flower stalk is paniculate and is typically cut when the plant is being grown for food production so the plant's energy goes to producing the edible petiole, not flowers. Naturally, if you want to collect seed, you'll have to leave some flower stalks intact.

RHUBARB SEED HEAD

ISOLATION REQUIREMENT: Rhubarb's flower structure favors insect cross-pollination. Recommended isolation distance is ½ mile.

SEED COLLECTION: It's fine to harvest a few early leaf stalks for eating. The seeds are achenes. Cut the flower stalk when the achenes are dry and flaky and before they shatter in the wind.

SEED CLEANING AND STORAGE: Separate seeds from the stalk by hand and spread to dry further.

GERMINATION: Rhubarb seeds will germinate in 7 to 21 days at soil temperatures between 68 and 86°F (20 and 30°C).

TRANSPLANTING: Rhubarb is generally propagated by crown division, and like other clonally propagated vegetables, plants resulting from collected seed will be quite diverse. Few if any will resemble the mother plant. If you decide to experiment with growing rhubarb from seed, and for seedling plants that are slow to bolt, select and sow seed directly in the garden.

RUTABAGA

Brassica napus (Napobrassica group)

Siberian kale

FAMILY: Brassicaceae; mustard family
PLANT TYPE: Biennial
SEED VIABILITY: 4 years
SPACING FOR SEED SAVING: 1'

FLOWERING: Rutabaga flowers are perfect. Plants flower after exposure to a cool period. The flowers are insect-pollinated.

OVERWINTERING RUTABAGA

RUTABAGAS GENERALLY WON'T FLOWER until after they've been exposed to winter cold. Overwintering rutabaga is similar to overwintering carrots.

YEAR 1

Rutabaga plants require a long season to mature; sow rutabaga seed as early as possible in the spring. In the fall, carefully dig and select those with the best root size and shape. Remove all but 1 to 2 inches of the tops and store the roots in a box (cover them with damp sand) over winter in a damp location at 33 to 40°F (1 to 4°C).

Rutabagas may also be sown in early August and left in place in areas of less severe winters. Thin your rutabaga stand during the eating season until the remaining plants are 1 foot apart. Mulch them over the winter.

YEAR 2

When spring arrives, select the stored roots that kept best and replant them. If you overwintered plants in the garden, remove the mulch in early spring. The plants will flower and set seed.

ISOLATION REQUIREMENT: Rutabaga varieties will cross-pollinate, and rutabagas will also cross with some varieties of turnip. If you have any concerns about cross-pollination, cage your rutabaga plants. Recommended isolation distance is 1 mile.

SEED COLLECTION: The siliques (pods) ripen from the bottom up on the seed stalk, turning from green to brown as they ripen. Hand-pick as they mature or harvest the entire stalk when enough seedpods have dried to satisfy you. Hang upside down out of direct sunlight over a cloth or bucket to catch the seeds as the pods shatter.

SEED CLEANING: Place any pods that don't shatter in a bag and flail. Winnow excess plant material.

SEED TREATMENT: Prior to planting, hold rutabaga seeds in 122°F (50°C) water for 20 minutes to control seedborne diseases, such as black rot, if they are issues in your garden.

GERMINATION: Seeds will germinate best at temperatures between 45°F (7°C) and 90°F (32°C). Don't plant before soils are at least 40°F (4°C).

TRANSPLANTING: Rutabagas are usually direct-seeded to the garden.

SALSIFY

Tragopogon porrifolius

Oyster plant, salsify, vegetable-oyster

FAMILY: Asteraceae; sunflower family
PLANT TYPE: Biennial
SEED VIABILITY: 1 year
SPACING FOR SEED SAVING: 1'

FLOWERING: Salsify's flowers are bluish purple and perfect and are open only in the morning. Salsify plants require exposure to winter cold to induce flowering. In mild locations, salsify may be mulched and left in the ground over winter. In more severe winter locations, dig the roots in late fall, trim the tops to 2 to 3 inches above the taproots, and store over winter in damp sand in a cool, dark location. Replant in the spring.

ISOLATION REQUIREMENT: The perfect flowers are mostly self-pollinating, but some crossing may occur because of insect activity. If you're concerned about crossing, you can isolate your salsify plants. Recommended isolation distance is ½ mile.

SEED COLLECTION: The fruit is an achene, and the seed head looks like a large dandelion head. An achene can be borne on the wind on its brownish pappus, like an oversize dandelion seed. Remove seeds from the head before the wind takes them or cut individual heads as they mature on a still morning.

SEED CLEANING: Flail to remove the pappus, then dry for another week or two out of direct sunlight. Seeds will snap in two easily when ready for storage.

GERMINATION: Optimum temperature for seed germination is approximately 70°F (21°C). Germination typically requires about 12 days.

TRANSPLANTING: Salsify is usually direct-seeded.

———————— CROP ALERT ————————

In some locations, salsify has escaped from cultivation and grows wild. These wild plants could cross with your garden plants, so take precautions if this is a problem where you live. There's also a yellow-flowered weedy relative of salsify, *Tragopogon pratensis*, but it will not cross with garden salsify.

SALSIFY SEED HEADS

Scorzonera

Scorzonera hispanica

Black oyster plant, black salsify, Spanish salsify, viper's grass

FAMILY: Asteraceae; sunflower family
PLANT TYPE: Perennial grown as annual
SEED VIABILITY: 2 years
SPACING FOR SEED SAVING: 1'

FLOWERING: Perfect yellow flowers occur the second season and are open only in the morning.

ISOLATION REQUIREMENT: Very few varieties are available, but if you grow more than one type, they can cross-pollinate because of insect activity. Also, scorzonera has escaped cultivation in some parts of the United States, and your garden scorzonera can cross with the wild type. Cage your plants if cross-pollination is a concern. Recommended isolation distance is ½ mile.

SEED COLLECTION: Watch for the seed capsules to compress and the fluffy pappus to form on the achenes. Collect ripe seeds daily during their ripening period.

OVERWINTERING SCORZONERA

ALTHOUGH SCORZONERA is a perennial, gardeners treat it as an annual, planting it in spring for harvest in fall or as a winter crop. When growing it for seed, treat it as a biennial, and do not harvest any shoots from plants from which you intend to save seed.

In temperate areas, mulch the plants and leave them in the ground over winter. *In areas where winters are severe,* dig the roots, trim the tops and any small side roots, and store the roots in damp sand in a damp location at 32 to 40°F (0 to 4°C). Replant in early spring of year 2.

SEED CLEANING: Dry in a warm spot for several days until brittle.

GERMINATION: Optimum soil temperature for germination is 68 to 86°F (20 to 30°C).

TRANSPLANTING: Scorzonera is typically direct-seeded.

SEA KALE

Crambe maritima

Scurvy grass

FAMILY: Brassicaceae; mustard family
PLANT TYPE: Perennial
SEED VIABILITY: 1 year
SPACING FOR SEED SAVING: 18"–36"

FLOWERING: Sea kale is grown for its blanched shoots, but it also will flower each year once it reaches maturity. Flowers are held in panicles and are insect-pollinated.

ISOLATION REQUIREMENT: Sea kale is insect-pollinated, but there are so few varieties available that isolation should be unnecessary.

GETTING STARTED WITH SEA KALE

SEA KALE TAKES 1 to 3 years to establish; once it reaches maturity, it will flower every summer. If you want to try growing sea kale, you may have trouble finding seed for sale, or it may be expensive.

Keep in mind that sea kale is hardy only to Zone 6. In areas where winters are severe, you'll need to dig the plants in fall and store them over the winter as you would carrots. See Overwintering Carrots, on page 102, for details.

SEED COLLECTION: Individual seeds are enclosed in a round, indehiscent silicle. Seed stalks will dry while the plant still remains green. When fully dry, remove the pods from the plant.

SEED CLEANING: Store entire seedpods over winter, then rub to separate the seeds before sowing.

GERMINATION: Optimum soil temperature for seed germination is around 75°F (24°C).

TRANSPLANTING: Sow seeds 6 to 8 weeks prior to setting in the garden.

SORREL

Rumex spp.

Garden sorrel, sour dock (*R. acetosa*)

Spinach dock, patience, monk's rhubarb (*R. patientia*)

Garden sorrel, French sorrel (*R. scutatus*)

FAMILY: Polygonaceae; buckwheat family
PLANT TYPE: Perennial
SEED VIABILITY: 4 years
SPACING FOR SEED SAVING: 1'

FLOWERING: Sorrel will overwinter in temperate regions. The 2- to 3-foot-tall seed stalk is formed the second year. Plants are monoecious with imperfect flowers or bisexual flowers in panicles. Flowers are insect-pollinated.

ISOLATION REQUIREMENT: Sorrel varieties will cross-pollinate. If needed, isolate varieties with alternate-day caging. Also be sure to control any weedy sorrel in your garden and prevent it from flowering (and cross your fingers that your neighbors will as well).

SEED COLLECTION: Select for slow-to-bolt plants.

SEED CLEANING AND STORAGE: Rub the achenes off the stalk when they are dry.

GERMINATION: Seed germination takes about 20 days.

TRANSPLANTING: Sorrel is typically grown from cuttings, but seeds may also be used. Direct-seed or transplant.

―――――――――― CROP ALERT ――――――――――

Sorrel plants are usually propagated by divisions and not by seed. Some species have escaped from cultivation and have become noxious weeds, so harvest all seeds from your plants.

SOYBEAN

Glycine max

Edamame, soya bean

FAMILY: Fabaceae; legume family
PLANT TYPE: Annual
SEED VIABILITY: Unknown
SPACING FOR SEED SAVING: 4–6"

FLOWERING: Soybean flowers are self-pollinating.

ISOLATION REQUIREMENT: Pollination occurs before the flowers open, so there is no concern about isolation with them.

SEED COLLECTION: The fruit is a dehiscent legume. Allow pods to dry on the plants; wear gloves to harvest them.

SEED CLEANING AND STORAGE: Flail as you would snap beans (see the Bean entry for details). Wear gloves if hand-shelling to avoid cuts from the sharp pods.

GERMINATION: Optimum soil temperatures for seed germination are 70 to 85°F (21 to 29°F).

TRANSPLANTING: Soybeans are direct-seeded.

SPINACH

Spinacia oleracea

FAMILY: Chenopodiaceae; goosefoot family
PLANT TYPE: Annual
SEED VIABILITY: 3 years
SPACING FOR SEED SAVING: 6–12"

FLOWERING: Most spinach plants are dioecious, but some may be monoecious. A 1:2 ratio of male to female plants should be maintained for best seed production, but the sex of a plant cannot be determined until the flower stalks have formed. There are two male-type plants: the "extreme male," a small, early-bolting type that should be rogued out, and the "vegetative male," which will produce much more foliage and provide good pollen. Spinach flowers are inconspicuous and have no petals. They are wind-pollinated.

ISOLATION REQUIREMENT: Different varieties of spinach will cross-pollinate one another. If your gardening neighbors are also saving spinach seed, you could propose growing the same variety. But if you have concerns about unwanted pollination, you'll need to bag the plants using spun-poly row cover, because spinach pollen is so small it will go right

MALE SPINACH

FEMALE SPINACH

through a fine-mesh screen. If bags are used, contain a minimum of two male and four female plants in each to ensure adequate pollination. Recommended isolation distance is 2 miles, but the pros separate spinach seed crops by 5 to 10 miles.

SEED COLLECTION: Select plants for seed production that have leaf characteristics you like (flat, savoyed, dark green, tender) and that are late to bolt. Some of the outer leaves may be harvested for eating before the plants bolt. A combination of warm temperatures and long days contributes to bolting, as does close spacing within the row; remove early bolters from your garden. Seeds are reaching maturity when the plants turn yellow. There are two types of seeds, smooth and prickly. Wear gloves to hand-harvest the prickly type. You can pull the plants after the seed has formed but when they are not yet dry, or you can leave the plants in the field to dry fully. Strip the seeds from a plant (you'll get leaves, too) by pulling the stalk through your hands from the base of the plant up.

SEED CLEANING: Winnow dry seeds and leaves immediately or allow them to dry out of the direct sun and then winnow. Removing the prickles from prickly seeded varieties does not increase germination.

SEED TREATMENT: Prior to planting, hold spinach seeds in 122°F (50°C) water for 25 minutes to control seedborne diseases if they are an issue in your garden.

GERMINATION: Minimum soil temperature for germination is 35°F (2°C), and optimum germination is achieved when soil is 45 to 75°F (7 to 24°C).

TRANSPLANTING: Spinach is usually direct-seeded.

FEMALE SQUASH FLOWER
AND DEVELOPING FRUIT

SQUASH

Cucurbita spp.

Pumpkins (*C. maxima, C. mixta, C. moschata, C. pepo* var. *pepo, C. pepo* var. *melopepo*)

FAMILY: Cucurbitaceae; squash family
PLANT TYPE: Annual
SEED VIABILITY: 4 years

FLOWERING: Squash and pumpkin plants are monoecious with imperfect flowers. The male flower stalk is straight and thin, whereas the female flower has an ovary resembling that of a squash. Squash and pumpkins are insect-pollinated.

ISOLATION REQUIREMENT: Cross-pollination among *Cucurbita* varieties and species can occur; see Sorting Out Squash and Pumpkin Varieties, on page 131, for details. In addition, cross-pollination of squash by cucumber pollen may occasionally result in parthenocarpic fruit. Recommended isolation distance for varieties that can cross-pollinate is 1½ to 2 miles; recommended isolation distance for other *Cucurbita* species is ¼ mile.

SEED COLLECTION: Squash and pumpkin fruits are indehiscent fleshy berries that have a hard rind. Harvest fully mature squash and hold for 3 to 4 weeks for the seeds to ripen. (Summer squashes are fully mature for seed collection when the rinds are hard and yellowish. Most winter squashes are fully mature

for seed collection when they have turned yellowish orange.) Cut the squash in half, then scoop out the seeds and pulp.

SEED CLEANING AND STORAGE: Rinse off the pulp. You can also mix the seeds and pulp with water. Viable seeds will sink. Drain the seeds and spread them on towels, paper towels, or cookie sheets to dry for about 2 weeks, stirring occasionally to be sure all seeds have dried evenly.

GERMINATION: Optimum range of soil temperature for germination is 70 to 95°F (21 to 35°C).

TRANSPLANTING: Sow seeds 4 to 5 weeks prior to setting in the garden. These species do not transplant well, so sow them in peat pots and transplant both plant and pot to the garden.

C. maxima   C. moschata   C. pepo

Examining the fruit stem of a squash can help identify which species it is. A *Cucurbita maxima* fruit stem is soft and rounded and appears to be inserted into the fruit. *C. moschata* stems have a flat plate at the point of stem attachment. *C. pepo* stems are hard, prickly, and enlarged at the point of attachment.

SORTING OUT SQUASH AND PUMPKIN VARIETIES

This chart groups commonly grown varieties and types of squash and pumpkins within four species. You can use the information here to compare the varieties you plan to grow to see which ones would be able to cross-pollinate. Keep in mind that all varieties will cross within their own species. Also, the various cucurbit species will cross with one another, with the exception that *C. maxima* will not cross with *C. pepo* or *C. mixta*. Thus, for example, 'Baby Boo' pumpkins will not cross-pollinate with 'Harvest Moon' pumpkins, but they will with 'Kentucky Field' pumpkins.

	CUCURBITA maxima	*CUCURBITA mixta*	*CUCURBITA moschata*	*CUCURBITA pepo*
PUMPKINS	Amish Pie, Atlantic Giant, Big Max, Big Moon, Burgess Giant, German Sweet Potato, Harvest Moon, King of Mammoths, Rouge Vif d'Etampes	Green Striped Cushaw, Jonathan, Pennsylvania Crookneck, White Crookneck, White Cushaw	Cheese, Dickinson, Kentucky Field, Long Island, Longfellow, Musquée de Provence, Quaker Pie, Tennessee Sweet Potato	Baby Boo, Baby Pam, Big Tom, California Sugar, Caserta, Cheyenne Bush, Cinderella, Connecticut Field, Cornfield Pumpkin, Cow, Halloween, Happy Jack, Howden, Jack-Be-Little, Jack-o'-Lantern, Lady Godiva, New England Pie, Pankow
SQUASH	Australian Butter, Banana, Boston Marrow, Buttercup, Galeux d'Eysines, Golden Delicious, Hokkaido, Hubbard, Kabocha, Lakota, Marina di Chioggia, Potimaron, Queensland Blue, Sibley (also called Pike's Peak), Turk's Turban	Cushaw	Butternut, Chirimen, Golden Cushaw, Kikuza, Musquée de Provence, Orange Cushaw, Pennsylvania Dutch Crookneck, Small Flat Cheese	Acorn, Cocozelle, Crookneck, Delicata, Patty Pan, Spaghetti, Straightneck, Sweet Dumpling, Thelma Sanders Sweet Potato, Tours Squash, Triamble (also called Shamrock, Triangle, or Tristar), Vegetable Marrow, Yugoslavian Finger Fruit, Zucchini

Sources: Gough and Moore-Gough, 2009; Ashworth, 2002; Whealy, 2010

TOMATILLO

Physalis ixocarpa

Mexican husk tomato, jamberry

FAMILY: Solanaceae; nightshade family
PLANT TYPE: Annual
SEED VIABILITY: 4 years
SPACING FOR SEED SAVING: 1–2'
SAVE SEED FROM: As many plants as possible

FLOWERING: Tomatillos are annual plants with prolific, perfect flowers; they are a mostly self-pollinated crop.

ISOLATION REQUIREMENT: There is little outcrossing with tomatillo, but if you want, you can plant a tall crop between varieties for extra protection or use alternate-day caging.

SEED COLLECTION: The fruit is a berry, enclosed in an expanding and often split calyx and should be picked fully ripe and ready to eat.

SEED CLEANING AND STORAGE: Remove the husk, then put berries in a blender with water to cover. Whir until completely blended, then dump into a bowl, adding the same amount of water as there is berry pulp. Stir, then allow the seeds to settle. Remove excess plant material and repeat until the seeds are clean. They are quite small, so do not use a strainer with large holes. Place on a paper towel or cookie sheet to dry.

GERMINATION: The optimum range of soil temperatures for germination of tomatillo seed is 75 to 90°F (24 to 32°C). If starting seeds indoors, use bottom heat if available.

TRANSPLANTING: Start seeds 4 to 6 weeks prior to setting transplants to the garden. Direct-seed where growing seasons are longer.

TOMATO

Lycopersicon esculentum (syn. *L. lycopersicum*)

FAMILY: Solanaceae; nightshade family
PLANT TYPE: Tender perennials usually grown as annuals
SEED VIABILITY: 4 years
SAVE SEED FROM: As many plants as possible

FLOWERING: Tomato plant types are determinate (bush), semideterminate (semierect), and indeterminate. Determinate plants flower during a single relatively short period, indeterminate throughout the life of a plant, and semideterminate somewhere in between. Flowers are perfect, with the pistil enclosed by a cone of anthers. Tomatoes are a mostly self-pollinated crop.

ISOLATION REQUIREMENT: In the flowers of older heirloom varieties, the pistil may protrude from the corolla, which allows cross-pollination to occur more readily. Plant a tall crop to separate such heirloom varieties from your other tomatoes for extra protection against crossing, if desired.

SEED COLLECTION: The fruit is a berry and should be picked fully ripe and ready to eat, usually 6 to 8 weeks from fruit set.

SEED CLEANING: There are two pretreatment options for tomato seeds. You can allow the fruit to sit after harvest until it softens and becomes very mushy, or even begins to rot, then squeeze the seeds out of the

TOMATOES

fruit. Or immediately after harvest, squeeze the pulp and seeds from the fruit into a container, add a small amount of water, and let ferment several days at or below 70°F (21°C), stirring several times a day to speed fermentation. The fermentation process, which takes about 4 days, protects against seedborne bacterial canker. Nonviable seeds will float, while seeds having the best chance of being viable will sink. Pour off the water and the floating seeds.

After either pretreatment procedure, spread the seeds in a single layer on a paper towel or screen and let them dry for a week or two.

SEED TREATMENT: Prior to planting, hold tomato seeds in 122°F (50°C) water for 25 minutes to control seedborne diseases if they are an issue in your garden.

GERMINATION: The optimum range of soil temperatures for germination of tomato seed is 65 to 75°F (18 to 24°C), with minimum soil temperature of about 50°F (10°C).

TRANSPLANTING: Direct-seed to the garden in longer-season growing locations or use transplants for an earlier crop. Start seeds about 8 weeks prior to setting transplants to the garden.

Turnip

Brassica rapa (Rapifera group)

FAMILY: Brassicaceae; mustard family
PLANT TYPE: Usually biennial
SEED VIABILITY: 4 years
SAVE SEED FROM: At least 6 plants
SPACING FOR SEED SAVING: 2'

FLOWERING: Turnip flowers are borne in a terminal raceme. The flowers are perfect and insect-pollinated; they are not usually self-pollinating. Turnip usually does not flower until after a cool period, but there are a few varieties that will produce seed within one growing season.

OVERWINTERING TURNIPS

IN MILD-WINTER AREAS you can overwinter turnips in the garden. In cold-winter locales, you'll need to dig the plants and store them in a protected area for the winter.

YEAR 1
Sow turnip seed for plants intended for seed saving later than you would for an eating crop. Your goal is to have the roots reach 2 to 3 inches in diameter at digging time in fall. *In cold-winter regions,* dig up the plants and trim the leaves, leaving 1 to 2 inches of stem. Select for roots with best color, shape, and size. Store these stecklings in a dark, humid location. If your storage area tends to be dry, pack the stecklings in a box with moist sand or sawdust. Maintain the temperature between 35 and 45°F (2 and 7°C).

In warmer locations, dig the roots and select those with the best color, shape, and size. Remove the tops, then replant the roots right away.

YEAR 2
Replant stecklings outdoors in early spring. The plants will flower and produce seed.

ISOLATION REQUIREMENT: Several common vegetable crops are varietas of *Brassica rapa*, and they can all cross-pollinate each other. Among them are broccoli raab, Chinese cabbage, mustard, rutabaga, and turnips. If you want to save seed from more than one of these crops simultaneously, you'll need to either bag the plants or plant a tall-growing, non-related crop between the different *Brassica* "cousins." In a small garden you can also cage the plants and introduce pollinators. A third option is to use an alternate-day caging method. Recommended isolation distance for *Brassica rapa* crops is ½ mile.

SEED COLLECTION AND CLEANING: Handle turnip seed as you would cabbage. See the Cabbage entry for instructions.

SEED TREATMENT: Prior to planting, hold turnip seeds in 122°F (50°C) water for 20 minutes to control seedborne diseases if they are an issue in your garden.

GERMINATION: Optimal soil temperature for germination of turnip seeds ranges from 60 to 105°F (16 to 41°C).

TRANSPLANTING: Turnips are usually direct-seeded.

Unicorn Plant

Proboscidea louisianica (syn. *Martynia proboscidea*)

Devil's claw, unicorn flower

IMMATURE PODS OF UNICORN PLANT

FAMILY: Martyniaceae; unicorn family
PLANT TYPE: Annual
SEED VIABILITY: 2 years
SPACING FOR SEED SAVING: 2'

FLOWERING: Unicorn plant flowers are held in loose racemes and are very ornamental. They are insect-pollinated.

ISOLATION REQUIREMENT: Varieties within each species of unicorn plant will cross, but crosses among species do not occur. If cross-pollination is a concern, use alternate-day caging. Recommended isolation distance is ½ mile.

SEED COLLECTION: The fruit is a pod that has a long, curved proboscis resembling a sharp claw (hence the descriptive genus name). Since unicorn plants are native to the southwestern United States and Mexico, in cooler or short-season areas, transplants should be used or the fruits will not ripen. The green fruits shed their hairy green casing as they mature, exposing the dark woody seedpod.

SEED CLEANING AND STORAGE: The seeds are ready to be removed when the pods split. Unicorn plant pods are quite sharp; take care when removing seeds and keep animals out of the garden when fruits are drying.

GERMINATION: Soil temperatures should be between 75 and 90°F (24 and 32°C) for best germination.

TRANSPLANTING: Unicorn plant is usually direct-seeded, but transplants may be started to lengthen the growing season. Supply bottom heat if possible to improve and speed germination, and allow about 8 weeks to grow the transplants.

—————— CROP ALERT ——————

Martynia annua is a noxious weed in some areas. Contact your local weed board for more information.

RIPENING WATERMELON

Watermelon

Citrullus lanatus

FAMILY: Cucurbitaceae; squash family

PLANT TYPE: Annual

SEED VIABILITY: 4 years

SPACING FOR SEED SAVING: 10–25', depending on the variety.

FLOWERING: Watermelon plants are usually monoecious, but there are andromonoecious varieties. Determine the female flowers by looking for a small, watermelon-shaped ovary under the flower. Male flower stems are straight and thin. There are more male flowers than female, and many of the female flowers won't set fruit, often due to inadequate pollination. Watermelon plants are insect-pollinated, and professional seed growers will bring in one or two honeybee hives per acre of watermelons.

ISOLATION REQUIREMENT: Watermelon varieties will cross-pollinate one another, and they will also cross with citron (*C. lanatus*). If any cross-pollination occurs between watermelons and melons (*Cucumis melo*), cucumbers, squash, or pumpkins, it will not result in formation of viable seed; any fruit that forms would be parthenocarpic. To ensure seed purity, hand-pollinate, following the instructions on page 29, Hand-Pollinating Squash. Recommended isolation distance for different varieties of watermelon is ½ mile.

SEED COLLECTION: A watermelon is technically a berry. Plants usually produce only one or two decent fruits per vine. If more fruits set, then all the fruits will be small and of poor quality. Remove any extra fruits that have set. Seeds are ripe when watermelons are good for eating. This may take from 70 to 110 days from planting, depending on the location, environment, fertility, water availability, and other variables. Check the ground spot — it should turn yellow when the fruit is ready to be picked. Also check the tendril closest to the fruit; when the fruit is ripe, this tendril will have died and blackened. Save the seed from particularly tasty melons, but do not save seed from melons with "hollow heart" or if the heart flesh is paler than that of the rest of the flesh.

SEED CLEANING AND STORAGE: Wash seeds lightly (especially if your helpers have been engaged in a spitting contest!), rinse, and spread on a screen, paper towel, or cookie sheet to dry. Dry for about 1 week.

GERMINATION: When starting plants indoors, use heating coils or a heat mat if possible, as watermelon seeds germinate best when soil temperatures are between 80 and 90°F (27 and 32°C). Seeds will germinate in 3 to 12 days, depending on the temperature.

TRANSPLANTING: Start watermelon seeds 4 weeks before you intend to set transplants to the garden. The optimum range of soil temperatures for transplanting is 70 to 85°F (21 to 29°C). As with other members of this family, watermelon does not transplant easily, so sow seeds in peat pots, and plant pot and plant to the garden.

CHAPTER 9

SAVING HERB SEEDS

CREATING A CHAPTER ON HERBS presents a challenge for anyone writing a gardening book, because so many kinds of plants can be called "herbs." We decided that the best focus for a book on seed saving would be culinary herbs, such as dill, parsley, and basil. We've also included some dual-purpose herbs, such as clary (*Salvia sclaria*), which has both culinary and medicinal uses. On the other hand, some culinary herbs, such as bee balm (*Monarda*), are also planted in perennial beds and borders simply for the beauty of their flowers, and we made the choice to include such herbs/flowers in chapter 10, which covers annual, biennial, and perennial flowers. Still other herbs are woody plants (lavender is one example), and we cover those in chapter 11.

Some herbs are very easy to propagate from seed, although many require more pregermination treatment than vegetable seeds typically do. Other types of herbs are normally propagated asexually by division or cuttings rather than by seed. Little is known about isolation requirements for herbs; this may be because most herbs are not sold as named varieties. We simply plant "dill" or "anise," so there is little chance of mixing genetic material. However, if you have planted more than one named variety of an herb and want to try

saving seeds from one variety, isolate it from other varieties by bagging or caging or staggering planting times, as explained in chapter 2.

Asexual propagation is the more reliable way to maintain named varieties of many herbs, such as apple mint and chives. In fact, seeds of many mint-family species will be nonviable and will produce no seedlings at all. So with some herbs, the fun of propagating them by seed is precisely to see what unique-looking or -tasting new plants will result. If your wish is to reproduce the parent plants exactly, stick with dividing your herbs or taking cuttings instead.

Botanically speaking, most of the herbs discussed in this chapter are members of Lamiaceae (the mint family) and Apiaceae (the parsley family). Because some herbs are known by multiple common names, we've elected to list them alphabetically by genus name. We've also done this as a reminder that it's important to make sure you identify a plant precisely if your goal is to collect seeds from it. Seeds of various herbs vary widely in their requirements for germination. If you misidentify an herb, you'll end up following the wrong instructions for getting its seeds to germinate, and you may end up with a flat full of nothing!

Most of the plants discussed in this chapter have dry fruits: the fruits of the mints are almost always nutlets, while those of the parsley family are almost always schizocarps. Moreover, the seed heads of the parsley family are notorious for uneven ripening and early shattering. You'll need to bag the seed heads to prevent seed loss by shattering, and you'll find specific directions for this technique on page 40.

The seeds of all herbs are considered orthodox and should be dried thoroughly, winnowed, then stored in sealed containers at 36 to 41°F (2 to 5°C). For most herbs, it's not known how long seeds will remain viable, so your best bet is to save them only for a short period. And unless we've indicated otherwise, herb seeds that need stratification should be treated as described on page 60.

A WORD OF CAUTION

MANY HERBS HAVE TOXIC PROPERTIES and/or the potential to become weeds — tansy is just one example. We've noted these characteristics in some of the plant entries in this chapter. By including such plants, we are not necessarily advocating growing them. Check with your local weed board to see which plants are viewed as noxious weeds in your region, and if you plan to grow any herbs with toxic properties, always take precautions if there are young children living nearby by informing your neighbors and/or fencing the area.

AGASTACHE

Giant hyssop, hummingbird mint, mosquito plant

A. aurantiaca (**orange hummingbird mint**), *A. barberi, A. cana* (**mosquito plant**), *A. foeniculum* (**anise hyssop, blue giant hyssop, fennel giant hyssop, fragrant giant hyssop, giant hyssop**), *A. nepetoides* (**yellow giant hyssop**), *A. pallidiflora* (**Texas hummingbird mint**), *A. pringlei, A. rugosum* (**Korean mint, wrinkled giant hyssop**), *A. rupestris* (**licorice mint, threadleaf giant hyssop**), *A. scrophulariifolia* (**purple giant hyssop**)

AGASTACHE FLOWER SPIKES

FAMILY: Lamiaceae; mint family
PLANT TYPE: Perennial
SEED VIABILITY: Unknown

FLOWERING: The small flowers, which are perfect, are arranged on long spikes.

SEED COLLECTION: Harvest the spikes by hand when they have browned.

SEED CLEANING: Spread the spikes on screens to dry, then rub to separate the nutlets. Each flower will bear four nutlets.

SEED TREATMENT: Stratify seeds of purple giant hyssop (*A. scrophulariifolia*) for 2 months at about 32°F (0°C).

GERMINATION: Sow the seeds when soil temperatures reach 68°F (20°C), and be patient. Some species need up to 4 weeks for germination to occur, but germination occurs in less than 2 weeks in others.

TRANSPLANTING: Seedlings need 6 to 8 weeks to reach transplanting size at 70°F (21°C).

Allium

Chives, garlic, leek, onion

A. acuminatum, A. aflatunense, A. ampeloprasum (**wild leek**), *A. amethystinum, A. amphibolum, A. atropurpureum, A. cernuum* (**nodding onion, wild onion, lady's leek**), *A. carinatum* ssp. *pulchellum, A. caeruleum, A. christophii* (**stars of Persia**), *A. cyathophorum* var. *farreri, A. fistulosum* (**ciboule, Japanese bunching onion, Spanish onion, two-bladed onion, Welsh onion**), *A. flavum, A. giganteum, A. hollandicum, A. karataviense, A. mairei, A. moly* (**lily leek**), *A. narcissiflorum, A. neapolitanum* (**daffodil garlic, flowering onion**), *A. nigrum, A. oreophilum, A. ostrowskianum, A. praecox, A. pulchellum, A. ramosum* (**fragrant-flowered garlic**), *A. rosenbachianum, A. schoenoprasum* (**chives, cive**), *A. senescens, A. sikkimensis, A. sphaerocephalum* (**round-headed garlic**), *A. splendens, A. stipitatum, A. stellatum* (**prairie onion**), *A. tanguticum* (**lavender globe lily**), *A. tuberosum* (**Chinese chive, garlic chive, Oriental garlic**), *A. ursinum* (**bear garlic, buckram, gypsy garlic, hog's garlic, ramsons**), *A. victorialis*

FAMILY: Liliaceae; lily family
PLANT TYPE: Biennial or perennial
SEED VIABILITY: About 1 year

FLOWERING: The flowers are small and perfect and are borne in an umbel.

SEED COLLECTION: The fruits are capsules, each containing shiny black, flat, angled or round seeds. Harvest the seed heads when they have begun to brown but before seed-shed.

SEED CLEANING: Spread the drying heads on screens to dry completely; then rub, thresh, or flail to remove the seeds.

SEED TREATMENT AND GERMINATION: For most of the species listed above, sow the seeds at 68°F (20°C) for a month, stratify them for another month, then move them to 50°F (10°C) for germination. Some species, though, require different treatment:

ALLIUM SEED HEADS WITH OPEN CAPSULES AND SEEDS

◗ Stratify seeds of lavender globe lily (*A. tanguticum*) and Chinese chive (*A. tuberosum*) for 6 weeks before moving them to 68°F (20°C) for germination.

◗ Sow seeds of chives (*A. schoenoprasum*), wild leek (*A. ampeloprasum*), *A. amphibolum, A. caeruleum, A. mairei, A. senescens, A. sikkimensis*, and *A. splendens* at 68°F (20°C) and allow about 2 weeks for germination.

◗ Sow the seeds of daffodil garlic (*A. neapolitanum*) and fragrant-flowered garlic (*A. ramosum*) at 70°F (21°C), and keep the relative humidity very high.

◗ Sow fresh seeds of *A. karataviense* in the garden right after harvest.

◗ Sow seeds of *A. amethystinum* and *A. hollandicum* at 39°F (4°C) and be patient — germination is erratic and takes a long time.

TRANSPLANTING: Grow the seedlings on at about 68°F (20°C) for about 3 months before transplanting to the garden in early spring. You may also direct-sow the seeds in very early spring.

——————— HERB ALERT ———————

Chives and some of other *Allium* species can become pesky weeds in some areas, self-sowing and growing out of bounds. To reduce the chance of this happening, pick off the flower heads before their seeds are mature (if you don't intend to save the seeds). Chives are usually propagated by division or bulbs.

Anethum

Dill

A. graveolens (**dill**)

FAMILY: Apiaceae; parsley family
PLANT TYPE: Annual, but may be biennial in some areas
SEED VIABILITY: Unknown

FLOWERING: The small yellow perfect flowers are borne in umbels. Dill is cross-pollinated.

SEED COLLECTION: The umbels shatter easily. As each umbel begins to turn brown, cover it with a paper bag to catch the seeds. Then snip off the entire seed head and leave it in the bag to dry further. Alternatively, harvest the umbels when the stem is still slightly green, and rub off the seeds at that time. Then spread them on a screen to dry. If you leave any seed heads intact in the garden, dill will self-sow readily; it can become weedy in some areas.

SEED CLEANING: After a couple of weeks of drying time, shake the bagged seed heads well and the seeds will separate. Rub any clinging seeds off the heads. Then winnow the seed to clean it.

GERMINATION: Sow the seeds in light at 60°F (16°C) and allow up to 4 weeks for full germination. Use peat pots so as not to disturb the roots during transplanting.

TRANSPLANTING: Grow the plants on for about 8 weeks at 60°F (16°C), then transplant them very carefully. Direct-sowing is also an option.

Angelica

Angelica, masterwort

A. archangelica (**angelica, masterwort**)

FAMILY: Apiaceae; parsley family
PLANT TYPE: Biennial
SEED VIABILITY: Very brief

FLOWERING: The small white or greenish perfect flowers are borne in umbels.

SEED COLLECTION: The fruits are compressed and winged. Allow umbels to dry in place, but harvest the seeds before the umbels shatter or the plants will self-sow.

SEED CLEANING: Spread the umbels on screens to dry further. Rub, thresh, or flail to remove the seeds.

SEED TREATMENT: Stratify the seeds for about a month before germination.

GERMINATION: Move the seeds to 68°F (20°C) in light, and allow about a month for them to germinate. Seeds of this species have a very low germination percentage.

TRANSPLANTING: Use only very fresh seeds to grow transplants, or, even better, sow the seeds in the fall and let them germinate in place in spring.

Artemisia

Dusty miller, tarragon, wormwood

A. absinthium (**absinthe, common wormwood**), *A. chamaemelifolia*, *A. dracunculus* (**estragon, tarragon**), *A. genipi*, *A. laxa* (**alpine wormwood**), *A. ludoviciana* (**white sage**), *A. stelleriana* (**beach wormwood, dusty miller, old woman**), *A. umbelliformis* (**alpine wormwood**), *A. vulgaris* (**felon herb, mugwort**)

FAMILY: Asteraceae; sunflower family
PLANT TYPE: Perennial
SEED VIABILITY: Unknown

FLOWERING: The disk flowers are perfect, the ray flowers pistillate.

SEED COLLECTION: The fruit are achenes. Harvest the spent flower heads when they have browned and begun to dry.

SEED CLEANING: Spread the heads on screens to dry, then rub to separate the achenes.

GERMINATION: Sow the seeds of most species at 41°F (5°C) and allow germination to proceed for several months. Germination is irregular. Seeds of absinthe (*A. absinthium*) and white sage (*A. ludoviciana*) germinate in a couple of weeks at 68°F (20°C).

TRANSPLANTING: The seeds are usually sown in place.

BORAGO

Borage, cool-tankard, tailwort

B. officinalis (**borage, cool-tankard, tailwort**)

FAMILY: Boraginaceae; borage family
PLANT TYPE: Annual
SEED VIABILITY: Unknown

FLOWERING: The small blue perfect flowers are borne in cymes.

BORAGE SEEDS

SEED COLLECTION: Each fruit contains four nutlets. Harvest the dried flower heads by hand.

SEED CLEANING: Spread the dried heads on screens to dry further; then thresh, flail, or rub to remove the seeds.

GERMINATION: Sow the seeds at about 68°F (20°C) in darkness. They'll take about 2 weeks to germinate.

TRANSPLANTING: Borage does not transplant well, so start the new plants in peat pots. Better yet, simply plant the seeds directly in the garden in spring.

——————— HERB ALERT ———————

Borage can self-sow and become weedy.

CARUM

Caraway

C. carvi var. *annua* (**annual caraway**), *C. carvi* var. *biennis* (**biennial caraway**)

FAMILY: Apiaceae; parsley family
PLANT TYPE: Annual or biennial, depending on species
SEED VIABILITY: Very short; use only fresh seeds

FLOWERING: The small white to pink flowers are perfect and borne in umbels.

SEED COLLECTION: The fruits are small, compressed schizocarps. Bag the umbels before they shatter and allow them to dry further in the bag. Biennial caraway can become weedy if allowed to self-sow.

SEED CLEANING: When seeds are fully dry, rub the seed heads to singularize the seeds.

GERMINATION: Sow seeds of annual caraway in light at about 68°F (20°C). Sow seeds of biennial caraway directly in the garden when soil temperature reaches about 68°F. Seeds of both should germinate in about 2 weeks.

TRANSPLANTING: The annual variety develops a heavy taproot and should be transplanted when quite young, perhaps after 4 to 6 weeks. Otherwise, direct-sow it in the garden in spring, as you would the biennial variety.

CHAMAEMELUM

Chamomile

> *C. nobile* (**chamomile, garden chamomile, Roman chamomile, Russian chamomile**)

FAMILY: Asteraceae; sunflower family
PLANT TYPE: Perennial
SEED VIABILITY: Unknown

FLOWERING: The flowers are borne in heads.

SEED COLLECTION: The fruit are achenes. Collect the dried heads by hand.

SEED CLEANING: Spread the heads on screens to dry; then rub, thresh, or flail to remove the fruit.

GERMINATION: Sow the seeds at 68°F (20°C), then allow 2 weeks for germination.

TRANSPLANTING: Grow plants on for 4 to 6 weeks at about 65°F (18°C). You may also direct-sow chamomile seeds to the garden in spring.

CHENOPODIUM

Epazote, good King Henry, quinoa

> *C. ambrosioides* (**American wormseed, epazote, Mexican tea, Spanish tea, wormseed**),
> *C. bonus-henricus* (**allgood, fat hen, good King Henry, goosefoot, Mercury, wild spinach**),
> *C. capitatum* (**Indian plant, strawberry blite**),
> *C. quinoa* (**quinoa, quinua**)

FAMILY: Chenopodiaceae; goosefoot family
PLANT TYPE: Annual but can be perennial in warmer areas
SEED VIABILITY: Unknown

FLOWERING: The small, mostly perfect flowers are borne in panicles. The fruits are utricles.

SEED COLLECTION: Harvest the panicles when they have turned brown, but handle them carefully so as not to encourage shattering.

SEED CLEANING: Spread the panicles on screens to dry, then rub or thresh to remove the seeds.

GERMINATION: Surface-sow seeds directly into the garden when the soil temperature has reached 68°F (20°C) at a depth of 2 inches, and allow about 2 weeks for germination. The seeds require light for germination.

——————— HERB ALERT ———————

Many of the *Chenopodium* species self-sow readily and can become quite weedy, so take care to keep them in bounds.

CHRYSANTHEMUM

Chrysanthemum

> **ANNUAL SPECIES:** *C. coronarium* (syn. *Leucanthemum coronarium*; **crown-daisy chrysanthemum, garland chrysanthemum, shungiku**)

> **PERENNIAL SPECIES:** *C. balsamita* (**alecost, costmary, mint geranium**)

FAMILY: Asteraceae; sunflower family
PLANT TYPE: Annual or perennial, depending on species
SEED VIABILITY: 9 to 10 years

FLOWERING: The disk flowers are perfect, the ray flowers pistillate.

ISOLATION REQUIREMENTS: All species are cross-pollinated, so isolate by at least ¼ mile.

SEED COLLECTION: The fruits are achenes. Harvest the plants as soon as the flower heads are mature. Any delay may result in great loss of seeds from shattering.

SEED CLEANING: Dry the plants further on a screen, then rub, flail or tumble to remove the seeds.

GERMINATION: For seeds of most species, germination takes 14 to 35 days at 65°F (18°C).

TRANSPLANTING: Chrysanthemums need about 2 months to reach transplanting size at 65°F (18°C).

——————— HERB ALERT ———————

Some folks develop an allergic skin reaction (a rash) from contact with the foliage of the plants in this genus. Wear gloves when handling the plants.

CORIANDRUM

Chinese parsley, cilantro, coriander

> *C. sativum* (**Chinese parsley, cilantro, coriander**)

FAMILY: Apiaceae; parsley family
PLANT TYPE: Annual
SEED VIABILITY: Unknown

FLOWERING: The perfect flowers are borne in umbels. The plants are naturally outcrossed by insects and will cross with other varieties. Fortunately, coriander is usually sold simply as "coriander," without varieties given, so there is little to worry about. If crowded in the plant bed, the plants will tend to bolt early, so give them plenty of room.

SEED COLLECTION: The seeds are oval shaped. Harvest the umbels as they dry but before they shatter. (When grown for its seed, this herb is called coriander. When leaves are harvested, they are called cilantro.)

SEED CLEANING: Spread the umbels on a screen to dry further; then rub, flail, or thresh to remove the seeds.

SEED TREATMENT: Soak seeds for 30 minutes in 127°F (53°C) water just prior to planting to control seedborne diseases if these are an issue in your garden.

CILANTRO (CORIANDER) FOLIAGE & SEEDS

GERMINATION: Sow the seeds in darkness at 60°F (16°C) and allow up to 3 weeks for germination.

TRANSPLANTING: Start the seeds in peat pots and transplant to the garden after about 4 weeks. These plants develop a large taproot, and transplanting more mature plants will be difficult. Plants that have been transplanted also tend to bolt earlier than those that were direct-seeded. You may direct-seed your coriander into the garden in early spring or in fall.

CRITHMUM

Rock samphire, samphire, sea fennel

> *C. maritimum* (**rock samphire, samphire, sea fennel**)

FAMILY: Apiaceae; parsley family
PLANT TYPE: Perennial
SEED VIABILITY: Seeds lose viability quickly

FLOWERING: The perfect flowers are arranged in umbels.

SEED COLLECTION: Take measures to prevent the seed heads from shattering and dispersing their seeds on the ground by bagging when the first seeds turn brown.

SEED CLEANING: Spread the heads on screens to dry, then rub or flail to remove the seeds.

GERMINATION: Sow the seeds at 68°F (20°C) and hold them there for a month. If none germinates, then stratify them for a month, and return them to 68°F (20°C) for germination.

TRANSPLANTING: Little is known about transplanting samphire because the usual method of propagation is by division in spring.

CUMINUM

Cumin

C. cyminum (**cumin**)

FAMILY: Apiaceae; parsley family
PLANT TYPE: Annual
SEED VIABILITY: Unknown

FLOWERING: The small white- or rose-colored flowers are borne in umbels.

SEED COLLECTION: Harvest the umbels as they brown but before they shatter. Seed ripening will be uneven, so you should bag the umbels to catch any seeds that shed early.

SEED CLEANING: Spread the heads on screens to dry, then rub to separate the seeds.

GERMINATION: Direct-sow to the garden when the soil temperature has reached 68°F (20°C) and allow about 2 weeks for germination.

ELETTARIA

Cardamom

E. cardamomum (**cardamom**)

FAMILY: Zingiberaceae; ginger family
PLANT TYPE: Perennial
SEED VIABILITY: Unknown

FLOWERING: The perfect flowers are borne on spikes.

SEED COLLECTION: The fruit is a capsule. Harvest the capsules when they have begun to turn brown.

SEED CLEANING: Spread the capsules on screens to dry, then rub or flail to remove the seeds.

GERMINATION: Sow the seeds in place outdoors when the soil temperature has reached 70°F (21°C). Plants from seeds may take up to 3 years to produce a good crop of capsules; because of this, cardamom is usually propagated by dividing its irislike rootstock.

ERYNGIUM

Culantro

E. foetidum (**culantro, eringo, fitweed, shado beni**)

FAMILY: Apiaceae; parsley family
PLANT TYPE: Biennial
SEED VIABILITY: Unknown

FLOWERING: The small perfect flowers are borne in heads.

SEED COLLECTION: Harvest the heads when they are drying but before they shatter.

SEED CLEANING: Spread the heads on screens to dry, then thresh or flail to remove the seeds.

GERMINATION: Sow the seeds in place when the soil temperature reaches 68°F (20°C) and allow about 2 weeks for germination, or fall-sow.

Melissa

Lemon balm, lemon mint

M. officinalis (**lemon balm, lemon mint**)

FAMILY: Lamiaceae; mint family
PLANT TYPE: Perennial
SEED VIABILITY: Unknown

FLOWERING: The flowers are whitish, perfect, and borne in cymes.

SEED COLLECTION: There are four single-seed nutlets per flower.

SEED CLEANING: Harvest the cymes as they dry and spread them on screens. When thoroughly dry, flail, thresh, or rub to separate the nutlets.

GERMINATION: Sow the seeds in light at 68°F (20°C), then allow about 2 weeks for germination.

TRANSPLANTING: Grow the plants on for about 8 weeks at 68°F (20°C). You may also sow the seeds directly into the garden in spring.

Mentha

Mint

M. aquatica (**water mint**), *M. longifolia* (**horsemint**), *M. ×piperita* (**peppermint**), *M. pulegium* (**pennyroyal**), *M. spicata* (**spearmint**), *M. suaveolens* (**apple mint**)

FAMILY: Lamiaceae; mint family
PLANT TYPE: Usually perennial but sometimes annual
SEED VIABILITY: Unknown

FLOWERING: The perfect flowers are produced in dense spikes.

SEED COLLECTION: Harvest the spikes when they brown.

SEED CLEANING: Spread the spikes on screens to dry; then flail, rub, or thresh to separate the nutlets.

GERMINATION: Sow seeds at 68°F (20°C) then allow a few weeks for germination. These species are usually propagated by cuttings or division of rhizomes.

TRANSPLANTING: If you do decide to transplant mint, place the dormant roots in water at 112°F (44°C) for 10 minutes just prior to planting to help control diseases.

————————— HERB ALERT —————————

The perennial mints are generally best propagated by division, as they hybridize readily and do not come true from seed. In fact, many hybrids are sterile and produce no viable seed at all. Of course, if you like to experiment, try seed propagation, but if you want to keep, for example, apple mint growing in your garden as apple mint, it's better to propagate it asexually. The mints tend to be weedy in many areas.

Myrrhis

Myrrh

M. odorata (**anise, myrrh, sweet chervil, sweet cicely**)

FAMILY: Apiaceae; parsley family
PLANT TYPE: Perennial
SEED VIABILITY: Less than 1 year.

FLOWERING: The small white flowers are perfect and borne in umbels.

SEED COLLECTION: Hand-pick the dry umbels before the fruits shatter.

SEED CLEANING: Spread the umbels on screens to dry further; then rub, thresh, or flail to remove the seeds.

GERMINATION: Sow the seeds at 68°F (20°C) and allow about a month for germination.

TRANSPLANTING: Grow the plants on at about 65°F (18°C) for 8 to 10 weeks before setting to the garden. The plants have a large taproot and do not transplant well, so start them in peat pots and transplant when they are no older than 10 weeks. You may also sow seeds in place in the fall or in the early spring. In fact, because of poor seed viability, it's probably best to sow the seeds outdoors in the fall soon after they ripen.

NEPETA

Catmint, catnip

N. cataria (**catmint, catnip**)

FAMILY: Lamiaceae; mint family
PLANT TYPE: Annual in cooler locales; perennial in warmer areas
SEED VIABILITY: Unknown

FLOWERING: The flowers may be perfect or imperfect and are borne in spikes.

SEED COLLECTION: Collect the dried flowers and spikes by hand.

SEED CLEANING: Spread the dried flowers and spikes on screens to dry further; then flail, rub, or thresh to remove the nutlets.

GERMINATION: Sow seeds at 68°F (20°C) and allow a week or two for germination.

TRANSPLANTING: Grow the plants on at about 68°F (20°C) for 6 to 8 weeks, then set to the garden in

NEPETA FLOWER SPIKES

spring. You may also sow the seeds directly into the garden in early spring.

OCIMUM

Basil

O. americanum (**sweet basil**), *O. basilicum* (**basil, Siam queen, Thai basil, Thai sweet basil, common basil**), *O. basilicum* var. *citriodorum* (**lemon basil, lime basil**), *O. basilicum* var. *minimum* (**spicy basil**), *O. basilicum* var. *purpurascens* (**purple leaf basil**), *O. bullatum, O. kilimandscharicum* (**hoary basil**), *O. micranthum, O. rotundifolium* (**creeping basil**), *O. tenuiflorum*

FAMILY: Lamiaceae; mint family
PLANT TYPE: *O. basilicum* is an annual; the other species can be annual or perennial
SEED VIABILITY: Unknown

FLOWERING: The small perfect flowers are borne in racemes or panicles. The plants tend to inbreed with the help of various insects.

SEED COLLECTION: Harvest the clusters when they brown. When the bottom fruits brown, harvest the entire raceme.

SEED CLEANING: Spread the clusters on screens to dry, then rub or thresh to remove the nutlets. Put the nutlets and chaff in a bowl and gently swirl them around. The very small seeds will sink to the bottom, then you can rake off the chaff with your hands.

GERMINATION: Seeds of all basils should be sown at 68°F (20°C) in light and will take about 10 days to germinate.

TRANSPLANTING: It takes 6 to 8 weeks at 68°F (20°C) to grow good transplants of these species.

herbs

ORIGANUM

Oregano, sweet marjoram

O. majorana (**sweet marjoram**), *O. vulgare* (**oregano**)

FAMILY: Lamiaceae; mint family

PLANT TYPE: Both species are annual, oregano may be perennial in warmer areas

SEED VIABILITY: About 1 year; plant only fresh seeds

FLOWERING: The perfect flowers are arranged in complex clusters.

SEED COLLECTION: Harvest the clusters when dry.

SEED CLEANING: Spread the dry clusters on screens to dry further, then thresh or rub to separate the nutlets.

GERMINATION: Sow seeds of both species in light at 68°F (20°C). It'll take about 14 days for germination to be complete.

TRANSPLANTING: You may grow plants on at about 68°F (20°C) for 6 to 8 weeks before transplanting to the garden. A much easier way to handle these species is simply to direct-sow to the garden in spring.

HERB ALERT

O. vulgare var. *hirtum* is the aromatic culinary oregano that cooks and gardeners love. In our experience, culinary oregano grown from seed doesn't have a lot of punch. If your goal is oregano to use in the kitchen, taking cuttings from a particularly flavorful *O. vulgare* var. *hirtum* plant is your best bet.

PETROSELINUM

Parsley

P. crispum (**parsley**), *P. crispum* var. *neapolitanum* (**Italian parsley**)

FAMILY: Apiaceae; parsley family

PLANT TYPE: Biennial

SEED VIABILITY: 3–5 years

FLOWERING: The perfect flowers are borne in umbels.

SEED COLLECTION: As with most members of the Apiaceae, parsley umbels tend to shatter. As the umbels brown and the seeds near maturity, place a paper bag over a head, tying it at its base, and allow the shattered seeds to fall into the bag. After a couple of weeks, snip the umbels off the plant, leaving them in the bag, and allow them to dry further.

SEED CLEANING: When the umbels are perfectly dry, shake the bag to remove more seeds, then rub the remaining seeds from the umbels. Winnow to clean the seeds.

GERMINATION: Seeds of parsley (*P. crispum*) can be sown at 68°F (20°C), and they will take about 21 days to germinate. Seeds of Italian parsley (*P. crispum* var. *neapolitanum*) should be soaked in water overnight, then sown at 68°F (20°C).

TRANSPLANTING: Allow about 8 weeks to grow good transplants of these species. Transplanting biennials will sometimes cause them to bolt rapidly, so you would be better off to sow the seeds directly into the garden in spring.

OREGANO FLOWER CLUSTERS

PIMPINELLA

Anise

P. anisum (**anise**)

FAMILY: Apiaceae; parsley family
PLANT TYPE: Annual
SEED VIABILITY: Unknown

FLOWERING: The perfect flowers are borne in umbels, which tend to shatter easily.

SEED COLLECTION: Harvest the seed heads as they begin to brown, or bag them to capture the seeds as they shed.

SEED CLEANING: Spread the umbels on screens to dry further, then rub or flail to remove the seeds. It may take a couple of weeks for a thorough harvest, as the seeds don't ripen evenly.

GERMINATION: Sow the seeds in flats at 68°F (20°C) and allow 2 weeks for germination.

TRANSPLANTING: Grow the plants on at 60°F (16°C) for about 8 weeks, then transplant to the garden. You may also direct-sow to the garden when soil temperature has reached 68°F (20°C).

POTERIUM

Burnet, salad burnet

P. sanguisorba (**burnet, salad burnet**)

FAMILY: Rosaceae; rose family
PLANT TYPE: Perennial
SEED VIABILITY: Unknown

——————— HERB ALERT ———————

Once established, this species will self-sow readily and can become weedy in some areas.

——————————————————————————

FLOWERING: Sex in this species is interesting. The lower flowers on the plant are staminate, the midlevel flowers are perfect, and the upper flowers are pistillate.

SEED COLLECTION: Harvest the perfect and pistillate flowers when they dry.

SEED CLEANING: Spread the spent flowers and fruits on screens to dry, then thresh to remove the seeds.

GERMINATION: Seeds take about 2 weeks to germinate at 68°F (20°C).

TRANSPLANTING: Burnet plants do not transplant well, so sow the seeds directly into peat pots and grow them on for about 8 weeks at 68°F (20°C). You can also simplify things by direct sowing them to the garden in spring.

RUTA

Rue

R. graveolens (**An Ruadh Lus, countryman's treacle, herb of grace, rue**)

FAMILY: Rutaceae; citrus family
PLANT TYPE: Perennial
SEED VIABILITY: Unknown

FLOWERING: The greenish yellow, perfect flowers are borne in terminal panicles.

——————— HERB ALERT ———————

This plant may be poisonous if prepared improperly. For some people, the oil and leaves, if rubbed on the skin, can cause severe blistering in full sunlight. Ingestion of rue oil can cause vomiting and death. The fruits do not contain rue oil.

——————————————————————————

SEED COLLECTION: The fruits are four- or five-lobed capsules. Harvest them when they are dried.

SEED CLEANING: Spread the capsules on screens to dry further, then flail or rub to remove the seeds.

GERMINATION: Sow the seeds at 68°F (20°C); they should germinate in 1 to 2 weeks.

SALVIA

Chia, clary, ramona, sage

S. elegans (**pineapple sage**), *S. officinalis* (**common salvia, garden salvia, salvia**), *S. sclarea* (**clary**)

FAMILY: Lamiaceae; mint family
PLANT TYPE: Perennial
SEED VIABILITY: About 1 year

FLOWERING: The flowers are perfect and borne in long clusters.

SEED COLLECTION: Harvest the flower clusters by hand as they dry.

SEED CLEANING: Spread the flowers on screens to dry; then flail or thresh to separate the fruits, three-angled nutlets. Each flower produces four fruits.

GERMINATION: Sow the seeds at 68°F (20°C) and allow up to 3 weeks for germination.

TRANSPLANTING: Grow the plants on at about 68°F (20°C) for 8 weeks before transplanting to the garden.

SATUREJA

Calamint, savory

ANNUAL SPECIES: *S. hortensis* (**summer savory**)

PERENNIAL SPECIES: *S. montana* (**winter savory**)

FAMILY: Lamiaceae; mint family
PLANT TYPE: Annuals and perennials
SEED VIABILITY: Unknown

FLOWERING: The perfect flowers are borne in cymes.

SEED COLLECTION: Harvest the clusters when they dry.

SEED CLEANING: Spread the clusters on screens to dry; then thresh to remove the fruits, which are four shiny nutlets per flower.

GERMINATION: Surface-sow nutlets in light at 68°F (20°C), and allow about 2 weeks for germination to be complete. Germination of winter savory (*S. montana*) seeds is erratic, and you may have to allow more time for germination to proceed.

TRANSPLANTING: Grow the plants on at around 60°F (16°C) for about 8 weeks before setting to the garden.

TANACETUM

Tansy

T. vulgare (**golden buttons, common tansy**)

FAMILY: Asteraceae; sunflower family
PLANT TYPE: Rhizomatous perennial
SEED VIABILITY: Unknown

FLOWERING: The small yellow perfect flowers are borne in clusters at the tops of the plants.

SEED COLLECTION: Harvest the heads when they brown but before they shatter.

— HERB ALERT —

This is a perennial that can become a noxious weed in some areas. The leaves and stems are poisonous if ingested in large quantities. Think carefully before planting this herb or using it as a remedy.

SEED CLEANING: Spread the heads on screens to dry, then rub to remove the achenes.

GERMINATION: Surface-sow the seeds at 68°F (20°C) in light. If after a month they have not germinated, stratify them for another month and return them to 68°F (20°C).

TANSY FLOWER HEADS

TRANSPLANTING: Grow the plants on at around 65°F (18°C) for about 8 weeks, then transplant to the garden. Tansy is usually sown directly into the garden.

THYMUS

Thyme

> *T. ×citriodorus* (**lemon thyme**), *T. herba-barona* (**caraway**), *T. pulegioides, T. serpyllum* (**creeping thyme, lemon thyme, wild thyme**), *T. vulgaris* (**common thyme, garden thyme**)

FAMILY: Lamiaceae; mint family
PLANT TYPE: Perennial
SEED VIABILITY: Unknown

FLOWERING: The perfect flowers are crowded on terminal heads.

SEED COLLECTION: Collect the flowers when they have dried.

SEED CLEANING: Spread the flowers on screens to dry further; then thresh or rub to separate the fruits, four shiny nutlets per flower.

GERMINATION: Sow nutlets at 68°F (20°C) and allow up to 4 weeks for germination.

TRANSPLANTING: Grow the plants on at about 65°F (18°C) for 6 to 8 weeks before transplanting.

TRIGONELLA

Fenugreek

> *T. foenum-graecum* (**fenugreek, sicklefruit**)

FAMILY: Fabaceae; legume family
PLANT TYPE: Annual
SEED VIABILITY: Unknown

FLOWERING: The flowers are borne singly or in pairs in the leaf axils and are probably self-pollinated.

SEED COLLECTION: The seeds are borne in pods or legumes that may or may not be dehiscent. Harvest the pods as they brown but before they release their seeds.

SEED CLEANING: Spread the pods on screens to dry, then flail to remove the seeds.

GERMINATION: Sow the seeds in place when the soil temperature has reached 68°F (20°C), then allow about 2 weeks for germination.

SAVING FLOWER SEEDS

FOLKS SAVE FLOWER SEEDS for a variety of reasons. Perhaps you want to save seed from your aunt Zoe's stocks, or you might want to become a specialist in developing new flower types of cosmos. Your grandma's sweet peas may have been the most fragrant around, and you'd like to enter them in the county fair. Saving seeds from rhizomatous plants and plants typically grown from perennial bulbs will allow you to develop brand-new types of plants that are unavailable in the market. Saving flower seeds can be a real adventure for the patient gardener.

As you use this chapter, remember that some familiar plants have been reclassified into new genera or species. For example, about 20 years ago, Shasta daisy was renamed *Leucanthemum ×superbum*. However, it has now been officially reclassified as *Chrysanthemum ×superbum*, but you'll still find both names in use in seed catalogs and on websites. In these cases, we include both botanical names in the plant entries to lessen confusion. Also, we've organized the chapter by genus name rather than common name, because many flowering plants go by an assortment of common names. We do list widely recognized common names for your reference as well, but be aware that some common names are used for more than one species. For example, the name dusty miller is used to refer to *Centaurea gymnocarpa*, but *Centaurea ragusina* is called dusty miller, too.

Each entry includes a listing of plant type (annual, biennial, or perennial). However, a plant that is perennial in its native environment is often grown as an annual in colder areas; Johnny jump up (*Viola tricolor*) is a good example. We have tried to provide the most accurate classification for various species, but a perfectly accurate description depends to a large degree on where the plant is being grown.

Some familiar annuals and perennials have escaped cultivation in some regions, and they can become problematic weeds in agricultural fields or displace native plants in natural areas. So before you save seeds from flowers in your garden, check to make sure that the plants are not considered invasive in your area (see page 38 for more about this.).

Less information has been recorded about the specific requirements for saving and germinating seeds of garden flowers than for seeds of vegetables and some woody plants. The most important things to remember in saving and storing flower seeds are to let them ripen on the plant for as long as possible, to dry them well, and to store them in a cool, dark place in sealed containers. Here are some other points to keep in mind as well:

SPACING. Flower plants grown especially for seeds often require wider spacing than do those grown for their flowers. These larger spacings provide for good air circulation and retard diseases. If your garden area is large enough, follow the recommendations in this chapter for wider spacings. If you don't have the room, don't worry too much about it. Most likely your plants still will produce viable seeds.

INBREEDING DEPRESSION. Some flowering plants will suffer from inbreeding depression if seeds are saved from only a few individuals each year (for an explanation of inbreeding depression, see page 20). Save seeds from as many plants as you can of a variety that still display the characteristics you are striving to maintain. At minimum, try to collect flower seeds from at least six plants.

CONTROLLING POLLINATION. Flowers are either insect- or wind-pollinated. Some species, such as sweet peas, are self-pollinating, whereas others are typically self-pollinated but can be cross-pollinated by insects and still others are only cross-pollinated. Self-sterility can occur as well.

Bagging and caging are good ways to avoid cross-pollination. Isolation can also be accomplished by planting different varieties of a plant at different times of the year where your season allows or by planting varieties with different days to maturity. We've also included recommended isolation distances for those plants where that information is available, but keep in mind that isolation distances vary by authority. Your best bet is to separate different varieties of the same flower species by as much distance as you can. If your space is limited, isolate different varieties by planting a tall plant of some other type in between. Hand-pollination is another good option to maintain pure seed. See chapter 2 for details.

SEED COLLECTION. The fruits of some species of flowers shatter easily and/or ripen unevenly and must be hand-picked as they ripen over time, then finished by drying in bags or on screens set up over a sheet to capture the seeds. Among these are columbines, delphiniums, salvias, pansies, petunias, and zinnias. In fact, when you make your first harvest of ripe seed

heads from these plants, there may still be flowers in bloom and "green" fruits on the plants.

Fruits of marigolds, phlox, and many other species ripen more or less evenly, which allows for an entire plant to be harvested, dried, and threshed. Pinks and alyssum, for example, ripen their fruit unevenly among, but not within, plants. For these plants you'll need to check your patch every few days and hand-harvest those plants with ripe fruits.

SEED EXTRACTION. Almost all flower species produce dry fruit, usually achenes, capsules, or follicles. Only a few genera, *Lantana* and *Convallaria*, for example, include herbaceous ornamentals that produce fleshy fruit. Therefore, drying and threshing, rubbing, and flailing are the predominant means of extracting the seeds.

DRYING AND STORAGE. Seeds of almost all flowers are orthodox; that is, they are best dried well and stored in sealed containers at 34 to 41°F (1 to 5°C). We specify exceptions to this where known.

GERMINATION REQUIREMENTS. The details are often contradictory, too, especially regarding requirement for light and the proper germination temperatures for flower seeds. Some flower seeds require special pretreatments before or during germination. Sometimes the special treatments are initiated only after the seeds have been sown and placed for some time in a certain environment. In the plant entries in this chapter, if there is no mention of special treatment for a particular plant, you can assume that the seeds will germinate well at a standard temperature range of 65 to 75°F (18 to 24°C).

One special treatment that is applicable to many types of herbaceous perennials is stratification of seed at temperatures between about 34 and 41°F (1 and 5°C). The seeds must be imbibed and kept moist in that temperature range for the duration of the stratification treatment. Since many flowers produce tiny seeds, you'll find it easiest to sow your flower seeds in flats or pots, then expose the sown containers to cold. Seeds of some flower species require warm stratification, and we provide details on that in the plant entries also.

SEED VIABILITY. We provide information on seed viability when it is known, but information is not available for some flower species and is contradictory for others. The numbers we provide are for seeds stored under good conditions that have maintained greater than 50 percent germination over the period of time indicated. Seed viability depends on how the seeds were harvested, cleaned, and stored and upon the age of the seeds. Our rule of thumb is to use seeds that are as fresh as possible. In the long run, it really doesn't matter too much if seeds of a particular type of flower will remain viable for 10 years. You are likely to use all of yours long before that anyway.

HEAT-TREATING SEEDS. Seeds of most flowers require no seed treatment to destroy seedborne pathogens, but we have noted those few that do.

RAISING TRANSPLANTS. Information for growing flower transplants is often lacking. Generally, transplants are best grown at temperatures from 5 to 10°F (3 to 6°C) cooler than those used to germinate their seeds. Be sure to add 2 weeks to the time given for growing transplants to allow for hardening your plants prior to setting them to the garden.

ABUTILON

Flowering maple

A. darwinii, A. theophrasti (**velvetleaf**), *A. vitifolium* (**parlour maple**)

FAMILY: Malvaceae; mallow family
PLANT TYPE: Perennial; *A. darwinii* can be annual or perennial
SEED VIABILITY: Unknown

———————— FLOWER ALERT ————————

Velvetleaf (*A. theophrasti*) is deemed a troublesome weed in some regions.

————————————————————————————

FLOWERING: The flowers are perfect. Flowers in F_1 hybrids are mostly sterile.

SEED COLLECTION: The fruits are schizocarps, with anywhere from five to many mericarps, each with two to nine seeds. Hand-pick the brown seed heads.

SEED CLEANING: Dry the flower heads on a screen, then rub or flail to remove the seeds.

GERMINATION: Germinate the seeds for 14 to 28 days at 75°F (24°C).

TRANSPLANTING: Seedlings need 6 to 8 weeks at 70°F (21°C), to reach transplanting size but seeds are usually sown in place.

ACHILLEA

Common yarrow, nose-bleed, yarrow

A. ageratum, A. clypeolata, A. erba-rotta,
A. filipendulina (**fernleaf yarrow**), *A. millefolium* (**thousandseal**), *A. nobilis, A. ptarmica,*
A. pyrenaica, A. sibirica, A. tomentosa

FAMILY: Asteraceae; sunflower family
PLANT TYPE: Perennial
SEED VIABILITY: 4 years

FLOWERING: The disk flowers are perfect and the ray flowers are pistillate. The plants are cross-pollinated.

SEED COLLECTION: The fruits are shiny achenes. Harvest the spent flowers before they shed their fruits.

SEED CLEANING: Spread the flowers on a screen to dry, then flail to remove the seeds.

GERMINATION: Sow most species at 68°F (20°C) in light. Germination will take 10 to 14 days. Germinate *A. erba-rotta* at 41°F (5°C); germination of this species is irregular and may require many months.

TRANSPLANTING: Seedlings need 8 to 10 weeks at 50 to 60°F (10 to 16°C) to reach transplanting size.

ACONITUM

ACONITUM

Monkshood

A. anthora, *A. ×cammarum*, *A. carmichaelli* (**violet monkshood**), *A. fischeri* (**azure monkshood**), *A. henryi* (**autumn monkshood**), *A. lamarckii*, *A. napellus* (**garden monkshood, garden wolfbane**), *A. septentrionale*, *A. variegatum*, *A. vulparia* (**wolfbane**)

FAMILY: Ranunculaceae; buttercup family
PLANT TYPE: Perennial
SEED VIABILITY: Very short-lived
SPACING FOR SEED SAVING: 9–12"

——————— FLOWER ALERT ———————
Monkshood plants are highly poisonous if ingested.

FLOWERING: The flowers are perfect.

SEED COLLECTION: The fruits are follicles, which you should harvest when they turn brown but before they shed their seeds.

SEED CLEANING: Spread the fruit on screens to dry, then rub or flail to remove the seeds.

SEED TREATMENT: Stratify the seeds for 6 to 12 weeks.

GERMINATION: Germinate at 68°F (20°C) for 3 weeks, move to 20°F (–7°C) for 5 weeks, then move to 50°F (10°C). Germination sometimes takes up to a year.

TRANSPLANTING: Seedlings need 6 to 8 weeks at 70°F (21°C), to reach transplanting size but seeds of this genus are usually sown in place. Seedlings have a 2- to 3-year juvenile period, so be patient.

AGASTACHE

Giant hyssop

A. aurantiaca, *A. barberi*, *A. cana* (**mosquito plant**), *A. cusickii*, *A. foeniculum* (**anise hyssop, blue giant hyssop, fragrant giant hyssop, giant hyssop**), *A. nepetoides* (**yellow giant hyssop**), *A. pallidiflora* (**Texas hummingbird mint**), *A. pringlei*, *A. rugosa* (**Korean mint, wrinkled giant hyssop**), *A. rupestris* (**threadleaf giant hyssop**), *A. scrophulariifolia* (**purple giant hyssop**)

FAMILY: Lamiaceae; mint family
PLANT TYPE: Perennial
SEED VIABILITY: Unknown

FLOWERING: The small flowers are perfect and are arranged on long spikes.

SEED COLLECTION: Harvest the spikes by hand when they have browned.

SEED CLEANING: Spread the spikes on screens to dry, then rub to separate the nutlets. Each flower will bear four nutlets.

SEED TREATMENT: Stratify seeds of purple giant hyssop (*A. scrophulariifolia*) for 2 months at about 32°F (0°C).

GERMINATION: Sow the seeds when soil temperatures reach 68°F (20°C), and be patient. Some species need up to 4 weeks for germination to occur but germination occurs in less than 2 weeks in others.

TRANSPLANTING: Seedlings need 6 to 8 weeks at 70°F (21°C) to reach transplanting size.

AGERATUM

Flossflower, pussyfoot

A. houstonianum

FAMILY: Asteraceae; sunflower family
PLANT TYPE: Annual
SEED VIABILITY: 4 years
SPACING FOR SEED SAVING: 10"

FLOWERING: Flower heads are arranged in a cymose cluster. The individual flowers are perfect.

SEED COLLECTION: The fruit is a five-angled achene. Harvest the flowers and/or the plants when the heads are spent.

SEED CLEANING: Spread the plants and flowers on screens to dry, then flail or tumble to remove the seeds.

GERMINATION: Sow seeds in the light at 75°F (24°C); germination takes about 3 weeks.

TRANSPLANTING: Grow the plants on at 65°F (18°C) for 6 to 8 weeks. Transplant when flower buds form on the seedlings.

AGROSTEMMA

Corn cockle, purple cockle

A. githago, *A. gracilis*

FAMILY: Caryophyllaceae; pink family
PLANT TYPE: Annual
SEED VIABILITY: 4 years

FLOWERING: The flowers are perfect.

SEED COLLECTION: The fruits are dehiscent capsules containing many black seeds. Collect the capsules before they shed their seeds.

SEED CLEANING: Dry the fruits on screens with a sheet beneath to capture the seeds. Then rub or flail to remove more seeds.

GERMINATION: Germinate at 55 to 65°F (13 to 18°C); allow about 21 days for complete germination.

TRANSPLANTING: Corn cockle is usually sown in place.

——————— FLOWER ALERT ———————

Corn cockle can become a weed in some locales.

ALCEA

Hollyhock

BIENNIAL SPECIES: *A. ficifolia.* (**Antwerp hollyhock**), *A. rosea* (**hollyhock**)

PERENNIAL SPECIES: *A. pallida, A. rugosa*

FAMILY: Malvaceae; mallow family
PLANT TYPE: Biennials and perennials
SEED VIABILITY: 9 years
SPACING FOR SEED SAVING: 12"

FLOWERING: Flowers are perfect and cross-pollinated by insects.

ISOLATION REQUIREMENT: To prevent cross-pollination, isolate by ¼ mile.

SEED COLLECTION: The fruit is a schizocarp. Each fruit has about 40 indehiscent, two-celled mericarps. Collect fruits when dry. If the fruits of biennial hollyhocks are left in place, plants will self-sow in some locations.

SEED CLEANING: Rub the fruits to separate the mericarp from the schizocarp, then winnow or sift to remove the chaff.

SEED TREATMENT: Stratify for 2 weeks.

GERMINATION: Sow seeds in light at 70°F (21°C); germination takes about 3 weeks. Germinate seeds of *A. pallida* and *A. rugosa* at 41°F (5°C) over several months.

TRANSPLANTING: Hollyhocks do not transplant well bare-rooted; start seeds in peat pots to minimize root disturbance. Grow seedlings on at about 60°F (16°C) for 6 to 8 weeks, then set out plants, pot and all, in the garden.

ALSTROEMERIA

Alstroemeria, lily-of-the-Incas, Peruvian lily

A. aurea (**yellow alstroemeria**), *A. hookeri*, *A. ligtu*, *A. psittacina* (**parrot alstroemeria**)

FAMILY: Alstroemeriaceae; alstroemeria family
PLANT TYPE: Perennial
SEED VIABILITY: Unknown

FLOWERING: The flowers are perfect.

SEED COLLECTION: The fruit is a three-celled capsule. Hand-pick the capsules when they are brown and dry.

SEED CLEANING: Dry the fruits on screens, then flail or thresh to release the seeds.

GERMINATION: Sow the seeds at 41°F (5°C). Germination is irregular and may take several months, so be patient.

TRANSPLANTING: Grow the plants on at about 68°F (20°C) for 10 weeks before setting out in spring.

ALYSSUM

Alyssum, basket of gold, madwort

ANNUAL SPECIES: *A. alyssoides* (**madwort**), *A. desertorum* (**desert madwort**)

PERENNIAL SPECIES: *A. argenteum* (syn. *A. murales*), *A. cuneifolium*, *A. montanum* (**mountain alyssum**), *A. petraea*, *A. repens*, *A. saxatilis* (**basket of gold**), *A. serpyllifolium*, *A. spinosum* (**spiny alyssum**), *A. wulfenianum*

ALYSSUM FLOWER HEADS

FAMILY: Brassicaceae; mustard family
PLANT TYPE: Perennials and annuals; madwort (*A. alyssoides*) may be biennial in some areas
SEED VIABILITY: 4 years

FLOWERING: The flowers are perfect.

SEED COLLECTION: The fruits are silicles that are dehiscent or indehiscent, depending upon species. Hand-pick the flower heads when they are brown and have begun to dry.

SEED CLEANING: Spread the heads on screens to dry further, with sheets beneath to capture shed seed. Rub, thresh, or flail gently to remove the seeds.

SEED TREATMENT AND GERMINATION: Sow seeds of most perennial species at 68°F (20°C). If seeds of *A. cuneifolium* do not germinate in a month, stratify them for a month and return to 68°F (20°C). Sow seeds of annual species in light at the same temperature.

TRANSPLANTING: Grow the plants on for 6 to 8 weeks at 60 to 65°F (16 to 18°C), or sow the seeds directly outdoors in fall or early spring, while the soil is still cool.

AMARANTHUS
Amaranth

A. caudatus (**love-lies-bleeding**), *A. chloro-stachys* (**slim amaranth**), *A. cruentus* (**prince's feather**), *A. gangeticus* (**elephant head amaranth**), *A. hypochondriacus* (**prince's feather**), *A. lividus* (**purple amaranth**), *A. tricolor* (**tampala**)

FAMILY: Amaranthaceae; amaranth family
PLANT TYPE: Annual
SEED VIABILITY: 4–5 years

FLOWERING: The inconspicuous flowers are imperfect.

SEED COLLECTION: The fruit is a utricle (single-seeded fruit). Collect the spent flower heads before they shatter.

SEED CLEANING: Spread the flower heads on a screen to dry; then thresh, tumble, or flail to remove the seeds.

GERMINATION: Germination takes 10 to 14 days at 75°F (24°C) in light.

TRANSPLANTING: Grow the plants on at 65°F (18°C). Seedlings are very sensitive to damping-off, so give them plenty of air circulation. Seedlings need 6 to 7 weeks to reach transplanting size.

——————PLANT BREEDING TIP——————

If you want to try your hand at breeding a new strain of love-lies-bleeding (*A. caudatus*), keep in mind that red leaf color is dominant over green.

AMARANTH FLOWER HEADS

AMMOBIUM
Everlasting

A. alatum (**winged everlasting**), *A. grandiflora* (**winged everlasting**)

FAMILY: Asteraceae; sunflower family
PLANT TYPE: Annual
SEED VIABILITY: 1–2 years

FLOWERING: The flowers are perfect.

SEED COLLECTION: The fruits are four-angled achenes. Collect the spent flower heads.

SEED CLEANING: Dry the heads on a screen, then thresh or flail to remove the seeds.

GERMINATION: Sow seeds of *A. alatum* at 60°F (16°C) and seeds of *A. grandiflora* at 70°F (21°C). Germination takes about 14 days.

TRANSPLANTING: Everlasting is difficult to transplant; sow seeds in place.

ANCHUSA
Bugloss

A. azurea, A. capensis (**bugloss, alkanet**), *A. officinalis*

FAMILY: Boraginaceae; borage family
PLANT TYPE: Perennial; alkanet (*A. capensis*) can be annual or biennial
SEED VIABILITY: Unknown
SPACING FOR SEED SAVING: 8"

FLOWERING: The flowers are perfect.

SEED COLLECTION: Fruits are composed of four rough nutlets. Collect the fruits as soon as the flower heads are dry.

SEED CLEANING: Spread the fruits and/or plants on screens to dry, then tumble or flail.

GERMINATION: Germinate for 7 to 28 days in light at 70°F (21°C). Germination can be erratic and very slow.

TRANSPLANTING: Six to 8 weeks are needed to grow transplants.

ANEMONE

Anemone, lily-of-the-field, windflower

A. baicalensis, A. baldensis (**moraine anemone**), *A. barbulata, A. canadensis* (**meadow anemone**), *A. caroliniana* (**Carolina anemone**), *A. cylindrica, A. drummondii, A. hortensis* (**garden anemone**), *A. hupehensis* (**dwarf Japanese anemone**), *A. ×lesseri, A. leveillei, A. lyallii, A. multifida, A. narcissiflora, A. nemorosa* (**European wood anemone**), *A. obtusiloba, A. parviflora, A. ranunculoides* (**yellow wood anemone**), *A. rivularis, A. rupicola, A. sylvestris* (**snowdrop anemone**), *A. tomentosa, A. vernalis, A. virginiana* (**thimbleweed, Virginia anemone**)

FAMILY: Ranunculaceae; buttercup family
PLANT TYPE: Perennial
SEED VIABILITY: Unknown

──────**PLANT-BREEDING HINT**──────

With anemones, blue flower color and red flower color are dominant over white.

FLOWERING: The mostly solitary flowers are perfect.

SEED COLLECTION: The fruits are achenes. Harvest the spent flower heads when they dry and are brown.

SEED CLEANING: Spread the heads on screens to dry further, then thresh or flail to remove the achenes.

SEED TREATMENT AND GERMINATION: Germinate the seeds of most species at 68°F (20°C). The germination of some species is quite fast, within 2 weeks of sowing, and for other species it's slow, so be patient

and take good notes. The following species require stratification:

- Sow seeds of moraine anemone (*A. baldensis*), Carolina anemone (*A. caroliniana*), *A. cylindrica*, garden anemone (*A. hortensis*), dwarf Japanese anemone (*A. hupehensis*), *A. leveillei, A. narcissiflora, A. tomentosa*, and Virginia anemone (*A. virginiana*) at 68°F (20°C) and wait a month. If none has germinated, stratify for a month, then return to 68°F for germination.
- Sow seeds of meadow anemone (*A. canadensis*), *A. multifida*, and *A. rivularis* at 68°F and hold for a month; stratify them for a month, then move them to 50°F (10°C) for germination.
- Seeds of yellow wood anemone should be sown at 68°F (20°C) and held for a month, moved to 23°F (−5°C) for a month, then moved to 50°F (10°C) for germination.
- Sow seeds of European wood anemone (*A. nemorosa*) and *A. ×lesseri* outdoors immediately after harvest. They will stratify over winter and germinate in spring.

TRANSPLANTING: You may Grow the plants on at 50°F (20°C) for about 8 weeks before setting out, or simply sow seeds of all species in place right after harvest or in the very early spring.

ANTIRRHINUM

Snapdragon

A. braun-blanquetti, A. majus (**common snapdragon**), *A. nuttallianum* (**violet snapdragon**), *A. siculum* (**common snapdragon**)

FAMILY: Scrophulariaceae; figwort family
PLANT TYPE: Annual; *A. braun-blanquetti* and common snapdragon (*A. majus*) are perennial in warmer areas
SEED VIABILITY: 3–4 years
SPACING FOR SEED SAVING: 12"

FLOWERING: The flowers are borne in racemes and are perfect and self-pollinated.

ISOLATION REQUIREMENT: Bumblebees cross-pollinate flowers. Isolate varieties by about 600 feet.

SEED COLLECTION: The seed capsules ripen from the base of the inflorescence upward. Hand-pick the pods when about 70 percent of them are brown and dry. Or cut the spikes when most of the pods are dry, and a second spike will form. Depending upon your season, you may then get another harvest.

SEED CLEANING: Put the pods on screens to dry, then shake or flail to extract the seeds.

SEED TREATMENT: Stratify for 1 to 2 weeks.

GERMINATION: Germinate the seeds for 7 to 14 days in light at 65°F (18°C).

TRANSPLANTING: After germination, Grow the plants on at 50°F (10°C). Seedlings need about 9 weeks to reach transplanting size.

--------PLANT-BREEDING HINT--------

Be forewarned that heredity is complex in snapdragons. In general, though, dark colors in flowers are dominant over the light colors. Red flower color is dominant over white. Smooth stems are dominant over hairy stems.

AQUILEGIA
Columbine

> *A. alpina, A. atrata, A. bertolonii, A. buergeriana, A. caerulea, A. canadensis* (**wild columbine**), *A. chrysantha, A. clematiflora, A. flabellata, A. flabellata* var. *pumila, A. formosa, A. glandulosa, A. laramiensis, A. longissima, A. nevadensis, A. olympica, A. rockii, A. saximontana, A. scopulorum, A. sibirica, A. transylvanica, A. viridiflora, A. vulgaris* (**garden columbine**)

FAMILY: Ranunculaceae; buttercup family
PLANT TYPE: Perennial
SEED VIABILITY: 2 years
SPACING FOR SEED SAVING: 12"

--------PLANT-BREEDING HINT--------

Blue flower color tends to be dominant in columbines, and wide petals are dominant over narrow petals. Midlength spurs are dominant over both long spurs and short spurs.

FLOWERING: Flowers are perfect and cross-pollinated by insects.

SEED COLLECTION: Fruit is a many-seeded follicle. Harvest the heads as soon as they are somewhat but not fully dry, as they will shatter at that later stage. The seeds should be dark green to almost black when harvesting begins.

SEED CLEANING: Spread the fruits on a screen to dry, then thresh.

SEED TREATMENT: Stratify for 3 weeks before sowing.

GERMINATION: Germinate most species for 21 to 28 days in light at 70°F (21°C). Sow seeds of *A. alpina, A. buergeriana, A. formosa, A. glandulosa, A. nevadensis, A. saximontana, A. transylvanica, A. viridiflora,* and garden columbine (*A. vulgaris*) at 41°F (5°C). Germinate seeds of *A. atrata* at 68°F (20°C) for 2 to 4 weeks. Then move them to 32°F (0°C) for 5 weeks, then to 50°F (10°C) and let them germinate. Germination is erratic and can take a long time.

TRANSPLANTING: Seedlings need 6 to 8 weeks to reach transplanting size.

COLUMBINE FLOWERS

ARCTOTIS

African daisy

A. acaulis (**African daisy**), *A. fastuosa* (**bitter gousblom**), *A. hirsuta* (**gousblom**), *A. venusta* (**blue-eyed African daisy**)

FAMILY: Asteraceae; sunflower family
PLANT TYPE: Half-hardy annual
SEED VIABILITY: 5 years
SPACING FOR SEED SAVING: 10"

FLOWERING: The disk flowers are perfect and the ray flowers are pistillate. This genus is usually self-fruitful.

SEED COLLECTION: The fruits are three- to five-winged achenes. Harvest when flowers have faded.

SEED CLEANING: Spread the flowers on a screen to dry, then flail.

GERMINATION: Germinate the seeds for 3 to 5 weeks at 60 to 70°F (16 to 21°C).

TRANSPLANTING: Transplants require 6 to 7 weeks to grow.

ASCLEPIAS

Butterfly milkweed, milkweed

A. amplexicaulis, A. curassavica (**bloodflower**), *A. exaltata, A. fascicularis, A. hirtella, A. incarnata, A. physocarpa, A. purpurescens, A. speciosa, A. sullivantii, A. tuberosa, A. verticillata, A. viridiflora*

FAMILY: Asclepiadaceae; milkweed family
PLANT TYPE: Perennial; bloodflower (*A. curassavica*) and *A. amplexicaulis* are annuals in cool areas
SEED VIABILITY: Unknown

FLOWERING: The flowers are perfect and often pollinated by butterflies.

SEED COLLECTION: Collect the follicles as they ripen and begin to split.

SEED CLEANING: Dry the fruits on screens, then open them by hand and remove the seeds. Each seed will have attached a long tuft of "silk."

SEED TREATMENT: Stratify for 6 to 10 weeks. *A. fascicularis, A. physocarpa,* and *A. speciosa* need no stratification.

GERMINATION: Germinate the seeds at about 70°F (21°C). Germination is generally slow in the milkweeds and can take from 2 weeks to several months. Be patient.

TRANSPLANTING: Seeds are usually sown in place.

ASTER

Aster, frost flower,
Michaelmas daisy, starwort

A. alpinus (**alpine aster**), *A. amellus* (**Italian aster**), *A. azureus, A. bigelovii, A. coloradoensis* (**Colorado aster**), *A. cordifolius* (**bluewood aster**), *A. delavayi, A. diplostephoides, A. divaricatus* (**white wood aster**), *A. ericoides* (**heath aster**), *A. farreri* (**farrer aster**), *A. ×frikartii, A. laevis* (**smooth aster**), *A. lateriflorus* var. *horizontalis* (**calico aster**), *A. linariifolius* (**savory leaf aster**), *A. linosyris* (**goldilocks aster**), *A. macrophyllus* (**bigleaf aster**), *A. novae-angliae* (**New England aster**), *A. novi-belgii* (**New York aster**), *A. oblongifolius, A. pilosus, A. ptarmicoides* (**white upland aster**), *A. puniceous* (**swamp aster**), *A. sagittifolius* var. *drummondii, A. sericeus, A. sibiricus, A. simplex, A. stracheyi, A. tongolensis* (**East Indies aster**), *A. turbinellus, A. umbellatus*

FAMILY: Asteraceae; sunflower family
PLANT TYPE: Perennial
SEED VIABILITY: 1–2 years

C

💧 Sow seeds of *C. fenestrellata* and *C. garganica* at 70°F (21°C) for 2 months, then move them to 32°F (0°C) for 8 weeks. After that, move them to 50°F (10°C) for germination. If the seeds do not germinate after 6 weeks, repeat the cycle.

──────PLANT-BREEDING HINTS──────

→ In bellflowers in general, dark blue flower color is dominant over light shades and white.

→ In Canterbury bells (*C. medium*), "hose-in-hose" type (a flower that looks as though it has one corolla within another) is partially dominant, and colored flowers are dominant over white.

TRANSPLANTING: Seedlings need about 10 weeks to reach transplanting size; set them outdoors after all danger of frost is past.

CAREX

Sedge

C. acuta, C. acutiformis, C. alba, C. arenaria, C. atrata, C. aurea, C. berggrenii, C. buchananii, C. comans, C. flacca, C. flava, C. grayi, C. lurida, C. macrocephala, C. muskingumensis, C. nigra, C. paniculata, C. pendula, C. pseudocyperus, C. remota, C. riparia, C. secta, C. sylvatica, C. tenuiculmis, C. testacea, C. umbrosa, C. vulpina

FAMILY: Cyperaceae; sedge family
PLANT TYPE: Perennial
SEED VIABILITY: Unknown

FLOWERING: The flowers are imperfect, and plants may be monoecious or dioecious. Flower spikes may (1) be composed entirely of staminate flowers, (2) be composed entirely of pistillate flowers, (3) have staminate flowers above and pistillate flowers below, or (4) have staminate flowers below and pistillate flowers above.

SEED COLLECTION: The fruits are achenes. Harvest the spikes when they have browned and dried but before fruit-shed. Only pistillate flowers will bear fruit.

SEED CLEANING: Spread the spikes on screens to dry further, then flail or thresh to remove the seeds.

SEED TREATMENT AND GERMINATION: Seeds of *C. flava* and *C. umbrosa* will germinate at 41°F (5°C). Those of *C. alba, C. comans, C. grayi, C. nigra, C. riparia,* and *C. testacea* should be sown and held at 68°F (20°C) for a month, stratified for a month, then moved to 50°F (10°C) for germination. Follow a similar protocol with seeds of the remaining species except after stratification, return them to 68°F (20°C). Germination should take place within 2 weeks after the necessary treatments.

TRANSPLANTING: Maintain seedlings at about 75°F (24°C) for 6 to 8 weeks, then transplant to the garden.

CATHARANTHUS

Periwinkle

C. roseus (**periwinkle**)

FAMILY: Apocynaceae; dogbane family
PLANT TYPE: Annual
SEED VIABILITY: Usually less than 1 year

FLOWERING: The flowers are perfect.

SEED COLLECTION: The fruit is a narrow cylindrical follicle containing 15 to about 30 seeds. Hand-pick the follicles when they are brown but before they shatter.

SEED CLEANING: Place the follicles on a screen to dry, then rub to remove the seeds.

GERMINATION: Sow the seeds, hold them for 3 days at 80°F (27°C) in the dark, then lower the temperature to 75°F (24°C). Germination will take 14 to 21 days.

TRANSPLANTING: Grow the plants on at 65°F (18°C) for about 10 weeks.

CELOSIA

Cockscomb

C. argentea (**cockscomb**), *C. cristata* (**crested cockscomb**), *C. plumosa*, *C. roripifolia*, *C. spicata*

FAMILY: Amaranthaceae; amaranth family
PLANT TYPE: Annual; *C. roripifolia* is biennial
SEED VIABILITY: 4 years
SPACING FOR SEED SAVING: 10"

FLOWERING: The flowers are small, perfect, and multicolored.

SEED COLLECTION: Harvest the flower heads when they are spent.

SEED CLEANING: Dry the heads on screens, then flail.

GERMINATION: Germinate the biennial at 68°F (20°C) and the other species at 75°F (24°C) for 7 to 14 days.

TRANSPLANTING: Grow the plants on at 65°F (18°C) for about 8 weeks.

———————— PLANT-BREEDING HINT ————————

If you self-pollinate red-flowered crested cockscomb (*C. cristata*), a mixture of red-flowered plants and plants with mosaic (red/yellow mixture) flowers will result. Red is dominant over mosaic.

CENTAUREA

Bachelor's button, cornflower

ANNUAL SPECIES: *C. americana* (**basket flower**), *C. cineraria* (**dusty miller**), *C. cyanus* (**bachelor's button, blue-bottle, cornflower**), *C. erythaea* (**European centaury**), *C. gymnocarpa* (**dusty miller**), *C. moschata* (**sweet sultan**), *C. ragusina* (**dusty miller**), *C. rothrockii* (**Rothrock's knapweed**), *C. venustum* (**charming centaury**)

PERENNIAL SPECIES: *C. alpestris*, *C. alpina*, *C. bella*, *C. dealbata*, *C. jacea*, *C. macrocephala*, *C. montana*, *C. nigra*, *C. orientalis*, *C. phrygia*, *C. pulcherrima*, *C. rupestris*, *C. ruthenica*, *C. uniflora*

FAMILY: Asteraceae; sunflower family
PLANT TYPE: Perennial and annual; bachelor's button (*C. cyanus*) is sometimes biennial
SEED VIABILITY: 7–10 years
SPACING FOR SEED SAVING: 1–2'

FLOWERING: The flowers are formed in heads 1 to 4 inches in diameter. Flowers are cross-pollinated by insects.

ISOLATION REQUIREMENTS: Isolate by at least ¼ mile to prevent cross-pollination.

SEED COLLECTION: The fruits are achenes. Harvest the plants when most have ceased flowering. Birds love the seeds and so can be troublesome. Protect your seed plants with bird netting.

SEED CLEANING: Place the plants on screens to dry, then thresh or flail as soon thereafter as you can to extract the seeds. The fruits shatter badly, so don't delay. If left in place in the garden, annual species will self-sow.

SEED TREATMENT: Stratify 5 days.

CENTAUREA FLOWER HEADS

GERMINATION: Germinate in darkness for 7 to 28 days at 75°F (24°C).

TRANSPLANTING: Seedlings need about 8 weeks to reach transplanting size.

CERASTIUM

Mouse-ear chickweed

C. alpinum (**alpine chickweed**), *C. arvense* (**field chickweed, starry grasswort**), *C. biebersteinii* (**taurus chickweed**), *C. candidissimum, C. grandiflorum, C. tomentosum* (**snow-in-summer**)

FAMILY: Caryophyllaceae; pink family
PLANT TYPE: Perennial
SEED VIABILITY: Several years

FLOWERING: The flowers are perfect.

SEED COLLECTION: The fruit is a cylindrical dehiscent capsule containing many seeds. Harvest the capsules by hand when they are brown but before they dehisce.

SEED CLEANING: Spread the capsules on screens to dry with a sheet beneath to capture the seeds. Then thresh or rub to extract the seeds.

MOUSE-EAR CHICKWEED FLOWERS

SEED TREATMENT AND GERMINATION: Surface-sow seeds of most species at 68°F (20°C) in light and allow a month for germination. If seeds have not germinated in a month, stratify them for another month and return them to 68°F (20°C). Germinate seeds of *C. grandiflorum* at 41°F (5°C) for several months.

––––––––––– FLOWER ALERT –––––––––––

Mouse-ear chickweeds can self-sow easily and become a terrible weed in some areas. Monitor them closely, and don't let seed escape from the garden.

TRANSPLANTING: Grow the plants on for 8 to 10 weeks at about 60°F (16°C), then transplant in early spring. Or sow the seeds in place in the fall.

CERINTHE

Honeywort

ANNUAL SPECIES: *C. major* (**honeywort**)

PERENNIAL SPECIES: *C. glabra*

FAMILY: Boraginaceae; borage family
PLANT TYPE: Annual and perennial
SEED VIABILITY: Unknown

FLOWERING: The flowers are perfect.

SEED COLLECTION: The fruits are nutlets. Collect the spent flower heads before they shed their fruit.

SEED CLEANING: Spread the heads on a screen to dry, then rub or flail to release the seeds.

SEED TREATMENT AND GERMINATION: Germinate at 60 to 65°F (16 to 18°C) for 14 days. If the seeds have not germinated in a month, stratify them by moving them to 32°F (0°C) for 4 weeks, then return them to the higher temperature. Perennial honeywort is likely to require this treatment.

TRANSPLANTING: Seedlings need about 8 weeks to reach transplanting size.

CHAENACTIS

Pincushion flower

C. douglasii (**Douglas' dusty maiden**), *C. fremontii* (**pincushion flower**), *C. glabriuscula* (**yellow chaenactis**), *C. xantiana* (**fleshcolor pincushion**)

FAMILY: Asteraceae; sunflower family
PLANT TYPE: Annual; Douglas' dusty maiden (*C. douglasii*) is biennial
SEED VIABILITY: Unknown

FLOWERING: The insect-pollinated flowers are tubular and solitary, or corymbose.

SEED COLLECTION: The fruit is an achene. Harvest the flower heads when they have browned.

SEED CLEANING: Place the spent heads on a screen to dry, then rub or flail to separate the seeds.

GERMINATION: Sow at 39°F (4°C) for 30 to 70 days.

TRANSPLANTING: Seedlings need about 10 weeks from germination to reach transplanting size.

CHAENORHINUM

Dwarf snapdragon

PERENNIAL SPECIES: *C. origanifolium* (**dwarf snapdragon**)

ANNUAL SPECIES: *C. minus* (**small snapdragon**)

FAMILY: Scrophulariaceae; figwort family
PLANT TYPE: Perennial and annual
SEED VIABILITY: Unknown

FLOWERING: The flowers are perfect.

SEED COLLECTION: The fruits are dehiscent capsules. Hand-pick them when they are brown but before they dehisce.

SEED CLEANING: Spread the capsules on a screen to dry, with a sheet beneath to capture the seeds. When dry, rub or gently flail or thresh the capsules to release the seeds.

GERMINATION: Sow seeds of the perennial species at 68°F (20°F).

—————— FLOWER ALERT ——————

Small snapdragon (*C. minus*) can become weedy in some areas.

TRANSPLANTING: Grow the plants on at about 75°F (24°C) for 6 to 8 weeks before spring planting. Sow seeds of the annual species outdoors in early spring and expect up to a month for germination.

CHEIRANTHUS

Wallflower

C. allionii (**Siberian wallflower**), *C. cheiri* (**wallflower**)

FAMILY: Brassicaceae; mustard family
PLANT TYPE: Annual or biennial
SEED VIABILITY: 5 years
SPACING FOR SEED SAVING: 12"

FLOWERING: Flowers are perfect.

ISOLATION REQUIREMENTS: Plants are cross-pollinated. Isolate by ¼ mile.

SEED COLLECTION: Fruit is a long silique. Collect the fruit when it has turned yellow-brown.

——————PLANT-BREEDING HINT——————

In wallflowers, dark red and yellow-brown are both dominant over bright yellow; dark red is favored over yellow-brown.

Your wallflower-breeding efforts will probably result in a few plants that exhibit the abnormal recessive form *C. cheiri* var. *gyanthus*, in which the stamens are replaced by additional carpels and the petals are strongly reduced.

SEED CLEANING: It is difficult to remove the seeds when the plants are absolutely dry, so flail or thresh when the plants are wet with dew. The moisture will help loosen the seeds. Be sure to dry the cleaned seeds well before storage.

GERMINATION: Germination takes about 14 days at 70°F (21°C).

TRANSPLANTING: Seedlings need 6 to 8 weeks to reach transplanting size.

CHRYSANTHEMUM
Chrysanthemum

ANNUAL SPECIES: *C. carinatum* (**tricolor daisy**), *C. coronarium* (**garland chrysanthemum or crown daisy chrysanthemum**), *C. multicaule* (**yellow buttons**), *C. ptarmiciflorum* now *Tanacetum ptarmiciflorum* (**dusty miller**), *C. segetum* (now *Glebionis segetum*; **corn chrysanthemum or corn marigold**)

PERENNIAL SPECIES: *C. alpinum* (now *Leucanthemopsis alpina*), *C. arcticum*, *C. cinerariifolium* (now *Tanacetum cinerariifolium*), *C. coccineum* (**pyrethrum**), *C. corymbosum* (now *Tanacetum corymbosum*), *C. frutescens* (now *Argyranthemum frutescens*), *C. indicum, C. koreanum, C. macrophyllum* (now *Tanacetum macrophyllum*), *C. maximum, C. parthenium* (now *Tanacetum parthenium*)

FAMILY: Asteraceae; sunflower family
PLANT TYPE: Annual and perennial
SEED VIABILITY: 9–10 years
SPACING FOR SEED SAVING: 12"

———————— FLOWER ALERT ————————

Corn marigold (*Glebionis segetum*) can become an invasive weed in some areas.

FLOWERING: The disk flowers are perfect, the ray flowers pistillate.

ISOLATION REQUIREMENTS: All species are cross-pollinated, so isolate by at least ¼ mile.

SEED COLLECTION: The fruits are achenes. Harvest the plants as soon as the flower heads are mature; any delay may result in great loss of seeds from shattering.

SEED CLEANING: Further dry the plants on a screen, then rub, flail, or tumble to remove the seeds.

SEED TREATMENT AND GERMINATION: For seeds of most species, germination takes 14 to 35 days at 65°F (18°C). Dusty miller (*Tanacetum ptarmiciflorum*) should be sown at 75°F (24°C). If seeds of *Leucanthemopsis alpina* do not germinate in a month, stratify them by moving them to 32°F (0°C) for about a month, then return them to germination temperature. Sow seeds of *C. arcticum* at 41°F (5°C); germination may take several months.

TRANSPLANTING: Chrysanthemums need about 2 months to reach transplanting size at 65°F (18°C).

———————— PLANT-BREEDING HINT ————————

Color inheritance is very complex in chrysanthemums. In addition, extreme double-flowered types may be sterile.

CLARKIA (GODETIA)
Clarkia, farewell-to-spring

C. amoena (**satinflower or godetia**), *C. bottae* (**Botta's fairyfan**), *C. concinna* (**red ribbons**), *C. deflexa* (**botta's fairyfan**), *C. elegans* (**elegant fairyfan**), *C. imbricata* (**vine hill fairyfan**), *C. pulchella* (**pinkfairies**), *C. purpurea* (**winecup fairyfan**), *C. rubicunda* (**rubychalice fairyfan**), *C. unguiculata* (**clarkia**)

FAMILY: Onagraceae; evening primrose family
PLANT TYPE: Annual
SEED VIABILITY: 3 years
SPACING FOR SEED SAVING: 8"

FLOWERING: The inflorescence is a leafy spike or a raceme. The perfect flowers are cross-pollinated by insects.

ISOLATION REQUIREMENTS: Isolate the plants by at least 200 feet.

If you want to try breeding clarkia, be sure you know which species you're working with.

→ In elegant fairyfan (*C. elegans*), purple-red flower color is most dominant, followed by salmon red, bright red, and white.

→ In pinkfairies (*C. pulchella*), purple is most dominant. Completely colored flowers are dominant over colored flowers with white margins, and colored flowers are dominant over white flowers.

→ In satinflower (*C. amoena*), spotted petals are dominant over nonspotted petals and double flowers are dominant over singles; the degree of "doubleness" is influenced by the petal-spot genes.

→ In *C. amoena* ssp. *whitneyi*, white color is dominant over yellow-margined white, red over lilac, rose-lilac over lilac, red over red spotted, large spot over small spot, and light-margined red over pure red. Single flowers are dominant over doubles.

SEED COLLECTION: Harvest the plants when the lower capsules have just begun to open.

SEED CLEANING: Place the plants on screens and let them dry for several more days, then flail or tumble to extract the seeds. Three shakings at 1-week intervals should remove just about all the seeds.

GERMINATION: The seeds are tiny (100,000 per ounce), so barely cover them when planting. Allow 7 to 14 days at 70°F (21°C) for seeds to germinate.

TRANSPLANTING: These species are usually sown in place in the garden as soon as the soil temperature reaches 70°F (21°C).

CLEOME

Spider flower

C. angustifolia (**spider flower**), *C. gynandra* (**spiderwisp**), *C. hasslerana* (**spider flower**), *C. lutea* (**yellow spider flower**), *C. serrulata* (**Rocky Mountain bee plant**), *C. viscosa* (**Asian spiderflower**)

FAMILY: Capparaceae; caper family
PLANT TYPE: Annual
SEED VIABILITY: Usually less than 1 year
SPACING FOR SEED SAVING: 12"

FLOWERING: Cleome flowers are spiderlike in appearance and are in indeterminate racemes. The plants are cross-pollinated by insects.

SEED COLLECTION: The fruit is a narrow capsule containing several to many seeds.

SEED CLEANING: Harvest the seed heads when the capsules have turned brown. Place the heads on a screen to dry, then tumble or flail to extract the seeds.

SEED TREATMENT: Stratify for 2 weeks.

GERMINATION: Germination takes 7 to 14 days in alternating 80°F day/70°F night (27°C/21°C) temperatures, with light during the day.

TRANSPLANTING: Seedlings need 6 to 8 weeks to reach transplanting size.

SPIDER FLOWER RACEME AND CAPSULES

CLITORIA

Butterfly pea

C. ternatea (butterfly pea)

FAMILY: Fabaceae; legume family
PLANT TYPE: Biennial or tender perennial
SEED VIABILITY: Unknown

FLOWERING: The perfect blue flowers are highly attractive to butterflies and bees.

SEED COLLECTION: The fruit is a legume pod. Hand-pick when the pods brown.

SEED CLEANING: Spread the fruits on a screen to dry, then rub or flail to release the seeds.

GERMINATION: Germination is erratic. Sow at 70 to 75°F (21 to 24°C).

TRANSPLANTING: Butterfly pea does not transplant well. Start the seeds in peat pots.

—————— PLANT-BREEDING HINT——————

Blue flower color is dominant over white in butterfly pea.

CONSOLIDA

Larkspur

C. ajacis (formerly rocket larkspur), C. regalis (formerly D. consolida)

FAMILY: Ranunculaceae; buttercup family
PLANT TYPE: Annual
SEED VIABILITY: 1–3 years
SPACING FOR SEED SAVING: 6"

FLOWERING: Infloresence is a raceme or a panicle. The corolla differs from that of delphinium in that the superior two petals are fused and the two flower petals are absent. The flowers are cross-pollinated by insects.

ISOLATION REQUIREMENTS: Isolate by 100 feet.

SEED COLLECTION: Harvest the plants when the lower follicles have browned and begun to open.

SEED CLEANING: Dry the plants on screens, then tumble or flail to extract the seeds.

SEED TREATMENT: Stratify for 14 days.

—————— FLOWER ALERT ——————

Larkspur plants are highly susceptible to crown rot and powdery mildew, so be sure your soil is well drained and that plants have adequate air circulation.

GERMINATION: Germinate for 21 to 28 days at 55 to 65°F (13 to 18°C). Dark is required.

TRANSPLANTING: Seedlings need 6 to 8 weeks to reach transplanting size.

CONVALLARIA

Lily of the valley

C. majalis

FAMILY: Liliaceae; lily family
PLANT TYPE: Perennial
SEED VIABILITY: Unknown

FLOWERING: The white or pink perfect flowers are borne in terminal clusters.

SEED COLLECTION: Fruits are many-seeded red or orange berries about ¼ inch in diameter. Plants do not produce many fruits. Harvest them by hand once they have turned bright red and have softened.

—————— FLOWER ALERT ——————

Propagating lily of the valley from seed can be quite difficult. These plants are usually propagated by clump division.

CAUTION: All plant parts are toxic if ingested.

SEED CLEANING: Macerate fruits in a blender, then collect and dry seeds. Or spread fruits on screens to dry, then rub them on wire mesh to separate seeds from dried pulp.

LILY OF THE VALLEY BERRIES

SEED TREATMENT AND GERMINATION: Sow the seeds and hold them at 68°F (20°C) for 1 month; then stratify them for 1 month. Then move them to 50°F (10°C) to germinate. Germination may take more than a year.

TRANSPLANTING: Grow the plants on for about 10 weeks at 60°F (16°C), then transplant them to their permanent location in spring.

COREOPSIS (CALLIOPSIS)

Tickseed

ANNUAL SPECIES: *C. basalis* (**golden wave**), *C. bigelovii* (**Bigelow's stickseed**), *C. grandiflora* (**large-flowered coreopsis**), *C. stillmanii* (**Stillman's tickseed**), *C. tinctoria* (**calliopsis, golden coreopsis**)

PERENNIAL SPECIES: *C. grandiflora, C. lanceolata, C. palmata, C. rosea, C. tripteris*

FAMILY: Asteraceae; sunflower family
PLANT TYPE: Perennial and annual; large-flowered coreopsis (*C. grandiflora*) can be perennial in warmer areas
SEED VIABILITY: 6 years
SPACING FOR SEED SAVING: 12"

FLOWERING: The ray flowers are sterile, the disk flowers perfect. They are cross-pollinated by insects.

SEED COLLECTION: The fruits are black achenes. Harvest the plants when the flowers are spent.

SEED CLEANING: Dry the flowers on a screen, then tumble or flail to extract the seeds.

SEED TREATMENT: Stratify seeds of most species for 3 to 4 weeks; stratify seeds of *C. palmata* for 2 months.

GERMINATION: Most species germinate in 14 to 28 days in light at 68°F (20°C). Sow seeds of *C. rosea* at 41°F (5°C). Germination may take several months.

TRANSPLANTING: Seedlings need about 8 weeks to reach transplanting size.

COSMOS

Cosmos

C. bipinnatus (**cosmos**), *C. sulphureus* (**yellow cosmos**)

FAMILY: Asteraceae; sunflower family
PLANT TYPE: Annual
SEED VIABILITY: 5 years
SPACING FOR SEED SAVING: 12"

FLOWERING: Flowers are heads with both ray and disk flowers.

SEED COLLECTION: The fruits are achenes. Harvest the entire plant when most of its seed heads are mature.

ISOLATION REQUIREMENTS: The plants are cross-pollinated by insects, so isolate by at least ¼ mile.

SEED CLEANING: Spread the plants or the seed heads on a screen for further drying, then flail or tumble to remove the seeds.

GERMINATION: Germination takes 14 to 28 days in light at 70°F (21°C).

TRANSPLANTING: Seedlings need about 5 weeks to reach transplanting size. Cosmos self-sow readily in garden beds.

CROCUS
Crocus

C. abantensis, C. asumaniae, C. tommasinianus, C. vernus (**common crocus**)

FAMILY: Iridaceae; iris family
PLANT TYPE: Perennial
SEED VIABILITY: Unknown

—————— FLOWER ALERT ——————

Most of the crocuses we grow are hybrid types that won't come true from seed. And note that even if you're growing one of the species crocuses named here, they are a challenge to grow from seed — the seeds take months to germinate. Crocuses are usually propagated by cormlets.

FLOWERING: The flowers are perfect.

SEED COLLECTION: The fruit is a three-valved capsule. Harvest the capsules when they are dry and brown. If you leave fruits in place, seed may self-sow.

SEED CLEANING: Spread the capsules on screens to dry, then rub or thresh to remove the seeds.

SEED TREATMENT AND GERMINATION: Sow and hold seeds at 68°F (20°C) for 1 month. Stratify for 1 month, then move to 50°F (10°C) for germination, which may take from several months to more than a year.

TRANSPLANTING: Grow the plants for about 10 weeks at 50°F (10°C), then transplant to the garden. You may also sow seeds in place in the fall and transplant the cormlets to their permanent location the following fall.

CUPHEA
Cigarflower

C. ignea (**cigarflower, firecracker plant**), *C. llavea, C. macrophylla, C. platycentra*

FAMILY: Lythraceae; loosestrife family
PLANT TYPE: Annual; perennial in warmer areas
SEED VIABILITY: Unknown

FLOWERING: The flowers are perfect.

SEED COLLECTION: The fruits are papery, dehiscent capsules. Hand-pick them before they split open.

SEED CLEANING: Place the capsules on a screen with a sheet beneath until they have dried, then rub or flail to release the seeds.

CIGARFLOWER BLOSSOMS AND CAPSULES

GERMINATION: Germinate at 70°F (21°C) for 12 to 15 days.

TRANSPLANTING: Seedlings need about 8 weeks to reach transplanting size.

CYCLAMEN

Alpine violet, Persian violet, sowbread

C. cilicium (**Sicily cyclamen**), *C. coum* (**Atkins' cyclamen**), *C. hederifolium* (**baby cyclamen, Neopolitan cyclamen**), *C. mirabile, C. persicum* (**florists' cyclamen**), *C. pseudibericum, C. purpurascens, C. repandum*

FAMILY: Primulaceae; primrose family
PLANT TYPE: Perennial
SEED VIABILITY: 4–5 years

FLOWERING: The flowers are perfect and borne on scapes.

SEED COLLECTION: The fruits are five-valved capsules. Harvest the entire scape or hand-pick the capsules when they brown and dry.

SEED CLEANING: Spread the scapes with capsules on screens to dry, then gently rub or flail to remove the seeds. The seeds are sticky and may be difficult to clean well.

SEED TREATMENT: Soak seeds of florists' cyclamen (*C. persicum* var. *giganteum*) in 80°F (27°C) water before sowing.

GERMINATION: Sow seeds of most species at 65°F (18°C) with light. Sow those of *C. purpurascens* at 41°F (5°C). Sow those of florists' cyclamen after treatment at 60°F (16°C) in darkness.

TRANSPLANTING: Grow the plants on at about 60 to 65°F (16 to 18°C) for 6 to 8 weeks for spring transplanting.

CYNOGLOSSUM

Houndstongue

BIENNIAL SPECIES: *C. amabile* (**Chinese forget-me-not**), *C. officinale* (**houndstongue**)

PERENNIAL SPECIES: *C. glochidatum,* *C. nervosum*

FAMILY: Boraginaceae; borage family
PLANT TYPE: Perennial and biennial
SEED VIABILITY: 2–3 years
PLANT SPACING FOR SEED: 12"

FLOWERING: Flowers are usually perfect and are cross-pollinated by insects.

SEED COLLECTION: The fruits are sets of four nutlets adhering together into a bur.

SEED CLEANING AND STORAGE: Harvest when the nutlets have browned. Dry on screens, then flail to release the seeds. The burs will stick together and stick to you, so be patient.

GERMINATION: Germination takes 14 to 21 days in the dark at 70°F (21°C). If after a month no seeds have germinated, move the flats into the light. Germinate seeds of *C. glochidatum* and *C. nervosum* at 41°F (5°C). Germination may take several months.

TRANSPLANTING: Seedlings need 7 to 8 weeks to reach transplanting size.

DAHLIA

Dahlia

D. ×*hybrida* (**garden dahlia**), *D. pinnata* (**pinnate dahlia**)

FAMILY: Asteraceae; sunflower family
PLANT TYPE: Perennial
SEED VIABILITY: 2–3 years

FLOWERING: The disk flowers are perfect, the ray flowers either pistillate or nonsexual.

SEED COLLECTION: The fruit is an achene. Gather the spent flower heads when they have begun to brown.

─────── **PLANT-BREEDING HINT** ───────

In dahlias, both yellow and ivory flower colors tend to be dominant over white. Both are undercolors colored over by purple, red, orange, or other anthocyanins.

SEED CLEANING: Spread the flowers on a screen to dry, then flail or thresh to separate the seeds.

GERMINATION: It takes 10 to 14 days for germination at 63°F (17°C).

TRANSPLANTING: Seedlings need 6 to 8 weeks to reach transplanting size.

DATURA

Angel's trumpet

> *D. ceratocaula, D. discolor* (**desert thornapple**), *D. innoxia* (**angel's trumpet**), *D. metel* (**horn of plenty**), *D. quercifolia* (**Chinese thornapple**)

FAMILY: Solanaceae; nightshade family
PLANT TYPE: Annual
SEED VIABILITY: Unknown

FLOWERING: The flowers are perfect and often nocturnal, and last only a day or so.

SEED COLLECTION: The fruit is a spiny, dehiscent capsule.

SEED CLEANING: Hand-pick the fruits, then spread them on a screen to dry. Rub or flail to remove the very hard seeds.

─────── **FLOWER ALERT** ───────

Angel's trumpet plants are toxic if ingested.

GERMINATION: Germinate at 80°F (27°C) for 7 to 42 days. Germination is highly erratic.

TRANSPLANTING: Seeds in this genus are usually direct-sown in place in the fall but could be started indoors in very cold regions. Allow 6 to 8 weeks to grow seedlings to transplant size.

DELOSPERMA

Delosperma

> *D. cooperi* (**delosperma**), *D. sutherlandii* (**delosperma**)

FAMILY: Aizoaceae; carpetweed family
PLANT TYPE: Biennial or perennial
SEED VIABILITY: Unknown

FLOWERING: The small, perfect flowers are borne singly or in clusters.

SEED COLLECTION: The fruit is a capsule. Hand-pick the fruits when they are dry and brown.

SEED CLEANING: Spread the fruits on a screen to dry, then rub or gently flail to remove the seeds.

GERMINATION: Seeds germinate in a week or two at 68°F (20°C).

TRANSPLANTING: Grow the plants on for 8 to 10 weeks at 60 to 65°F (16 to 18°C).

DELOSPERMA FLOWERS AND CAPSULES

DELPHINIUM

Larkspur

ANNUAL SPECIES: *D. ×belladonna* (**garden delphinium**) *D. grandiflorum* (**delphinium, larkspur, Siberian larkspur**); *D. menziesii* (**Menzi's larkspur**)

PERENNIAL SPECIES: *D. belladonna, D. cardinale, D. cashmirianum, D. chinense, D. elatum, D. exaltatum, D. glareosum, D. glaucum, D. grandiflorum, D. nudicaule* (**red larkspur**)*, D. patens, D. requinni, D. semibarbatum, D. tatsienense*

FAMILY: Ranunculaceae; buttercup family
PLANT TYPE: Perennial and annual; annual species may be perennial in warmer climates
SEED VIABILITY: 1 year
SPACING FOR SEED SAVING: 12"

FLOWERING: Flowers are perfect and cross-pollinated by insects.

SEED COLLECTION: Fruit has two to four carpels and forms three many-seeded follicles. The seed is triangular and wrinkled. Harvest the racemes as soon as they are dry.

SEED CLEANING: Spread the follicles on a screen, then rub or thresh to remove seeds. You may also flail the follicles. Hybrid seed is very perishable and will keep for only 1 year. Store in a plastic bag in the refrigerator.

SEED TREATMENT: Before sowing, soak seeds in 122 to 127°F (50 to 53°C) water for 10 minutes to control seedborne diseases. Stratify seeds for 2 to 4 weeks.

——————— PLANT-BREEDING HINT ———————

Double flowers are recessive to single. The "wild" color of *Delphinium* petals (red-violet) is dominant over others.

GERMINATION: Germinate for 14 to 20 days at 75°F (24°C). Red larkspur (*D. nudicaule*) has very erratic germination. Germinate its seeds at 39°F (4°C) for 1 month to 2 years. Germinate seeds of *D. exaltatum* at 41°F (5°C) for several months.

TRANSPLANTING: Transplant to a second container when the second true leaves appear and provide with partial shade until planting out. Seedlings need 8 to 10 weeks to reach transplanting size.

DIANTHUS

Pink carnation

ANNUAL SPECIES: *D. armeria* (**Deptford pink**), *D. barbatus* (**sweet William**), *D. caryophyllus* (**carnation or clove pink**), *D. chinensis* (**rainbow pink, garden pink**), *D. deltoides* (**maiden pink**), *D. plumarius* (**garden pink**)

PERENNIAL SPECIES: *D. allwoodii, D. alpinus, D. amurensis, D. arenarius, D. ×arvernensis, D. barbatus, D. carthusianorum, D. caryophyllus, D. cruentus, D. deltoides, D. glacialis, D. gratianopolitanus, D. knappii, D. microlepis, D. monspessulanus, D. myrtinervius, D. nardiformis, D. nitidus, D. pavonius, D. petraeus, D. pinifolius, D. plumarius, D. pontederae, D. pyrenaicus, D. seguieri, D. siberica, D. strictus* var. *bebius, D. subacaulis, D. superbus, D. sylvestris*

FAMILY: Caryophyllaceae; pink family
PLANT TYPE: Annual and perennial
SEED VIABILITY: 8–10 years
SPACING FOR SEED SAVING: 12"

——————— FLOWER ALERT ———————

Most of the annual species of *Dianthus* may behave as perennials when grown in warmer climates. Sweet William (*D. barbatus*) and some carnations can also be biennials.

FLOWERING: The perfect flowers are cross-pollinated by insects.

SEED COLLECTION: Fruit is a two-valved capsule that will shatter easily. Harvest the entire plant or hand-pick the capsules.

SEED CLEANING: Spread the plants and capsules on screens to dry, then rub or flail to remove the seeds.

SEED TREATMENT AND GERMINATION: Germinate for 14 to 21 days at 60 to 70°F (16 to 21°C). Germinate rainbow pinks (*D. chinensis*) at 70 to 75°F (21 to 24°C). Germinate carnations (*D. caryophyllus*) at 39°F (4°C) for 2 weeks, then move to 60 to 70°F. This species shows a very high percentage of germination.

If seeds of *D. alpinus*, and *D. pavonius* have not germinated in a month, stratify them for a month, then move them back to 68°F (20°C). Sow seeds of *D. glacialis* at 68°F, wait a month, stratify them for 6 weeks, then move them to about 50°F (10°C) for germination. Sow seeds of *D. strictus* var. *bebius* at 41°F (5°C), and they will germinate over several months.

———————PLANT-BREEDING HINT———————

Be aware that flowers of sweet William (*D. barbatus*) usually change color over the lifetime of the flower (color change is a dominant trait). In carnations (*D. caryophyllous*), white flowers are dominant over yellow and red-yellow flowers are dominant over red.

TRANSPLANTING: Seedlings need 8 to 10 weeks to reach transplanting size. Rainbow pinks and annual and perennial carnations should be grown on at 50 to 55°F (10 to 13°C).

DICENTRA

Bleeding heart

D. cucullaria (**Dutchman's breeches**), *D. eximia* (**fringed bleeding heart**), *D. formosa* (**Pacific bleeding heart**), *D. peregrina*, *D. scandens*, *D. spectabilis* (**bleeding heart**)

FAMILY: Fumariaceae; fumitory family
PLANT TYPE: Perennial
SEED VIABILITY: 1–2 years

FLOWERING: The flowers are perfect.

SEED COLLECTION: The fruit is a capsule. Hand-pick the capsules as they turn brown and begin to dry.

SEED CLEANING: Spread the capsules on screens to dry further, then thresh or flail to remove the seeds.

SEED TREATMENT AND GERMINATION: Sow and hold seeds at 68°F (20°C) for a month, then move to 50°F (10°C). Sow the seeds immediately after their harvest. If seeds of *D. scandens* do not germinate in a month at 68°F (20°C), stratify them for a month and return them to 68°F (20°C).

TRANSPLANTING: Grow the transplants on for about 10 weeks before transplanting. The easier method is to sow the seeds outdoors in late summer and provide a winter mulch. Transplant the small sets in spring.

DICTAMNUS

Gas plant

D. albus (**dittany, gas plant**)

FAMILY: Rutaceae; citrus family
PLANT TYPE: Perennial
SEED VIABILITY: Very poor

FLOWERING: The flowers are perfect.

SEED COLLECTION: The fruit is a deeply lobed-five-celled capsule. Harvest the seed heads or hand-pick the capsules when they turn brown but before they burst and shed their seeds.

SEED CLEANING: Place the capsules and heads on a screen to dry; then shake, thresh, flail, or rub to separate the seeds.

——————— FLOWER ALERT ———————

All parts of gas plants, especially the seeds, are poisonous.

SEED TREATMENT AND GERMINATION: Sow and hold the seeds at 75°F (24°C) for 2 months, move them to 39°F (4°C) for 2 months, then to 50°F (10°C) by slowly raising the temperature over a week or two. If they do not germinate in a few months, repeat the protocol. Germination may take more than 2 years.

TRANSPLANTING: Grow the plants on at about 55°F (13°C) or, better yet, sow the seeds directly outdoors in the fall. Then transplant the seedlings to their permanent location.

DIERAMA

Wandflower

D. pendulum (**angel's fishing rod, grassy bell**), *D. pulcherrimum*

FAMILY: Iridaceae; iris family
PLANT TYPE: Perennial
SEED VIABILITY: Unknown

FLOWERING: The flowers are perfect and appear in spikes in summer.

SEED COLLECTION: The fruit is a small, membranous capsule. Harvest the fruits carefully by hand when they are brown and dry.

SEED CLEANING: Dry the fruits further on screens; then rub, roll, or flail gently to remove the seeds.

GERMINATION: Sow seeds of these species at 68°F (20°C). If none germinates in a month, then stratify for a month and return to 68°F (20°C).

TRANSPLANTING: You can transplant outdoors in spring or, even better, sow the seeds outdoors in fall in a protected seedbed and provide a winter mulch. Seeds will germinate in spring.

DIGITALIS

Foxglove

BIENNIAL SPECIES: *D. ferruginea* (**rusty foxglove**), *D. grandiflora* (**yellow foxglove**), *D. lanata* (**Grecian foxglove**), *D. lutea* (**straw foxglove**), *D. purpurea* (**foxglove**)

FOXGLOVE FLOWERS AND CAPSULES

PERENNIAL SPECIES: *D. eriostachya* (syn. *D. lutea*), *D. laevigata*, *D. obscura*, *D. parviflora*, *D. thapsi*

FAMILY: Scrophulariaceae; figwort family
PLANT TYPE: Biennial and perennial; some biennial species may be perennial in warmer areas
SEED VIABILITY: 2 years
SPACING FOR SEED SAVING: 12"

FLOWERING: The perfect flowers are cross-pollinated by insects.

———————— FLOWER ALERT ————————

Foxglove plants are toxic if ingested.

SEED COLLECTION: The fruit is a capsule. Collect the capsules by hand as they mature.

SEED CLEANING: Dry the capsules on screens, then thresh or flail to remove the seeds.

SEED TREATMENT: Soak seeds in 130°F (54°C) water for 15 minutes before sowing to control soilborne diseases if they are an issue in your garden.

flowers

GERMINATION: Surface-sow and germinate 14 days in light at 70°F (21°C).

─────────**PLANT-BREEDING HINT**─────────

Purple petal color is dominant over white, and purple spotting is over brown spotting in *D. purpurea* ssp. *purpurea*. Nonpeloric is dominant over peloric. If you make crosses of foxglove and straw foxglove (*D. lutea*), the hybrid offspring will resemble the maternal parent in calyx and corolla. This may influence your choice of species to use as the pollen source.

Various hybrid combinations in foxgloves may be sterile — they may produce flowers but no seeds.

TRANSPLANTING: Grow the plants on at 55°F (13°C). It takes 8 to 10 weeks to form good transplants.

DIMORPHOTHECA

Cape marigold

D. pluvialis (**weather prophet**), *D. sinuata* (**Cape marigold**)

FAMILY: Asteraceae; sunflower family
PLANT TYPE: Annual
SEED VIABILITY: 1 year
SPACING FOR SEED SAVING: 8"

FLOWERING: The disk flowers are perfect; the ray flowers are pistillate.

SEED COLLECTION: The fruit are achenes. Harvest the seed heads when they are dry.

SEED CLEANING: Spread the seed heads on a screen to dry further, then flail.

GERMINATION: Germinate for 14 to 21 days at 60 to 70°F (16 to 21°C).

TRANSPLANTING: Seedlings need 6 to 7 weeks to reach transplanting size. Do not wet the leaves.

DOLICHOS

Hyacinth bean

D. lablab (syn. *Lablab purpureus;* **hyacinth bean**)

FAMILY: Fabaceae; legume family
PLANT TYPE: Annual
SEED VIABILITY: Unknown

FLOWERING: Flowers are perfect.

SEED COLLECTION: The fruit is a flat, beaked legume pod. Hand-pick the fruits when they yellow.

SEED CLEANING: Spread the fruits on a screen to dry, then flail or thresh to remove the seeds.

GERMINATION: Germinate at 65 to 70°F (18 to 21°C) for 10 to 20 days.

TRANSPLANTING: Seeds are best sown in place outdoors, as they don't transplant well.

─────────**PLANT-BREEDING HINT**─────────

Flower color, seed-coat color, and plant color are closely correlated in hyacinth bean; purple is slightly dominant over white. Climbing habit is dominant over bush habit.

ECHINACEA

Coneflower

E. angustifolia, E. pallida, E. paradoxa, E. purpurea (**purple coneflower**), *E. tennesseensis*

FAMILY: Asteraceae; sunflower family
PLANT TYPE: Perennial
SEED VIABILITY: Unknown

FLOWERING: The flowers are perfect and are cross-pollinated by insects.

SEED COLLECTION: The fruits are four-angled achenes. Harvest the flower heads when they have dried.

CONEFLOWER SEED HEAD

SEED CLEANING: Spread the spent heads on screens to dry, then flail or thresh to remove the seeds.

SEED TREATMENT: Stratify for 5 weeks.

GERMINATION: Seeds will germinate after 20 to 30 days in the dark at 60 to 65°F (16 to 18°C). Sow, then hold seeds of *E. angustifolia*, *E. paradoxa*, and *E. tennesseensis* at 68°F (20°C) for a month, move to 41°F (5°C) for 10 weeks, then germinate at 50°F (10°C).

TRANSPLANTING: Seedlings need 8 to 10 weeks to reach transplanting size at about 65°F (18°C).

ECHINOPS
Globe thistle

E. bannaticus, *E. ritro* (**small globe thistle**), *E. sphaerocephalus* (**great globe thistle**)

FAMILY: Asteraceae; sunflower family
PLANT TYPE: Biennial and perennial
SEED VIABILITY: Unknown

FLOWERING: The flowers are perfect and borne in a globular head.

SEED COLLECTION: The fruit are achenes. Hand-pick the flower heads when they have dried.

SEED CLEANING: Spread the spent heads on a screen, then thresh or flail to separate the fruit.

GERMINATION: Sow the seeds at 68°F (20°C). Germination should take 14 days or less.

TRANSPLANTING: Grow the plants on at about 65°F (18°C) for about 8 weeks, then transplant to the garden.

ERODIUM
Heron's bill, stork's bill

E. acaule, *E. manescavii* (**Pyrenees heronbill**), *E. pelargoniflorum* (**geraniumleaf heronbill**)

FAMILY: Geraniaceae; geranium family
PLANT TYPE: Perennial
SEED VIABILITY: Unknown

FLOWERING: The flowers are perfect and arranged in umbels.

SEED COLLECTION: Harvest the umbels as they dry. The fruit are capsules. Not all capsules ripen at once, so harvest may extend over a period of about a week.

SEED CLEANING: Spread the fruit capsules on a screen to dry further, then flail or rub to separate the seeds.

SEED TREATMENT: Grow on for 6 to 8 weeks at 60 to 65°F (16 to 18°C) before transplanting outdoors, or sow directly outdoors in early spring.

GERMINATION: Sow seeds at 68°F (20°C). If those of *E. acaule* do not germinate in a month, stratify for a month, then return to 68°F (20°C).

TRANSPLANTING: Grow the plants on at about 65°F (18°C) for about 10 weeks, then transplant outdoors. Otherwise, sow the seeds directly in the garden.

ESCHSCHOLZIA

California poppy

E. caespitosa (**tufted poppy**), *E. californica* (**California poppy**)

FAMILY: Papaveraceae; poppy family
PLANT TYPE: Annual
SEED VIABILITY: 10 years
SPACING FOR SEED SAVING: 6"

FLOWERING: The perfect flowers are cross-pollinated by insects.

SEED COLLECTION: The fruit is a slender capsule containing several seeds. Harvest the individual fruits when they brown and the seeds rattle in them.

SEED CLEANING: Spread the capsules on a screen to dry. Flail the dried capsules to remove the seeds.

——————PLANT-BREEDING HINT——————

In California poppy (*E. californica*), white flower color is dominant over yellow and orange; complete orange color is dominant over orange base color on yellow petals.

SEED TREATMENT: In autumn, sow seeds directly into the garden. Prior to sowing, soak seeds for 30 minutes in 125°F (52°C) water to control seedborne disease.

GERMINATION: Germination takes 7 to 21 days at 70°F (21°C).

TRANSPLANTING: This plant will not transplant well, so direct-sow in place.

EUPATORIUM

Joe-pye weed

BIENNIAL SPECIES: *E. fistulosum* (**Joe-pye weed**), *E. maculatum* (**spotted pye weed**)

PERENNIAL SPECIES: *E. ageratoides, E. altissimum, E. cannabinum, E. chinense* var. *simplicifolia, E. coelestinum, E. greggii, E. maculatum, E. perfoliatum, E. rugosum, E. sessilifolium*

FAMILY: Asteraceae; sunflower family
PLANT TYPE: Biennial and perennial
SEED VIABILITY: Unknown

FLOWERING: The flowers are perfect.

SEED COLLECTION: The fruit is an achene. Harvest the drying heads before they shatter.

SEED CLEANING: Spread the heads on screens to dry further, then thresh to remove the seeds.

SEED TREATMENT: Hold the seeds for a month at 68°F (20°C), then 6 weeks at 39°F (4°C). Stratify seeds of *E. altissimum* at 32°F (0°C) for 3 months.

GERMINATION: Surface-sow and germinate for 10 to 20 days at 70°F (21°C) in light. Seeds of *E. cannabium, E. perfoliatum, E. rugosum*, spotted pye weed (*E. maculatum*), and *E. sessilifolium* need no stratification. Sow directly at 68°F (20°C).

TRANSPLANTING: Seedlings need about 8 weeks to reach transplanting size.

EUSTOMA (LISIANTHUS)

Prairie gentian

E. exaltatum (**catchfly prairie gentian**), *E. russellianus* (**lisianthus, showy prairie gentian**)

FAMILY: Gentianaceae; gentian family
PLANT TYPE: Annual or biennial
SEED VIABILITY: Less than 3 years

FLOWERING: The flowers are perfect.

SEED COLLECTION: The fruit is an ellipsoid, many-seeded capsule. Hand-pick the fruits before they shed.

SEED CLEANING: Spread the capsules on a screen to dry, then rub or flail to separate the seeds.

GERMINATION: Surface-sow the seeds in light at 77°F (25°C). Germination takes 10 to 20 days.

TRANSPLANTING: Grow the plants on at 60°F (16°C) for 8 to 10 weeks.

FRITILLARIA

Fritillary

F. imperialis (**crown imperial, imperial fritillary**), *F. lanceolata* (**checker lily, narrow-leaved fritillary, rice root fritillary**), *F. meleagris* (**checkered lily, guinea hen flower, guinea hen tulip, snake's head**), *F. michailovskyi, F. pallidiflora, F. persica, F. pontica, F. pudica* (**yellow fritillary**), *F. pyrenaica, F. raddeana, F. tubiformis*

FAMILY: Liliaceae; lily family
PLANT TYPE: Perennial
SEED VIABILITY: Unknown

FLOWERING: The flowers are perfect.

SEED COLLECTION: The fruits are three-valved, sometimes six-angled capsules, each containing many flat seeds. Harvest the capsules when they brown and dry.

SEED CLEANING: Spread the capsules on a screen to dry, then rub or thresh to remove the seeds.

GERMINATION: Sow all seeds at 41°F (5°C), then wait several months for germination.

TRANSPLANTING: You may sow the seeds in place outdoors or grow the transplants at around 60°F (16°C) for about 10 weeks. Plants in this genus are usually propagated by offsets, not by seeds.

GAILLARDIA

GAILLARDIA

Blanket flower

ANNUAL SPECIES: *G. amblyodon* (**maroon blanket flower**), *G. pulchella* (**blanket flower**)

PERENNIAL SPECIES: *G. aristata, G. lanceolata, G. lorenziana, G. pinnatifida, G. pulchella*

FAMILY: Asteraceae; sunflower family
PLANT TYPE: Annual and perennial; *G. lanceolata* can be an annual in cooler areas
SEED VIABILITY: 2–4 years
SPACING FOR SEED SAVING: 12"

FLOWERING: The perfect flowers are cross-pollinated by insects.

SEED COLLECTION: The fruits are achenes. Harvest the plants when the flowers brown.

SEED CLEANING: Further dry the flowers on screens, then flail to separate the seeds.

GERMINATION: Germinate the seeds for 14 to 21 days in light at 70 to 75°F (21 to 24°C) day and 60 to 65°F (16 to 18°C) night.

TRANSPLANTING: Seedlings need 6 to 8 weeks to reach transplanting size.

GAZANIA

Treasure flower

G. hybrida, *G. krebsiana*, *G. linearis* (**treasure flower**)

FAMILY: Asteraceae; sunflower family
PLANT TYPE: Annual; perennial in warmer areas
SEED VIABILITY: Unknown

FLOWERING: The disk flowers are tubular and perfect, the ray flowers sterile.

SEED COLLECTION: The fruits are achenes. Harvest the flower heads when they are spent and brown.

SEED CLEANING: Spread the heads on a screen to dry, then rub or flail to remove the achenes.

GERMINATION: Germinate at 60 to 65°F (16 to 18°C) in the dark for 8 to 20 days.

TRANSPLANTING: Grow the plants on at about 60°F (16°C) for about 8 weeks.

GERANIUM

Geranium, cranesbill

ANNUAL SPECIES: *G. columbinum* (**long-stalked cranesbill**), *G. dissectum* (**wrinkled-seeded cranesbill**), *G. lucidum* (**shining geranium**), *G. molle* (**dove-foot geranium**), *G. pusillum* (**small geranium**)

PERENNIAL SPECIES: *G. asphodeloides*, *G. bohemicum*, *G. endressii* (**Pyrenean cranesbill**), *G. himalayense* (**lilac cranesbill**), *G. macrorrhizum* (**bigroot geranium**), *G. maculatum* (**alumroot, spotted cranesbill, wild cranesbill, wild geranium**), *G. maderense*, *G. nepalense*, *G. nodosum*, *G. oxonianum*, *G. palmatum*, *G. palustre*, *G. phaeum*, *G. pratense*, *G. psilostemon*, *G. pyrenaicum*, *G. renardii*, *G. robertianum* (**herb Robert**), *G. sanguineum*, *G. swatense*, *G. sylvaticum*, *G. viscosissimum*

FAMILY: Geraniaceae; geranium family
PLANT TYPE: Annual and perennial; *G. maderense* is sometimes perennial and herb Robert (*G. robertianum*) is sometimes biennial
SEED VIABILITY: Unknown

FLOWERING: The flowers are perfect and are borne solitary or clustered.

SEED COLLECTION: Collect the capsules when they have dried on the plants or hand-pick them when they have begun to dry, then dry them further in a paper bag or on screens in a protected location.

SEED CLEANING: Remove the seeds by rubbing the dried capsules on screens or by flailing them.

——————— FLOWER ALERT ———————

Herb Robert (*G. robertianum*) is an invasive weed in some areas.

SEED TREATMENTS AND GERMINATION: Annual species are best sown outdoors. If you choose to grow transplants of these, sow the seeds in flats or pots and Grow the transplants on at 70°F (21°C). Germination may take from 6 weeks to 13 months, so be patient. For the perennials, sow seeds of *G. bohemicum*,

GERANIUM BLOSSOM AND FLOWER BUDS

alumroot (*G. maculatum*), *G. nepalense*, *G. nodosum*, *G. palustre*, *G. phaeum*, *G. psilostemon*, *G. renardii*, *G. sanguineum*, *G. swatense*, and *G. viscosissimum* at 41°F (5°C); germination may take several months. Sow seeds of *G. pratense* and *G. pyrenaicum* at 68°F (20°C). If they do not germinate in a month, stratify them for 6 weeks, then move them to 50°F (10°C) for germination. Sow seeds of *G. asphodeloides*, lilac cranesbill (*G. himalayense*), *G. oxonianum*, and *G. palmatum* at 55°F (13°C); germination should take 10 to 14 days. Sow seeds of Pyrenean cranesbill (*G. endressii*), bigroot geranium (*G. macrorrhizum*), and *G. sylvaticum* at 68°F (20°C). Germination is slow and may take several months.

TRANSPLANTING: Grow seedlings on at 65°F (18°C) for 8 to 10 weeks before transplanting to the garden.

GEUM

Avens

G. 'Bulgaricum', *G. chiloense*, *G. coccineum*, *G. hybridum*, *G. macrophyllum*, *G. montanum*, *G. pyrenaicum*, *G. reptans*, *G. rivale*, *G. triflorum*, *G. urbanum*

FAMILY: Rosaceae; rose family
PLANT TYPE: Perennial
SEED VIABILITY: 2 years
SPACING FOR SEED SAVING: 12"

FLOWERING: Perfect flowers are cross-pollinated by insects.

SEED COLLECTION: Fruit are achenes with persistent styles. Collect the seed heads by hand as they mature.

SEED CLEANING: Spread the heads on screens to dry, then rub or flail to remove the seeds.

SEED TREATMENT AND GERMINATION: Germinate for 21 days at 70°F (21°C).

If seeds of *G.* 'Bulgaricum', *G. rivale*, *G. triflorum*, or *G. urbanum* do not germinate after a month, stratify them for 4 weeks, then repeat the cycle.

Sow seeds of *G. coccineum*, *G. montanum*, *G. pyrenaicum*, and *G. reptans* at 41°F (5°C), and they will germinate over several months.

TRANSPLANTING: Grow the plants on for about 8 weeks.

──────────PLANT-BREEDING HINT──────────

Red flower color is dominant over non-red in *Geum* and yellow is dominant over non-yellow. Large flowers are dominant over small flowers.

GILIA

Gilia

G. achilleifolia (**California gilia**), *G. capitata* (**bluehead gilia**), *G. latiflora* var. *davyi* (**hollyleaf gilia**), *G. leptantha* (**fineflower gilia**), *G. thurberi* (**El Paso skyrocket**), *G. tricolor* (**bird's eye**)

FAMILY: Polemoniaceae; phlox family
PLANT TYPE: Annual
SEED VIABILITY: 4–5 years

FLOWERING: The flowers are perfect.

SEED COLLECTION: The fruit is a three-celled capsule that eventually ruptures. Harvest the spent heads as they brown and before the capsules rupture.

SEED CLEANING: Spread the heads on screens to dry, then rub or flail to remove the seeds.

GERMINATION: Sow the seeds at 70°F (21°C); they will germinate in 14 to 21 days.

TRANSPLANTING: Seeds are usually direct-sown in the garden.

GOMPHRENA

Globe amaranth

G. canescens, G. decumbens (**arrasa con todo**), *G. dispersa* (**arrasa con todo**), *G. globosa* (**globe amaranth**), *G. haageana* (**Rio Grande globe amaranth**)

FAMILY: Amaranthaceae; amaranth family

PLANT TYPE: Annual; *G. canescens* may be perennial in warmer areas

SEED VIABILITY: 2–3 years

FLOWERING: The perfect flowers are borne in dense, chaffy heads.

SEED COLLECTION: The fruit is a utricle. Harvest the heads when they brown but before they shatter.

SEED CLEANING: Dry the heads on screens, then thresh or flail to remove the seeds.

SEED TREATMENT: Never store the seeds of these species below 60°F (16°C).

GERMINATION: Germinate at 70 to 75°F (21 to 24°C) in light for 7 to 14 days.

TRANSPLANTING: Seedlings need 8 to 12 weeks to reach transplanting size. Alternate the temperatures between 75°F (24°C) day and 65°F (18°C) night.

GYPSOPHILA

Gypsophila, baby's breath

ANNUAL SPECIES: *G. elegans* (**baby's breath**)

PERENNIAL SPECIES: *G. cerastoides*, *G. muralis* (**low babysbreath**) *G. pacifica*, *G. paniculata* (**baby's breath**), *G. repens*, *G. tenuifolia* var. *gracilipes*

FAMILY: Caryophyllaceae; pink family

PLANT TYPE: Annual and perennial

SEED VIABILITY: 2 years for baby's breath (*G. elegans*) and low babysbreath (*G. muralis*); 4 years for perennial species

SPACING FOR SEED SAVING: 12"

FLOWERING: Perfect flowers are cross-pollinated by insects.

SEED COLLECTION: Harvest the crop when most of the one-celled capsules have turned brown. If the plants have "gotten away" from you and become very mature, harvest only in the early morning when dew is on the plants. That will reduce shattering.

SEED CLEANING: Place entire plants on screens to dry further, then tumble or flail to extract the seeds.

—————— FLOWER ALERT ——————

Some species of baby's breath have escaped cultivation and become invasive in some areas.

GERMINATION: Germinate for 14 to 21 days at 70°F (21°C). Germinate seeds of *G. cerastoides* at 41°F (5°C), and they will germinate over several months.

TRANSPLANTING: It takes about 6 weeks for seedlings of annual species to reach transplanting size and about 10 weeks for perennial species.

HELIANTHUS

Sunflower

ANNUAL SPECIES: *H. annuus* (**sunflower**), *H. argophyllus* (**silver leaf sunflower**), *H. debilis* (**cucumberleaf sunflower**)

HELIANTHUS SEED HEAD

PERENNIAL SPECIES: *H. angustifolius, H. deca-petalus, H. grosseserratus, H. ×laetiflorus, H. maximiliani, H. mollis, H. occidentalis, H. rigidus, H. strumosus*

FAMILY: Asteraceae; sunflower family
PLANT TYPE: Annual and perennial
SEED VIABILITY: 2–3 years
SPACING FOR SEED SAVING: 12"

FLOWERING: The plants are cross-pollinated by insects. The disk flowers are perfect and the ray flowers pistillate.

SEED COLLECTION: Harvest the heads when achenes (seeds) are dark colored and begin to rub off easily.

SEED CLEANING: Hang the heads upside down to dry in the garage for another 2 to 3 weeks, or until the achenes rub off easily. Dry the achenes on a screen, then store dried in sealed containers.

SEED TREATMENT: Stratify seeds of *H. grosseserratus, H. laetiflorus, H. mollis,* and *H. strumosus* for 12 weeks before moving them to germination temperature.

─────── FLOWER ALERT ───────

Swamp sunflower is one of those plants that is often grown as an annual, but it can behave as a biennial or even a perennial, too.

GERMINATION: Germinate for 14 to 21 days at 70°F (21°C). Sow seeds of *laetiflorus, H. decapetalus, H. maximiliana, H. occidentalis,* and *H. rigidus* at 41°F (5°C), and they will germinate over several months.

TRANSPLANTING: Seedlings need 3 weeks to reach transplanting size, although seeds are usually direct-sown. If you want to start them indoors, do so in peat pots, and transplant the pots to the garden, as sunflowers do not transplant well bare-rooted.

HELICHRYSUM

Strawflower, everlasting

ANNUAL SPECIES: *H. bracteatum* (**strawflower**), *H. lindleyii, H. subulifolium*

BIENNIAL SPECIES: *H. foetidum* (**stinking strawflower**)

PERENNIAL SPECIES: *H. arenarium, H. bellidioides, H. italicum, H. plicatum, H. thianschanicum*

FAMILY: Asteraceae; sunflower family
PLANT TYPE: Annual, biennial, and perennial
SEED VIABILITY: 2 years
SPACING FOR SEED SAVING: 8"

FLOWERING: The flowers are perfect, though sometimes the outer rows are pistillate. Flowers are cross-pollinated by insects.

SEED COLLECTION: The fruits are five-angled achenes. Harvest the fruits when the pappus becomes fluffy. You can pick individual heads or cut an entire plant. Harvest the heads and plants only when they are very dry.

SEED CLEANING: Dry the plants and fruits on a screen with a sheet beneath, then flail or tumble to separate the seeds. The seed is bulky and is easily blown away by a moderate wind gust.

GERMINATION: Germinate for 14 to 21 days in light at 75°F (24°C).

TRANSPLANTING: Grow on for 8 weeks at 65°F (18°C).

───────PLANT-BREEDING HINT───────

In strawflower, sulfur and orange flower color are dominant over white. If both dominant genes occur together, the plant will have gold flowers.

Helipterum

Everlasting, strawflower

H. roseum (**paper daisy, everlasting, straw-flower**), *H. muellerii* (**immortelle;** also known as *Acroclinium roseum*)

FAMILY: Asteraceae; sunflower family
PLANT TYPE: Annual
SEED VIABILITY: 3 years
SPACING FOR SEED SAVING: 8"

FLOWERING: The flowers are perfect, except those in the outer rows, which are sometimes pistillate. They are cross-pollinated by insects.

SEED COLLECTION: The fruits are nearly cylindrical achenes. Harvest the seeds when the flowers have dried.

SEED CLEANING: Spread the flowers on a screen for further drying, then flail or tumble to remove the seeds.

GERMINATION: Germinate 14 to 40 days at 72°F (22°C).

TRANSPLANTING: Seedlings need 12 weeks to reach transplanting size.

Hemerocallis

Daylily

H. ×hybrida, H. liliasphodelus (**lemon daylily, lemon lily, yellow daylily**), *H. middendorffii* (**Middendorf daylily**), *H. minor* (**grass leaf daylily, small daylily**)

FAMILY: Liliaceae; lily family
PLANT TYPE: Perennial
SEED VIABILITY: Unknown

FLOWERING: The flowers, which are perfect, are insect-pollinated.

SEED COLLECTION: The fruit is a three-valved capsule. Harvest the capsules by hand when they have browned and dried.

SEED CLEANING: Further dry the capsules on screens, then flail or thresh to remove the few seeds they contain.

GERMINATION: Lightly cover the seeds, then germinate them at 41°F (5°C) over several months.

TRANSPLANTING: Grow the plants on for 8 to 10 weeks at 55 to 60°F (13 to 16°C), or sow the seeds directly outdoors in early spring, while the soil is still cool. The juvenile period may take 2 to 3 years, so be patient. These plants are usually propagated by clump division.

Heuchera

Heuchera, alumroot

H. americana (**American alumroot, rock geranium**), *H. ×brizoides, H. cylindrica, H. micrantha, H. nicratum, H. pulchella, H. richardsonii, H. sanguinea* (**coral bells**), *H. villosa* var. *macrorrhiza* (**hairy alumroot**)

FAMILY: Saxifragaceae; saxifrage family
PLANT TYPE: Perennial
SEED VIABILITY: Less than 3 years

FLOWERING: The flowers are perfect and arranged in panicles.

SEED COLLECTION: The fruit is a dehiscent capsule. Hand-pick the capsules as they brown and dry but before they dehisce, or harvest entire panicles when dried and before the capsules dehisce.

SEED CLEANING: Spread the panicles or capsules on screens to dry further; then rub, thresh, or flail to extract the seeds.

SEED TREATMENT: Stratify seeds of *H. richardsonii* for 3 months, then sow as for other species.

GERMINATION: Surface-sow at 68°F (20°C) in light the seeds of American alumroot (*H. americana*), *H. ×brizoides, H. cylindrica, H. micrantha, H. nicratum, H. pulchella,* coral bells (*H. sanguinea*), *H. versicola,* and hairy alumroot (*H. villosa* var. *macrorrhiza*). If the seeds have not germinated in a month, stratify for a month and repeat the cycle.

TRANSPLANTING: Grow the plants on at 60 to 65°F (16 to 18°C) for 6 to 8 weeks. Or sow seeds outdoors in spring, and their plants will form flowers the following year. Heucheras are drought-tolerant, but do best planted in a cool, shady spot.

Hibiscus

Mallow, giant mallow, rose mallow

H. coccineus (**scarlet rosemallow**), *H. manihot,* *H. moscheutos* (**common rosemallow**), *H. syriacus* (**rose of Sharon, shrub althea**)

FAMILY: Malvaceae; mallow family
PLANT TYPE: Perennial
SEED VIABILITY: 3–4 years

FLOWERING: The flowers are perfect.

SEED COLLECTION: The fruit is a five-celled capsule, each cell containing three seeds. Harvest the capsules by hand when they brown and dry.

SEED CLEANING: Further dry the capsules on screens, then rub or flail to remove the seeds.

———————— FLOWER ALERT ————————

Some species of mallow are classified as weeds in some areas.

GERMINATION: Seeds will germinate readily at 75°F (24°C).

TRANSPLANTING: Grow the plants on for 6 to 8 weeks at 75°F (24°C) for spring transplanting.

HOSTA CAPSULES AND SEEDS

Hosta

Hosta, lily-plantain

H. elata, H. minor, H. montana, H. sieboldiana, *H. ventricosa*

FAMILY: Liliaceae or Hostaceae; lily family or hosta family
PLANT TYPE: Perennial
SEED VIABILITY: Unknown

FLOWERING: The flowers appear in terminal clusters.

SEED COLLECTION: The fruits are three-valved capsules containing many seeds. Harvest the capsules by hand when they brown.

SEED CLEANING: Spread the capsules on screens to dry, then roll or flail to separate the small, winged seeds.

GERMINATION: Sow seeds of these species at 68°F (20°C). If they do not germinate in a month, stratify for a month and return them to 68°F (20°C).

TRANSPLANTING: Grow the plants on for 6 to 8 weeks at 75°F (24°C).

SWEET ALYSSUM FLOWERS

SEED COLLECTION: The fruits are silicles (pods), which ripen unevenly. Harvest the seeds when most pods have turned brown. Pods shatter easily, so don't delay the harvest. Harvest the plants in the early morning while they are still damp with dew; this will reduce the shattering.

SEED CLEANING: Dry the plants in thin layers with their pods on screens and a sheet beneath them to catch the seeds; then flail or tumble to extract the remaining seeds.

GERMINATION: Sprinkle seeds on the soil surface and press them in with a board or by hand. Germination takes 14 to 21 days at 80°F (27°C) in light.

TRANSPLANTING: Grow the plants on at 50 to 55°F (10 to 13°C) for about 9 weeks.

LUPINUS

Lupine

ANNUAL SPECIES: *L. albus* (**white lupine**), *L. angustifolius* (**narrowleaf lupine**), *L. benthamii* (**spider lupine**), *L. bicolor* (**bicolor lupine**), *L. albus* (**whitewhorl lupine**), *L. hartwegii* (**Hartweg's bluebonnet**), *L. hirsutissimus* (**stinging annual lupine**), *L. luteus* (**European yellow lupine**), *L. nanus* (**sky lupine**), *L. sparsiflorus* (**Mojave lupine**), *L. subcarnosus* (**Texas bluebonnet**), *L. texensis* (**Texas bluebonnet**), *L. subvexus* (**valley lupine**), *L. succulentus* (**hollowleaf annual lupine**), *L. truncatus* (**collared annual lupine**)

PERENNIAL SPECIES: *L. albicaulis* (**sicklekeel lupine**), *L. arboreus*, *L. lepidus*, *L. littoralis*, *L. perennis*, *L. podophyllus*, *L. polyphyllus*, *L.* Russell hybrids, *L. sericeus*, *L. texensis*

FAMILY: Fabaceae; legume family
PLANT TYPE: Annual and perennial
SEED VIABILITY: 2 years
SPACING FOR SEED SAVING: 9"

FLOWERING: Perfect flowers are held in terminal racemes. As with other legumes, they are papilionaceous and are self-pollinated.

SEED COLLECTION: The fruit is a legume (pod), often showing constrictions between the seeds. Harvest the pods as soon as they turn brown, before they dehisce, when fully mature.

SEED CLEANING: Spread the pods on a screen to dry, then flail to extract the seeds.

SEED TREATMENT: Nick the seeds and soak in warm water for 24 hours, or pour hot water over the seeds and let them soak for about 3 days, or until there is noticeable swelling.

GERMINATION: It will take 14 to 56 days to germinate seeds at 68°F (20°C). If seeds of *L. lepidus* and *L. perennis* do not germinate in a month, stratify them for 3 months, then return to 68°F (20°C).

TRANSPLANTING: Seedlings need 6 to 8 weeks to reach transplanting size. The plants do not transplant well bare-root, so start them in peat pots.

LYCHNIS

Campion

ANNUAL SPECIES: *L. coeli-rosa* (**rose of Heaven**), *L. haageana*

PERENNIAL SPECIES: *L. alpina*, *L.* ×*arkwrightii*, *L. chalcedonica*, *L. coronaria* (**rose campion**), *L. flos-cuculi*, *L. flos-jovis*, *L. miqueliana*, *L. sieboldii*, *L. viscaria*, *L. yunnanensis*

FAMILY: Caryophyllaceae; pink family

PLANT TYPE: Annual, biennial, and perennial

SEED VIABILITY: 2–3 years

FLOWERING: The flowers are often perfect but may be imperfect at times. The older types are normally dioecious.

SEED COLLECTION: The fruits are five-toothed capsules. Hand-pick the capsules before they dry fully.

———————— PLANT-BREEDING HINT ————————

Purple campion flowers are dominant over white and broad leaves are dominant over narrow ones.

SEED CLEANING: Spread the capsules on fine screens to dry, then rub or flail to remove the tiny black seeds.

GERMINATION: Sow seeds of *L. ×arkwrightii* at 70°F (21°C) in light and seeds of the other species at 60 to 65°F (16 to 18°C). All seeds germinate in 7 to 14 days. If seeds of *L. sieboldii* do not germinate in a month, stratify them for another month, then return them to 68°F (20°C).

———————— FLOWER ALERT ————————

With campions, plants don't always behave according to type. Many of the annual species may become perennials in warmer areas. And two perennial species, rose campion and *L. miquelina*, sometimes behave as biennials.

TRANSPLANTING: Seeds are usually sown in place, but if you want, you may Grow the transplants on at about 65°F (18°C) for about 10 weeks.

Malva

Hollyhock mallow

M. alcea (**hollyhock mallow**), *M. dendromorpha*, *M. moschata* (**musk mallow**), *M. sylvestris* (**high mallow**)

FAMILY: Malvaceae; mallow family

PLANT TYPE: Perennial

SEED VIABILITY: Unknown

FLOWERING: The flowers are perfect and borne either solitary or in clusters.

SEED COLLECTION: The fruit is a disk-shaped schizocarp with mericarps, each containing one seed. Harvest the spent brown seed heads.

SEED CLEANING: Dry the spent flowers on a screen, then flail or rub to separate the seeds.

GERMINATION: Sow seeds of hollyhock mallow (*M. alcea*) and musk mallow (*M. moschata*) at 41°F (5°C). Sow those of high mallow (*M. sylvestris*) in light at 68°F (20°C).

TRANSPLANTING: Grow transplants at about 75°F (24°C) for 6 to 8 weeks before setting them outdoors in early spring.

HOLLYHOCK MALLOW FLOWERS
AND SCHIZOCARPS

MATTHIOLA

Stock

BIENNIAL SPECIES: *M. incana* (**evening stock, gilliflower**), *M. sinuata*

ANNUAL SPECIES: *M. longipetala* (**night-scented stock**)

FAMILY: Brassicaceae; mustard family
PLANT TYPE: Biennial and annual
SEED VIABILITY: 10 years
SPACING FOR SEED SAVING: 4"

FLOWERING: The perfect flowers are cross-pollinated by insects.

ISOLATION REQUIREMENTS: The crop is mostly self-pollinated, so there is no need to isolate varieties.

SEED COLLECTION: The fruit is a flattened, long, narrow silique. Fruits from single flowers contain 30 to 60 seeds. The double flowers contain no sex organs and thus no seeds. Harvest the plants as soon as the fruits turn yellow-brown.

SEED CLEANING: Dry the entire plants on screens, with a sheet beneath them to capture the seeds, then thresh to extract the seeds.

SEED TREATMENT: Soak seeds in 130°F (54°C) water for 10 minutes to control seedborne disease if they are an issue in your garden.

——————— **PLANT-BREEDING HINT** ———————

Purple and red flower colors are dominant over white for stock. Double flowers are recessive to single. Earliness in blooming is dominant over lateness, and open growth habit is dominant over compactness.

GERMINATION: Germination takes 7 to 21 days in light at 70°F (21°C).

TRANSPLANTING: Seedlings need 7 to 8 weeks to reach transplanting size.

MESEMBRYANTHEMUM

Ice plant

M. criniflorum (**ice plant**), *M. crystallinum* (**common ice plant**), *M. occulatum*

FAMILY: Aizoaceae; carpetweed family
PLANT TYPE: Annual
SEED VIABILITY: 3–4 years

FLOWERING: The flowers are perfect.

SEED COLLECTION: The fruit is a four- to five-valved capsule. Hand-pick the fruits when they brown.

SEED CLEANING: Dry the fruits on screens, then rub or flail to extract the seeds.

GERMINATION: Germinate in light for 7 to 21 days at 65°F (18°C).

TRANSPLANTING: Grow the plants on for about 8 weeks before transplanting.

MIMULUS

Monkeyflower

ANNUAL SPECIES: *M. brevipes* (**Bolander's monkeyflower**), *M. hybridus* (**monkeyflower**)

PERENNIAL SPECIES: *M. cardinalis*, *M. cupreus*, *M. guttatus* (**seep monkeyflower**), *M. lewisii*, *M. luteus*, *M. minima*, *M. ringens*, *M. tilingii*

FAMILY: Scrophulariaceae; figwort family
PLANT TYPE: Annual and perennial; seep monkey-flower (*M. guttatus*) can be biennial
SEED VIABILITY: More than 3 years

FLOWERING: The flowers are perfect.

SEED COLLECTION: The fruits are capsules. Hand-pick them when they brown but before they shatter.

SEED CLEANING: Dry the capsules on screens, then rub or flail to release the seeds.

GERMINATION: Surface-sow the seeds of most species in light at 65°F (18°C). They germinate in 7 to 14 days. Sow seeds of *M. ringens* at 41°F (5°C).

--------------------PLANT-BREEDING HINT--------------------

In monkeyflower, distribution of spots over the entire petal surface is dominant over spots on only part of the petal surface. Single flowers are dominant over double. The trait of terminal flowers differing from flowers lower on the plant (peloric trait) is dominant over nonpeloric.

TRANSPLANTING: Grow the plants on at 55 to 60°F (13 to 16°C) for 8 to 10 weeks.

MIRABILIS
Four-o'clock

M. jalapa (four-o'clock)

FAMILY: Nyctaginaceae; four-o'clock family
PLANT TYPE: Annual but may be perennial in warmer areas
SEED VIABILITY: Seeds remain viable for years.
SPACING FOR SEED SAVING: 12"

FLOWERING: Perfect flowers are cross-pollinated by insects.

SEED COLLECTION: Fruits are smooth or ribbed oval achenes with persistent calyces. Collect the seed heads by hand as they mature.

--------------------PLANT-BREEDING HINT--------------------

Flower color inheritance in four-o'clocks is complex. Striped varieties are all heterozygous and segregate to self colors and striped in F_2. Therefore, if you plant the seeds from F_1 striped varieties, some of the offspring will bear self colors and some will be striped. Tall is dominant over semidwarf and dwarf. Semidwarf is dominant over dwarf.

SEED CLEANING: Dry the heads on screens, then flail or thresh to remove the seeds.

GERMINATION: Germinate at 72°F (22°C) in light. Germination takes 7 to 14 days.

TRANSPLANTING: Seedlings need about 6 weeks to reach transplanting size.

MISCANTHUS
Miscanthus

M. transmorisonensis

FAMILY: Poaceae; grass family
PLANT TYPE: Perennial
SEED VARIABILITY: Unknown

FLOWERING: The perfect flowers are borne on long spikes.

SEED COLLECTION: The fruit is a grain. Harvest the seed heads when they brown and before they shatter.

SEED CLEANING: Further dry the heads on screens; then thresh, flail, or rub to remove the seeds.

--------------------FLOWER ALERT--------------------

Miscanthus is classified as an invasive plant in some areas.

GERMINATION: Sow seeds at 68°F (20°C). If they do not germinate in a month, stratify for a month, then return them to 68°F.

TRANSPLANTING: It takes 6 to 8 weeks at 75°F (24°C) for seedlings to reach the transplanting stage.

MOLUCELLA
Bells of Ireland

M. laevis (bells of Ireland)

FAMILY: Lamiaceae; mint family
PLANT TYPE: Annual
SEED VIABILITY: Unknown

FLOWERING: The flowers are perfect.

BELLS OF IRELAND FLOWERS

SEED COLLECTION: Each flower produces four three-angled, shiny nutlets. Harvest the nutlets or pull the entire plant.

SEED CLEANING: Place the plants on a screen to dry, then thresh or flail to remove the seeds.

GERMINATION: Surface-sow and hold for 2 weeks at 39°F (4°C), then germinate at 68°F (20°C) in light for 10 to 20 days.

TRANSPLANTING: Grow the plants on at about 65°F (18°C) for 6 to 8 weeks.

Monarda

Bee balm

> *M. astromontana, M. bradburiana, M. citriodora* (**lemon mint**)*, M. didyma* (**bee balm**)*, M. fistulosa, M. hybrida, M. punctata*

FAMILY: Lamiaceae; mint family
PLANT TYPE: Perennial; lemon mint *(M. citriodora)* and bee balm *(M. didyma)* are annual in some areas
SEED VIABILITY: Unknown

FLOWERING: The perfect flowers are cross-pollinated and borne in long clusters.

SEED COLLECTION: Each flower produces four nutlets. Harvest the clusters when they have browned.

SEED CLEANING: Collect the spent flowers, dry them on a screen, then thresh or flail to remove the seeds.

SEED TREATMENT: Stratify seeds of lemon mint for about 4 weeks.

GERMINATION: Germination takes up to a month. Sow seeds in light at 68°F (20°C). If seeds of lemon mint and *M. fistulosa* have not germinated in a month, stratify for a month, then return to 68°F (20°C).

TRANSPLANTING: Seedlings need 8 to 10 weeks to reach transplanting size.

Myosotis

Forget-me-not

> **ANNUAL SPECIES:** *M. arvensis* (**field forget-me-not**)*, M. dissitiflora*

> **PERENNIAL SPECIES:** *M. alpestris, M. australis, M. palustris, M. pulvinaris, M. sylvatica* (**forget-me-not**)

FAMILY: Boraginaceae; borage family
PLANT TYPE: Annual, biennial, and perennial
SEED VIABILITY: 2 years
SPACING FOR SEED SAVING: 12"

FLOWERING: Perfect flowers are cross-pollinated by insects.

SEED COLLECTION: Each flower contains four smooth, shiny nutlets. Collect the seed heads by hand as they mature.

SEED CLEANING: Spread the seed heads on screens to dry, then gently rub to separate the seeds. The seeds are easily damaged, so use caution when cleaning.

PLANT-BREEDING HINT

In forget-me-nots, blue flower color is dominant over pink; pink is dominant over white.

GERMINATION: Field forget-me-not and *M. dissitiflora* are sown in the dark at about 70°F (21°C). *M. alpestris*, *M. palustris*, and forget-me-not are sown at 68°F (20°C) in light. Germination takes about 14 days. Sow seeds of *M. australis* in light at 68°F (20°C), and if they don't germinate in a month, stratify for a month, then return them to 68°F (20°C). Sow seeds of *M. pulvinaris* at 41°F (5°C).

FLOWER ALERT

Some forget-me-nots aren't content to be just annuals or just perennials. Field forget-me-not (*M. arvensis*) is an annual that sometimes behaves as a biennial. *M. dissitiflora* is an annual, but it acts like a perennial in some regions. *M. sylvatica* sometimes acts like a perennial, sometimes like a biennial.

TRANSPLANTING: Seedlings need 8 to 10 weeks to reach transplanting size. Forget-me-not is usually sown directly to the garden.

Narcissus

Daffodil, narcissus

N. poeticus (**pheasant's-eye, poet's narcissus**), *N. serotinus*

FAMILY: Amaryllidaceae; amaryllis family
PLANT TYPE: Perennial
SEED VIABILITY: Unknown

FLOWERING: The flowers are perfect.

SEED COLLECTION: The fruit is a capsule. Hand-pick the fruits when they have browned and dried.

SEED CLEANING: Spread the capsules on screens to dry further; then flail, thresh, or rub to separate the black seeds.

GERMINATION: Sow seeds of pheasant's-eye (*N. poeticus*) at 68°F (20°C) for a month, then stratify them for 6 weeks, then move them back to 50°F (10°C) for germination. Sow seeds of *N. serotinus* at 68°F (20°C) and wait a month. If they do not germinate, stratify them for another month and repeat the cycle.

TRANSPLANTING: Grow the plants on for about 10 weeks at 60°F day/50°F night (16°C/10°C) temperatures.

FLOWER ALERT

Most daffodils grown in gardens are hybrids that will not come true from seed. If you want to experiment with saving seed from daffodils, make sure you're growing one of the species listed above. Even these species are usually propagated by bulbs, not by seed.

Nemesia

Nemesia

N. cheiranthus, *N. floribunda*, *N. foetens*, *N. fruticans*, *N. strumosa*, *N. versicolor*

FAMILY: Scrophulariaceae; figwort family
PLANT TYPE: Annual
SEED VIABILITY: Less than 1 year

FLOWERING: The flowers are perfect, but *N. strumosa* is self-sterile.

SEED COLLECTION: The fruits are capsules. Harvest the individual capsules, or pull entire plants when the capsules are brown and dried.

SEED CLEANING: Place the plants or capsules on screens to dry, then flail or thresh to separate the seeds.

GERMINATION: Germinate at 55 to 60°F (13 to 16°C) for 7 to 21 days.

TRANSPLANTING: The plants do not transplant well, so start them in peat pots and allow 6 to 8 weeks to grow the transplants.

NICOTIANA

Flowering tobacco

N. alata (**flowering or jasmine tobacco**),
N. acuminata (**manyflowered tobacco**), *N. attenuata* (**coyote tobacco**), *N. bigelovii* (**Bigelow's tobacco**), *N. knightiana*, *N. langsdorfii* (**Langsdorf tobacco**), *N. rustica* (**wild tobacco**),
N. ×sanderae (**Sander's tobacco**), *N. sylvestris* (**South American tobacco**), *N. tabacum* (**cultivated tobacco**), *N. trigonophylla*

FAMILY: Solanaceae; nightshade family
PLANT TYPE: Annual; *N. knightiana* can be perennial and cultivated tobacco (*N. tabacum*) can be biennial
SEED VIABILITY: 3–4 years
SPACING FOR SEED SAVING: 10"

FLOWERING: Perfect flowers are cross-pollinated by insects.

SEED COLLECTION: Harvest the fruit, a two- or sometimes four-valved capsule, as soon as it is dry.

SEED CLEANING: Place the fruits and flower heads on screens to dry further, then flail or rub them between your hands to separate the very tiny seeds.

GERMINATION: It takes 14 to 21 days in light at 73°F (23°C) for seeds to germinate.

TRANSPLANTING: Sow outside after danger of frost, or start transplants under lights 6 to 8 weeks before setting out. Grow the transplants at 65°F (18°C). The very small, dustlike seeds need light to germinate and so should be simply sprinkled on the soil surface. Culture is easy, but you must keep the young plants growing rapidly.

———————**PLANT-BREEDING HINT**———————

In flowering tobacco, dark flower colors are dominant over the lighter colors.

NIGELLA

Nigella

N. arvensis, *N. ciliaris*, *N. damascena* (**love-in-a-mist**), *N. hispanica* (**Spanish fennel**), *N. garidella*, *N. orientalis*, *N. sativa* (**black cumin**)

FAMILY: Ranunculaceae; buttercup family
PLANT TYPE: Annual
SEED VIABILITY: 5 years
SPACING FOR SEED SAVING: 6"

FLOWERING: The perfect flowers are cross-pollinated by insects.

SEED COLLECTION: The fruit consists of 2 to 14 united follicles. Harvest when they have begun to dry.

SEED CLEANING: Spread the fruits on screens to dry further, then rub or flail to release the seeds.

GERMINATION: Germination takes 7 to 14 days at 70°F (21°C).

TRANSPLANTING: Direct-sow the seeds in the fall or allow 6 to 8 weeks for seedlings to reach transplant size. Plants in this genus do not transplant well bare-rooted, so start the seeds in peat pots.

———————**PLANT-BREEDING HINT**———————

In *Nigella*, long stems are dominant over dwarf forms. Single flowers are dominant over normal doubles, but there is an abnormal double type that is dominant over single. In petal color, all colors are dominant over ivory white, and blue is dominant over purple.

OENOTHERA

Evening primrose

O. acaulis (**dandelion sundrop**), *O. argilicola*,
O. berlandieri, *O. biennis* (**evening primrose**),
O. caespitosa (**twisted sundrop**), *O. elata*,
O. erythrosepala, *O. fruticosa* (**sundrop**),
O. glazioviana, *O. kunthiana*, *O. laciniata*,

O. macrocarpa, O. missouriensis (**Ozark sundrop**), *O. odorata, O. pallida, O. perennis, O. pilosella, O. rhombipetala, O. rosea, O. speciosa* (**white evening primrose**), *O. syrticola, O. tetragona, O. versicolor*

FAMILY: Onagraceae; evening primrose family
PLANT TYPE: Perennial
SEED VIABILITY: 2 years

FLOWERING: The flowers are perfect.

SEED COLLECTION: The fruits are capsules. Harvest the drying flower clusters.

SEED CLEANING: Dry the clusters on screens, then flail or rub to separate the capsules and seeds.

————————— FLOWER ALERT —————————

Some species of evening primrose are considered invasive in some regions.

GERMINATION: Seeds of most species will germinate in a couple of weeks at 68°F (20°C). Germinate seeds of *O. erythrosepala* at 41°F (5°C). If seeds of *O. macrocarpa*, Ozark sundrops (*O. missouriensis*), and *O. syrticola* do not germinate at 68°F (20°C) within a month, stratify them for a month and return them to 68°F (20°C).

TRANSPLANTING: Grow the seedlings on for 6 to 8 weeks at 68°F (20°C) before transplanting outdoors in early spring.

PAEONIA

Peony

P. anomala (**Ural peony**), *P. delavayi, P. lactiflora* (**bai shao yao, Chinese peony, white peony**), *P. lutea* (**Tibetan peony**), *P. mascula, P. mlokosewitschii* (**Caucasian peony**), *P. officinalis* (**common peony**), *P. peregrina, P. suffruticosa* (**tree peony**), *P. tenuifolia* (**fern-leaved peony**), *P. veitchii, P. wittmanniana* (**Irangold peony**)

FAMILY: Paeoniaceae; peony family
PLANT TYPE: Perennial
SEED VIABILITY: Usually less than 3 years

FLOWERING: The flowers are perfect.

SEED COLLECTION: Fruits are clusters of horizontally spreading follicles. Hand-pick the fruits when they are fully ripe and dry.

SEED CLEANING: Place fruits on screens to dry, then rub or flail to remove the seeds.

SEED TREATMENT AND GERMINATION: Seeds have an impervious seed coat. Shake in sharp sand or rub with sandpaper to scarify the seeds. In the fall, immediately after harvest, sow seeds in flats of moist sand outdoors in a shady spot to stratify over winter. In the spring, check the seeds for signs of sprouting, but note that seeds will take up to a year to germinate. Plant the sprouted seeds outdoors as soon as the radicles have emerged.

TRANSPLANTING: You can plant sprouted seeds directly in the garden, but seedlings may do better if planted in a nursery bed for a year or two before you plant out to their permanent garden location.

OPEN PEONY FOLLICLES

Papaver

Poppy

ANNUAL SPECIES: *P. aculeatum* (**poppy**), *P. argemone* (**long pricklyhead poppy**), *P. californicum* (**California poppy or western poppy**), *P. commutatum, P. dubium* (**blindeyes**), *P. glaucum* (**tulip poppy**), *P. laciniatum, P. nudicaule* (**Iceland poppy**), *P. paeoniflorum, P. rhoeas* (**corn poppy, field poppy, Flanders poppy, or Shirley poppy**), *P. somniferum* (**common poppy or opium poppy**), *P. triniifolium*

BIENNIAL SPECIES: *P. triniifolium*

PERENNIAL SPECIES: *P. alboroseum, P. atlanticum, P. burseri, P. kerneri, P. lateritium, P. nudicaule* (**Iceland poppy**), *P. orientale* (**oriental poppy**), *P. pilosum, P. radicatum, P. rhaeticum, P. rupifragum, P. sendtneri, P. spicatum*

FAMILY: Papaveraceae; poppy family

PLANT TYPE: Annual, biennial, and perennial; Iceland poppy (*P. nudicaule*) can be biennial

SEED VIABILITY: Corn poppy (*P. rhoeas*) 10 years; most other species, less than 3 years

SPACING FOR SEED SAVING: 8"–12"

FLOWERING: Flowers are perfect and cross-pollinated by insects.

SEED COLLECTION: The fruit is a cylindrical to almost globose capsule. Harvest the fruits when many of the capsules have begun to open. You can harvest individual fruits or entire plants. Harvest the seed heads of Iceland poppy slightly before they mature, as mature heads are readily eaten by birds.

SEED CLEANING: Dry the plants or fruits on screens, then tumble or shake the capsules to extract the seeds. You may have to shake the fruits two or three times at intervals of several days to extract all of the seeds.

───────**PLANT-BREEDING HINT**───────

White margins on poppy petals are dominant over no margins. In common poppy (*P. somniferum*), the basal spot on petals is dominant over its absence, color is dominant over white, and purple is dominant over red. Nonstriped petals are dominant over striped. Single flowers are dominant over double, and laciniated (split) petals are dominant over whole. The single-flower trait dominates over the double-flowered trait. In Flanders poppy (*P. rhoeas*), flowers eventually revert to single reds, and in common poppy they revert to single mauve. (See page 87 for more information on reversion.) Large plant size is dominant over small size.

GERMINATION: Germination takes about 14 days at 60°F (16°C) in the dark in annual species. Germinate most perennial poppies in light at 68°F (20°C). If seeds of *P. burseri, P. radicatum, P. rupifragum, P. sendtneri,* or *P. triniifolium* have not germinated in a month, stratify them for a month and return them to 68°F (20°C). Germinate seeds of *P. spicatum* at 41°F (5°C).

TRANSPLANTING: Poppies are best sown directly outdoors. If you start them in the house, do so in peat pots, and allow 6 to 8 weeks for them to reach transplant size.

Pelargonium

Geranium

ANNUAL SPECIES: *P. chamaedrifolium, P. grossularoides* (**gooseberry geranium**), *P. ×hortorum* (**geranium**), *P. nanum, P. peltatum* (**ivy geranium**), *P. senecioides*

PERENNIAL SPECIES: *P. endlicherianum*

FAMILY: Geraniaceae; geranium family

PLANT TYPE: Annuals and perennials

SEED VIABILITY: 1–2 years

FLOWERING: The flowers are perfect.

SEED COLLECTION: The fruit is a five-valved capsule that dehisces. Hand-pick the capsules when they brown but before they dehisce.

────── FLOWER ALERT ──────

Although most geraniums die down completely in the winter, gooseberry geranium (*P. grossularoides)* and *P. senecioides* can successfully overwinter in warmer areas and resprout in the spring. *P. chamaedrifolium* sometimes acts as a biennial, producing no flowers until its second year of growth.

SEED CLEANING: Spread the fruits on screens to dry, then rub or flail to remove the seeds.

GERMINATION: Germinate at 73°F (23°C) for 7 to 14 days. Sow seeds of *P. endlicherianum* at 41°F (5°C). They may take several months to germinate.

────── PLANT-BREEDING HINT ──────

In at least some species of geraniums, single flowers are dominant. Rose-pink color is dominant, salmon-pink recessive.

TRANSPLANTING: Grow the plants on for 6 to 8 weeks at about 75°F (24°C).

PENSTEMON

Penstemon, beard-tongue

P. barbatus (**beardlip penstemon**), *P. hartwegii* (**Hartweg's penstemon**)

FAMILY: Scrophulariaceae; figwort family
PLANT TYPE: Annual and perennial
SEED VIABILITY: 2 years

FLOWERING: The flowers are perfect.

SEED COLLECTION: The fruit is a capsule containing many seeds. Hand-pick the capsules as they dry.

SEED CLEANING: Spread the capsules on a screen to dry, then rub or flail to release the seeds.

PENSTEMON

────── A PROLIFIC GENUS ──────

There are hundreds of species in the genus *Penstemon,* including many fine garden plants. The two species we've mentioned here can act as either annuals or perennials. We've singled them out because they best represent the type of seed-collecting protocols used for penstemons.

GERMINATION: There are hundreds of protocols for germinating seeds of various species of penstemons. However, in most cases your best bet is to surface-sow the seeds to germinate at 40°F (4°C), under light. Seeds of most species germinate in 8 to 12 weeks, but some of those may take up to 2 years. In general, seeds of species native to the eastern United States germinate quite readily but those from species adapted to harsher climates are much more difficult to germinate. Try to replicate the conditions to which the seeds normally would be exposed in their native habitat. One trick is to sow the seeds in flats, then leave the flats outdoors until the next season, exposed to all the climatic conditions they normally would be but protected from animals. Try different strategies with difficult seed lots, and be patient. You'll find one that works for you.

TRANSPLANTING: Because of the long germination period, seeds of these species are usually direct-sown in the garden.

flowers

PETUNIA

Petunia

P. axillaris (**large white petunia**), *P. ×hybrida grandiflora* (**common garden petunia**), *P. multiflora* (**common garden petunia**), *P. violacea* (**violet-flowered petunia**)

FAMILY: Solanaceae; nightshade family
PLANT TYPE: Annual
SEED VIABILITY: 5 years
SPACING FOR SEED SAVING: 12"

FLOWERING: Perfect flowers are cross-pollinated by insects.

ISOLATION REQUIREMENTS: There is some crossing, so isolate by at least ⅛ mile. Single forms generally will come true from seed, but doubles and semi-doubles will not.

——————PLANT-BREEDING HINT——————

Most of the petunias sold in garden centers are complex hybrids, so will not come true to seeds. It may be fun, however, to save seeds from these plants to see what you get.

In species petunias rather than hybrids, violet petal color is dominant over red and lilac. Uniform color is dominant over green-edged petals. Doubleness is caused by stamens becoming petals. In the cross singles × singles, all progeny are single; in double × double, the double trait shows in the progeny by three to one. Large white petunia (*P. axillaris*) and violet-flowered petunia (*P. violacea*) are self-sterile, and there are several grades of self-sterility.

SEED COLLECTION: The fruits are two-celled capsules, each containing from 100 to 300 seeds. The seeds mature 3 to 8 weeks after pollination. As soon as the flowers mature, harvest the individual heads or cut the entire plants.

SEED CLEANING: Dry the heads or plants on a screen, then flail or tumble to remove the seeds.

GERMINATION: Germinate the seeds for 7 to 14 days in light at 77°F (25°C).

TRANSPLANTING: Seedlings need 10 to 12 weeks to reach transplanting size at 55 to 60°F (13 to 16°C). Double-flowered varieties do not produce seeds. Eventually, flowers will revert to the wild, white-flowered forms.

PHLOX

Phlox

ANNUAL SPECIES: *P. drummondii* (**annual phlox**)

PERENNIAL SPECIES: *P. ×arendsii, P. divaricata* (**wild blue phlox, woodland phlox**), *P. glaberrima, P. maculata* (**wild sweet William**), *P. multiflora* (**small alpine phlox**), *P. paniculata* (**garden phlox, summer phlox**), *P. pilosa, P. subulata* (**moss pink**)

FAMILY: Polemoniaceae; phlox family
PLANT TYPE: Annual and perennial
SEED VIABILITY: 3 years
SPACING FOR SEED SAVING: Annual phlox, about 8"; perennial phlox, about 12"

FLOWERING: The perfect flowers are cross-pollinated by insects.

SEED COLLECTION: The fruit is a three-valved capsule that ruptures at maturity. Harvest the capsules when they turn brown but before they shed their seeds.

SEED CLEANING: Spread the capsules on a screen to dry further, with a sheet beneath, then rub or flail to separate the seeds.

SEED TREATMENT AND GERMINATION: Annual species require 14 to 21 days in dark and perennial species 30 to 49 days at 68°F (20°C) for germination. If seeds of *P. ×arendsii* and garden phlox (*P. paniculata*) have not germinated in a month, stratify for a month, then return to 50°F (10°C) for germination. Sow seeds of *P. glaberrima* var. *interior* and *P. pilosa* outdoors immediately after harvest.

TRANSPLANTING: Seedlings of annual species need 7 to 8 weeks to reach transplanting size; seedlings of perennial species need 10 to 12 weeks.

──────PLANT-BREEDING HINT──────

The corolla of phlox flowers can have different shapes, and salver shape is dominant to funnel shape in annual phlox. Entire petals are dominant over deeply cut.

In annual phlox (*P. drummondii*), cream-yellow flower color is recessive to white.

If you make a cross between garden phlox and annual phlox, you will be able to harvest viable seed, but the plants that grow from those seeds (the F_1 progeny) will be infertile.

PLATYCODON

Balloon flower

P. apoyama, P. grandiflorus (**balloon flower**)

FAMILY: Campanulaceae; bellflower family
PLANT TYPE: Perennial
SEED VIABILITY: 2–3 years

FLOWERING: The perfect flowers are cross-pollinated by insects.

SEED COLLECTION: The fruit is a dehiscent, five-valved capsule. Harvest the spent flowers as they brown but before the fruits dehisce.

SEED CLEANING: Spread the fruits on a screen to dry further, then rub or flail to remove the seeds.

GERMINATION: Sprinkle the seeds on the medium surface. Germination will take 10 to 30 days in light at 65 to 70°F (18 to 21°C).

TRANSPLANTING: Seedlings need 6 to 8 weeks to reach transplanting size. Because this genus does not transplant well bare-root, start the seeds in peat pots.

PORTULACA

Portulaca

P. grandiflora (**moss rose, portulaca**), *P. pilosa* (**kiss me quick**)

FAMILY: Portulacaceae; portulaca family
PLANT TYPE: Annual
SEED VIABILITY: 3 years
SPACING FOR SEED SAVING: 6"

FLOWERING: The perfect flowers are cross-pollinated by insects.

SEED COLLECTION: The fruit is a capsule. Harvest the capsules by hand, or pull the entire plant when the capsules brown.

SEED CLEANING: Dry the plants or capsules on screens with a sheet beneath; then thresh, rub, or flail to remove the tiny seeds.

SEED TREATMENT: Stratify 15 days.

──────PLANT-BREEDING HINT──────

Single-flower form is dominant over double flowers in portulaca, and normal plant size is dominant over dwarfness. Branch length is a genetic trait as well, but it's not known whether long branches are dominant over short ones. Dwarf plants can have long branches, and normal-size plants can have short branches.

GERMINATION: Germination will take 14 to 21 days in light at 75°F (24°C).

TRANSPLANTING: Seedlings need about 6 weeks to reach transplanting size. The genus does not transplant well bare-root, so start seeds in peat pots.

PORTULACA FLOWERS AND CAPSULES

PRIMULA
Primrose

P. alpicola, P. auricula (**auricula primrose**),
P. beesiana, P. ×bulleesiana, P. bulleyana,
P. burmanica, P. candelabra hybrids, *P. capitata,*
P. chungensis, P. ×chunglenta, P. cockburniana,
P. denticulata, P. elatior (**oxslip**), *P. farinosa,*
P. floribunda (**buttercup primrose**), *P. florindae,*
P. frondosa, P. halleri, P. hirsuta, P. integrifolia,
P. japonica, P. juliae, P. luteola, P. malacoides
(**fairy primrose**), *P. nutans, P. pedemontana,*
P. poissonii, P. polyantha (**polyanthus primrose**),
P. polyneura, P. prolifera (formerly *P. smithiana*),
P. ×pubescens, P. pulverulenta, P. rosea, P. saxa-
tilis, P. scotica, P. secundiflora, P. sieboldii, P. sik-
kimensis, P. sinopurpurea, P. veris (**cowslip prim-**
rose), *P. verticillata, P. vialii, P. vulgaris* (**English**
primrose), *P. waltonii*

FAMILY: Primulaceae; primrose family
PLANT TYPE: Perennial
SEED VIABILITY: Usually less than 1 year

——————————— FLOWER ALERT ———————————

Although primroses are perennials, many types that
are popular in gardens are grown as annuals. If primroses don't overwinter well for you, saving seeds can
be a good way to avoid the expense of buying plants
every year.

FLOWERING: The perfect flowers are cross-
pollinated.

SEED COLLECTION: The fruit is a 5- or 10-valved
capsule.

SEED CLEANING: Harvest the fruits as they brown
and spread them on screens to dry, then rub or flail
to extract the seeds.

——————— PLANT-BREEDING HINT———————

White flower color is recessive in primroses. Small flower
"eyes" are dominant over large eyes. A large greenish eye
is recessive to normal. Doubleness is recessive to normal.
In fairy primrose (*P. malacoides*), the white-margined leaf
is recessive.

SEED TREATMENT AND GERMINATION: Soak primrose seeds in water overnight to leach germination
inhibitors from the seed coat. For most species, the
next step is to stratify the seeds for 3 weeks, then sow
the seeds at about 68°F (20°C) in light. Germination
takes 2 to 4 weeks.

Some species require different treatment:

◆ Sow the following species without initial strati-
fication at 68°F (20°C) for a month, stratify them
for a month, then move them to 50°F (10°C) for
germination: auricula primrose (*P. auricula*),
P. burmanica, P. chungensis, P. ×chunglenta,
P. elatior, P. hirsuta, P. integrifolia, P. japonica,
P. pedemontana, P. poissonii, P. rosea, P. sieboldii,
P. sinopurpurea, P. veris.

◆ Sow the following species at 41°F (5°C): *P. fron-*
dosa, P. luteola, P. scotica, P. secundiflora, P. ver-
ticillata.

◆ Sow *P. juliae* at 68°F (20°C) for 2 months, stratify
for 2 more months, then move them in 50°F
(10°C) to germinate. If the seeds do not germinate in 2 months, run the entire cycle repeatedly
until germination.

TRANSPLANTING: Seedlings need 8 to 10 weeks to
reach transplanting size.

PRUNELLA
Self-heal

P. grandiflora, P. laciniata, P. vulgaris (**self-heal**
or heal-all), *P. ×webbiana*

FAMILY: Lamiaceae; mint family
PLANT TYPE: Perennial
SEED VIABILITY: Unknown

FLOWERING: The perfect flowers are arranged in
spikes. The fruit is a nutlet.

SEED COLLECTION: Harvest the spikes when they
have browned.

SEED CLEANING: Dry the spikes on screens, then
thresh or flail to remove the seeds.

GERMINATION: If seeds do not germinate at 68°F (20°C) in a month, then stratify for a month and return to 68°F.

TRANSPLANTING: Grow on for 6 to 8 weeks at 75°F (24°C) before setting outdoors in spring.

RANUNCULUS

Buttercup

ANNUAL SPECIES: *R. asiaticus* (**buttercup**)

PERENNIAL SPECIES: *R. aconitifolius, R. amplexicaulis, R. buchananii, R. fascicularis, R. flammula, R. glacialis, R. gramineus, R. haastii, R. insignis, R. lyallii, R. nivicola, R. pachyrrhizus, R. parnassifolius, R. pyrenaeus, R. romboideus, R. sericophyllus*

FAMILY: Ranunculaceae; buttercup family
PLANT TYPE: Annual and perennial
SEED VIABILITY: 5–6 years

FLOWERING: The flowers are perfect.

SEED COLLECTION: The fruit is a head of achenes. Hand-pick the heads.

SEED CLEANING: Spread the heads on screens to dry, then thresh or flail to remove the seeds.

SEED TREATMENT AND GERMINATION: For the annual species, which is the type commonly grown as an ornamental, seeds germinate in 14 to 21 days at 68°F (20°C). For perennial species, sow the seeds at 68°F (20°C) for a month, then move them to 23°F (–5°C) for a month and back to 50°F (10°C) until they germinate.

TRANSPLANTING: Seedlings need 8 to 10 weeks to reach transplanting size at 60°F (16°C) day temperature and 50°F (10°C) night temperature.

RESEDA

Mignonette

ANNUAL SPECIES: *R. alba* (**white upright mignonette**), *R. luteola* (**Dyer's rocket, weld**)

BIENNIAL SPECIES: *R. lutea* (**yellow mignonette**), *R. odorata* (**common mignonette**)

FAMILY: Resedaceae; mignonette family
PLANT TYPE: Annual and biennial; annual types may be perennial in warmer areas
SEED VIABILITY: 2–4 years
SPACING FOR SEED SAVING: 8"

FLOWERING: The flowers are usually perfect and are borne in terminal racemes. They are cross-pollinated by insects.

SEED COLLECTION: The fruit is a capsule. Harvest the plants when the lowest capsules are fully mature.

SEED CLEANING: Place the plants on screens to dry, then flail or tumble to extract the seeds.

GERMINATION: Germination takes 14 to 21 days in light at 70°F (21°C).

TRANSPLANTING: Mignonette does not transplant well; seeds are usually sown in place.

RUDBECKIA

Black-eyed Susan, coneflower

R. fulgida, R. hirta (**black-eyed Susan**), *R. laciniata, R. maxima, R. nitida, R. occidentalis* (**western coneflower**), *R. purpurea* (now *Echinacea purpurea*), *R. subtomentosa, R. triloba*

FAMILY: Asteraceae; sunflower family
PLANT TYPE: Perennial; black-eyed Susan (*R. hirta*) is grown as an annual in some locales
SEED VIABILITY: Unknown

FLOWERING: The disk flowers are fertile and the ray flowers are sterile. The flowers are cross-pollinated.

SEED COLLECTION: The fruits are four-angled achenes. Collect the spent seed heads by hand.

SEED CLEANING: Place the seed heads on a screen for drying, then flail to remove the seeds.

SEED TREATMENT: Stratify for 3 weeks.

———————— FLOWER ALERT ————————

Black-eyed Susan can become invasive. If it is a problem in your area, be sure to deadhead blossoms before they set seed.

GERMINATION: Germinate the seeds for 7 to 21 days in light. Light is especially required for *R. fulgida.* Sow *R. subtomentosa* and *R. triloba* at 41°F (5°C), and the seeds will germinate over several months.

TRANSPLANTING: Seedlings need 6 to 8 weeks at 70°F (21°C) to reach transplanting size.

SALPIGLOSSIS

Salpiglossis

S. sinuata (**salpiglossis, painted tongue**)

FAMILY: Solanaceae; nightshade family
PLANT TYPE: Annual
SEED VIABILITY: 6–7 years
SPACING FOR SEED SAVING: 12"

FLOWERING: The flowers are perfect.

SEED COLLECTION: The fruit is a two-valved capsule. Harvest fruits as they turn brown.

SEED CLEANING: Spread the capsules on a screen to dry, with a sheet beneath, then thresh gently in a paper bag to capture the minute seeds.

GERMINATION: Germination takes 14 to 28 days in the dark at 70°F (21°C).

TRANSPLANTING: Seedlings need about 12 weeks to reach transplanting size.

SALVIA STALKS THAT HAVE SHED THEIR NUTLETS

SALVIA

Salvia

S. argentea (**silver sage**), *S. carduacea* (**thistle sage**), *S. coccinea* (**scarlet sage, scarlet salvia**), *S. columbariae* (**chia**), *S. farinacea* (**mealycup sage**), *S. forskaohlei* (**indigo woodland sage**), *S. hispanica* (**Spanish sage**), *S. nemerosa*, *S. patens*, *S. reflexa* (**Rocky Mountain sage**), *S. sclarea* (**clary sage**), *S. splendens* (**scarlet sage, scarlet salvia**), *S. tiliifolia* (**lindenleaf sage**), *S. viridis*

FAMILY: Lamiaceae; mint family
PLANT TYPE: Annual, biennial, and perennial
SEED VIABILITY: 1 year
SPACING FOR SEED SAVING: 8–12"

FLOWERING: The plants are cross-pollinated by insects.

SEED COLLECTION: The fruit are ovoid, three-angled nutlets. Harvest the flowers by hand as they dry, or pull entire plants when the calyces on the stems have turned brown.

SEED CLEANING: Place flowers or plants on screens to dry, then tumble or flail to extract the seeds.

SEED TREATMENT: Stratify seeds for 3 weeks.

GERMINATION: Germinate seeds for 7 to 21 days in light at 68°F (20°C).

TRANSPLANTING: Seedlings need about 8 weeks at 68°F (20°C) to reach transplanting size.

──────── FLOWER ALERT ────────

Many ornamental salvias are reliable perennials. However, thistle sage (*S. carduacea*), chia (*S. columbariae*), Spanish sage (*S. hispanica*), Rocky Mountain sage (*S. reflexa*), and scarlet salvia (*S. splendens)* are annuals. Scarlet salvia (*S. coccinea*), mealycup sage (*S. farinacea*), and lindenleaf sage (*S. tiliifolia*) can be either annual or perennial; *S. viridis* can be either an annual or a biennial; and silver sage (*S. argentea*) and indigo woodland sage (*S. forskaohlei*) are biennial. Clary sage (*S. sclarea*), which may become a weed in some areas, can be either a biennial or a perennial.

SCABIOSA

Scabiosa

ANNUAL SPECIES: *S. prolifera* (**carmel daisy**), *S. stellata* (**starflower pincushions**)

BIENNIAL SPECIES: *S. japonica* var. *alpina*

PERENNIAL SPECIES: *S. atropurpurea* (**mourning bride, scabiosa),** *S. caucasica*, *S. columbaria*, *S. graminifolia*, *S. japonica*, *S. lucida*, *S. ochroleuca*, *S. silenifolia*, *S. speciosa*, and *S. variifolia*

FAMILY: Dipsacaceae; teasel family
PLANT TYPE: Annual, biennial, and perennial
SEED VIABILITY: 5 years
SPACING FOR SEED SAVING: 10"

FLOWERING: The flowers are perfect.

SEED COLLECTION: The fruit is an achene. Harvest the flower heads when they have browned and begun to dry.

SEED CLEANING: Further dry the heads on a screen, then thresh to remove the seeds.

GERMINATION: Germination takes 14 to 21 days at 70°F (21°C). Sow seeds of *S. columbaria*, *S. graminifolia*, *S. japonica*, *S. lucida*, and *S. ochroleuca* at 41°F (5°C). Germination will take several months.

TRANSPLANTING: It takes about 5 weeks to grow good annual transplants and about 10 weeks to grow good perennial transplants. Grow the plants at about 65°F (18°C).

SCHIZANTHUS

Butterfly flower, poor man's orchid

S. candidus (**butterfly flower**), *S. hookeri* (**butterfly flower),** *S. grandiflorus*, *S. papilionaceus*, *S. pinnatus* (**poor man's orchid**), *S. porrigens*, *S. wisetonensis*

FAMILY: Solanaceae; nightshade family
PLANT TYPE: Annual
SEED VIABILITY: 4–5 years

FLOWERING: The flowers are perfect and borne in long clusters.

SEED COLLECTION: The fruit is a two-valved capsule containing many small seeds. Harvest the spent flower heads when they have dried and browned.

SEED CLEANING: Spread the heads on screens to dry further, with a sheet beneath to capture fallen seeds; then flail, rub, or thresh to remove the very small seeds.

GERMINATION: Surface-sow seeds of all species at about 68°F (20°C) and keep them in the dark. Germination will occur in about 2 weeks.

TRANSPLANTING: Seedlings need about 8 weeks at 65°F (18°C) to reach transplanting size.

SEDUM

Stonecrop, green cockscomb, widowscross

S. acre (**goldmoss stonecrop**), *S. aizoon* (**aizoon stonecrop**), *S. album* (**white stonecrop**), *S. anacampseros* (**evergreen orpine**), *S. ewersii*, *S. floriferum*, *S. kamtschaticum* (**orange stonecrop**), *S. lanceolatum* (**spearleaf stonecrop**), *S. oreganum* (**Oregon stonecrop**), *S. populifolium*, *S. pulchellum* (**widowscross**), *S. reflexum* (**Jenny's stonecrop**), *S. rupestre* (now *S. reflexum*), *S. spathulifolium*, *S. spurium* (**two row stonecrop**), *S. telephium* (**live forever, orpine**), *S. ternatum* (**woodland stonecrop**)

FAMILY: Crassulaceae; orpine family
PLANT TYPE: Perennial
SEED VIABILITY: Less than 3 years

FLOWERING: The perfect flowers are borne in cymes.

SEED COLLECTION: The fruits are follicles. Harvest the inflorescences when they have browned and dried.

SEED CLEANING: Spread the spent flower heads on screens to dry, then rub to separate the seeds.

SEED TREATMENT AND GERMINATION: Sow seeds of most species at 68°F (20°C). Sow seeds of white stonecrop (*S. album*), *S. ewersii*, *S. spathulifolium*, and woodland stonecrop (*S. ternatum*) at 41°F (5°C). Sow seeds of *S. populifolium* and orpine (*S. telephium*) at 68°F (20°C) for a month; then stratify them for a month; then move them to 50°F (10°C) for germination. Do not cover seeds of any of the species.

TRANSPLANTING: Grow the plants on at 60 to 65°F (16 to 18°C) for about 8 weeks.

STACHYS

Lamb's ears

S. byzantina (**lamb's ears, woolly betony**), *S. coccinea* (**scarlet betony, scarlet hedge nettle, Texas betony**), *S. grandiflora* (**big sage**), *S. lanata* (now *S. byzantina*), *S. monieri*, *S. officinalis* (**betony, common hedge nettle**), *S. olympica* (now *S. byzantina*), *S. palustris*

FAMILY: Lamiaceae; mint family
PLANT TYPE: Perennial
SEED VIABILITY: Unknown

—————— **NAME NOTES** ——————

Here's a genus that has collected a wide and entertaining set of common names. Lamb's ears is probably the most widely known name for *Stachys*, but among others are Excellency's supreme reflection, betony, big-sage, hedge nettle, mouse's-ear, staggerweed, wood mint, and woundwort.

FLOWERING: The perfect flowers are born in a spike or head.

SEED COLLECTION: Each flower produces four nutlets. Harvest the flower heads when they are brown and dry.

SEED CLEANING: Spread the heads on screens to dry further, then rub or flail to separate the seeds.

SEED TREATMENT: Stratify seeds of *S. palustris* for 4 months, then move them to 68°F (20°C).

GERMINATION: Sow seeds of wooly betony (*S. byzantina*), scarlet hedge nettle (*S. coccinea*), and *S. lanata* at 68°F (20°C). Sow seeds of big sage (*S. grandiflora*), *S. monieri*, and common hedge nettle (*S. officinalis*) at 41°F (5°C).

TRANSPLANTING: Grow the plants on at 60 to 65°F (16 to 18°C) for 8 to 10 weeks before setting out. Better still, simply sow the seeds in place in early spring.

MARIGOLD FLOWERS SHOWING SEEDS

TAGETES
Marigold

T. erecta (**African, Aztec, big marigold**), *T. fili-folia*, *T. minuta* (**muster John Henry**), *T. patula* (**French marigold**), *T. pusilla* (**marigold**), *T. tenuifolia* (**Mexican, signet marigold**), *T. tubiflora*

FAMILY: Asteraceae; sunflower family

PLANT TYPE: Annual; *T. tubiflora* can be perennial in warm areas

SEED VIABILITY: 5 years

SPACING FOR SEED SAVING: African marigold 12"; French marigold 10"; Mexican marigold 8"

FLOWERING: Flowers are perfect and cross-pollinated by insects. The double-flower form dominates over the single.

ISOLATION REQUIREMENTS: The flowers are normally cross-pollinated and should be isolated by about ¼ mile between varieties of the same species. There is little or no cross-pollination among species.

SEED COLLECTION: The fruits are club-shaped, elongated achenes. Pull the plants when they stop flowering.

SEED CLEANING: Place the plants on a screen to dry, then flail or tumble to extract the seeds.

GERMINATION: Germinate in light at 74°F (23°C) for 3 to 14 days. Sow the seeds very shallow and vertically into the medium. The literature is mixed on the light requirement. If your lot does not germinate with light, try moving it to darkness.

TRANSPLANTING: Seedlings need 3 to 7 weeks to reach transplanting size.

TRACHYMENE
Blue lace flower

T. coerulea (**blue lace flower**)

FAMILY: Apiaceae; parsley family

PLANT TYPE: Annual

SEED VIABILITY: Unknown

SPACING FOR SEED SAVING: 8"

FLOWERING: The flowers are perfect.

SEED COLLECTION: Harvest the plants when the flowers are spent and the umbels are brown.

SEED CLEANING: Place the flowers on screens to dry further, then thrash or flail to remove the seeds.

GERMINATION: Seeds germinate in 14 to 28 days in darkness at 70°F (21°C).

TRANSPLANTING: This species does not transplant well. Seedlings need about 10 weeks to reach transplanting size; grow them in peat pots.

TRICYRTIS
Toad lily

T. hirta (**hairy toad lily**), *T. latifolia*, *T. macropoda* (**speckled toad lily**), *T. maculata*

FAMILY: Liliaceae; lily family

PLANT TYPE: Perennial

SEED VIABILITY: Unknown

FLOWERING: The flowers are perfect and small and resemble those of lily.

SEED COLLECTION: Harvest the individual seed heads or the entire clusters as they turn brown.

SEED CLEANING: Dry on screens, then thresh or flail to remove the seeds.

TOAD LILY FLOWERS

GERMINATION: Sow seeds of speckled toad lily (*T. macropoda)* at 41°F (5°C) for several months. Sow seeds of the other species at 68°F (20°C) for a month, stratify for a month, then move them to 50°F (10°C) to germinate.

TRANSPLANTING: Grow on for 6 to 8 weeks at about 60°F (16°C) before setting outdoors. This genus is usually propagated by division, as you would daylilies.

TROLLIUS
Globeflower

T. acaulis, *T. altaicus*, *T. asiaticus* (**Siberian globeflower**), *T. chinensis*, *T. ×cultorum*, *T. europaeus* (**common globeflower**), *T. laxus* (**American globeflower, spreading globeflower**), *T. pumilus* (**dwarf globe flower**), *T. yunnanensis*

FAMILY: Ranunculaceae; buttercup family
PLANT TYPE: Perennial
SEED VIABILITY: Unknown

FLOWERING: Flowers are perfect and borne singly.

SEED COLLECTION: The fruits are follicles. Harvest the follicles or entire seed heads when they are dry and brown.

SEED CLEANING: Spread the follicles and seed heads on screens to dry; then rub, thresh, or flail to separate the seeds.

GERMINATION: Sow the seeds at 68°F (20°C) for a month, move them to 20°F (−7°C) for another month, then move them to 50°F (10°C) to germinate.

TRANSPLANTING: Grow the plants on for 6 to 8 weeks at 70 to 75°F (21 to 24°C) for spring transplanting.

TROPAEOLUM
Nasturtium

ANNUAL SPECIES: *T. majus* (**garden or tall nasturtium, Indian cress**), *T. minus* (**dwarf nasturtium**), *T. peltophorum* (**shield nasturtium**), *T. peregrinum* (**canary bird flower**), *T. polyphyllum*

PERENNIAL SPECIES: *T. ciliatum*, *T. speciosum*

FAMILY: Tropaeolaceae; nasturtium family
PLANT TYPE: Annual and perennial; canary bird flower (*T. peregrinum*) and *T. polyphyllum* can be perennial
SEED VIABILITY: 10 years
SPACING FOR SEED SAVING: Dwarf species 7"; semi-tall species 14"; tall species 24"

ISOLATION REQUIREMENTS: Varieties cross-pollinate readily, so separate them by at least ¼ mile.

FLOWERING: The perfect flowers are cross-pollinated by insects.

SEED COLLECTION: The fruit is wrinkled and three-lobed and separates into three one-seeded sections. Harvest the plants when the fruits have browned and begun to dry.

SEED CLEANING: Further dry the plants on screens with sheets beneath to catch the seeds, then flail or tumble to extract any remaining seeds.

SEED TREATMENT: Before planting, soak seeds for 30 minutes in 125°F (52°C) water to control seedborne diseases. Especially with perennial species, scarify the seeds in sharp sand or on sandpaper, then soak until they are noticeably swollen (about 36 hours).

GERMINATION: For annuals, germination takes 14 to 21 days at 68°F (20°C). For perennials, sow at 68°F (20°C) for a month, stratify for a month, then move back into 50°F (10°C) for germination, which can take many months. Plant in the dark.

———————PLANT-BREEDING HINT———————

Nasturtium flowers have large anthers, and the stamens form a tight cluster around the pistil. They are easily removed if you want to make controlled crosses. (Keep in mind that the flowers are insect-pollinated, so you'll need to protect them from insect visits, too.)

In nasturtium, variegated flowers are dominant over solid-colored. Climbing plant habit is dominant over dwarfness. Flower doubleness and fragrance are both recessive. "Super-double" is dominant over both normal double and single. If you make a cross between tall nasturtium (*T. majus*) and dwarf nasturtium (*T. minus*), red flower color will be dominant over yellow, green leaf color over variegated, and tall plants over dwarf.

TRANSPLANTING: Annuals are often sown in place. It's easier to sow perennials in place, too. If you want to try transplants, it'll take 4 to 6 weeks at 65°F (18°C) for them to reach garden size.

TULIPA

Tulip

T. kaufmanniana (**waterlily tulip**), *T. pulchella*, *T. sprengeri*, *T. sylvestris* (**Florentine tulip**), *T. tarda*, *T. turkestanica*, *T. urumiensis*

FAMILY: Liliaceae; lily family
PLANT TYPE: Perennial
SEED VIABILITY: Unknown

——————————— FLOWER ALERT ———————————

Most of the tulip bulbs planted in gardens are hybrid types and will not come true from seed. And even if you grow the species tulips listed above, note that they are usually propagated by bulbs, not by seed.

FLOWERING: The usually solitary flowers are perfect.

SEED COLLECTION: The fruit is a three-valved capsule with many flat seeds. Hand-pick the capsules when they are dry, brown, and soft.

SEED CLEANING: Spread the capsules on screens to dry further; then thresh, flail, or rub to separate the seeds.

SEED TREATMENT AND GERMINATION: Sow the seeds at 68°F (20°C) for a month, stratify them for another month, then move them to 50°F (10°C) for germination.

TRANSPLANTING: Grow the plants on at about 60°F (16°C) until they are ready to transplant in early spring.

VERBASCUM

Mullein

V. blattaria (**moth mullein**), *V. bombyciferum*, *V. chaixii*, *V. densiflorum*, *V. hybridum*, *V. longifolium*, *V. nigrum*, *V. olympicum*, *V. phoeniceum* (**purple mullein**), *V. roripifolium*, *V. thapsus* (**common mullein, flannel plant**)

FAMILY: Scrophulariaceae; figwort family
PLANT TYPE: Perennial
SEED VIABILITY: Unknown

MULLEIN FLOWER SPIKES

FLOWERING: The flowers are perfect and arranged in spikes.

SEED COLLECTION: The fruits are capsules. Harvest entire spikes when they are dry.

SEED CLEANING: Dry the spikes on screens; then flail, thresh, or rub to remove the capsules and the seeds.

SEED TREATMENT AND GERMINATION: Seeds of most species germinate readily at 68°F (20°C). Sow seeds of *V. nigrum* at 68°F (20°C) for a month, stratify them for a month, then return them to 68°F (20°C) for germination. Cover all seeds only very lightly.

TRANSPLANTING: Grow on at about 60°F (16°C) for 8 to 10 weeks prior to setting outdoors, or simply sow in place outdoors in fall or early spring.

VERBENA

Verbena, vervain

ANNUAL SPECIES: *V. hortensis* (**trailing verbena**), *V. ×hybrida* (**garden vervain**),

PERENNIAL SPECIES: *V. bonariensis* (**purpletop vervain**), *V. canadensis* (**rose vervain**), *V. hastata*, *V. rigida*, *V. stricta*, *V. tenuisecta*

FAMILY: Verbenaceae; verbena family
PLANT TYPE: Annual and perennial; purpletop vervain (*V. bonariensis*) can be perennial in warm areas
SEED VIABILITY: 3–5 years
SPACING FOR SEED SAVING: 12"

FLOWERING: The flowers are perfect.

SEED COLLECTION: The dry fruit is enclosed in a persistent calyx. Harvest the heads when they have dried.

SEED CLEANING: Further dry on a screen, then thresh or flail to remove the fruits from the calyces. The fruits are nutlets. There is no need to remove the seeds from the nutlets.

SEED TREATMENT: Stratify seeds of purpletop vervain, *V. rigida*, and *V. stricta* for 4 weeks.

-------------- **PLANT-BREEDING HINT** --------------
In verbena, light variegation in flower color is dominant over heavy variegation, and variegation is dominant over dilute (watered-down) self-color. Light self-color (color with white added) is dominant to dark.

GERMINATION: Germinate at 75°F (24°C) for 7 to 21 days. Seeds of purpletop vervain should be germinated in darkness. Sow seeds of *V. hastata* at 41°F (5°C).

TRANSPLANTING: Seedlings need 6 to 8 weeks at about 70°F (21°C) to reach transplanting size.

-------------- FLOWER ALERT --------------
Purpletop vervain can become weedy in many areas.

VERONICA

Speedwell

V. austriaca, *V. fruticans*, *V. fruticulosa*, *V. gentianoides*, *V. gutheriana*, *V. incana* (now *V. spicata* ssp. *incana*), *V. longifolia*, *V. officinalis* (**common speedwell**), *V. prostrata*, *V. repens*, *V. saturejoides*, *V. schmidtiana*, *V. spicata*, *V. subsessilis*, *V. teucrium* (now *V. austriaca* ssp. *teucrium*), *V. urticifolia*, *V. virginica* (now *Veronicastrum virginicum*)

FAMILY: Scrophulariaceae; figwort family
PLANT TYPE: Perennial
SEED VIABILITY: Usually less than 1 year

FLOWERING: The flowers are small and perfect.

SEED COLLECTION: The fruit is a flattened capsule. Harvest entire seed spikes when the flowers have browned.

SEED CLEANING: Dry the spikes on screens; then thresh, flail, or rub to separate the seeds from capsules.

SPEEDWELL FLOWERS AND FRUITS

GERMINATION: Sow seeds of *V. spicata* ssp. *incana*, *V. longifolia*, *V. spicata*, *V. subsessilis*, *V. austriaca* ssp. *teucrium*, and *V. urticifolia* at 68°F (20°C) in light. Those of *V. repens* need no light. Sow seeds of the other species at 68°F (20°C), and if they don't germinate in a month, stratify them for a month and return them to 68°F (20°C). They also need light.

TRANSPLANTING: Grow on at 70 to 75°F (21 to 24°C) for 6 to 8 weeks prior to setting outdoors.

──────── PLANT-BREEDING HINT ────────

In some species of *Veronica*, the short-styled flower form is dominant over the long-styled, and white flowers are recessive to colored flowers.

VIOLA

Pansy, violet

V. elatior, V. jooi, V. labradorica, V. lanceolata, V. odorata, V. obliqua, V. papilionacea, V. pedata, V. pumila, V. reichenbachiana, V. rupestris, V. sagittata, V. sororia, V. tricolor (**Johnny jump up**), *V. triloba*

FAMILY: Violaceae; violet family
PLANT TYPE: Annual, biennial, and perennial
SEED VIABILITY: 5 years
SPACING FOR SEED SAVING: Pansies, 6"; violets, 5"

──────── NAME NOTE ────────

The genus *Viola* contains far too many species for us to list them all here, but we have included germination instructions for a handful of species. Have fun propagating the violets that you find in your garden or neighborhood. Keep in mind that many *Viola* sold at garden centers are hybrid types that will not come true from seed.

FLOWERING: The plants are perfect-flowered and cross-pollinated by insects. Some plants are self-pollinating before the flowers open.

ISOLATION REQUIREMENTS: There may be some cross-pollination between varieties and among species, so separate by at least 100 feet, with a tall crop or structure between. Isolation by at least 600 feet is even better.

SEED COLLECTION: The fruit is a dehiscent capsule that shatters easily. The plants bloom continuously, so hand-pick the capsules as they become brown. The fruits will shatter quickly after the seeds ripen.

SEED CLEANING: Dry the fruits on a screen with a sheet beneath to capture the shed seeds. Place a cloth bag over the seeds, as they are ejected rather forcefully from the ripening fruits and thus may be lost. Rub the dry fruits over the screen to remove the seeds.

SEED TREATMENT: Stratify the seeds of most species for 2 weeks. Stratify seeds of *V. lanceolata, V. papilionacea*, and *V. sagittata* for 3 months.

flowers

V

GERMINATION: *Viola* seeds are very small and should be sprinkled only on the medium surface. In general, germinate them for 14 to 21 days in darkness at 70°F (21°C). Sow seeds of sweet violet (*V. odorata*), *V. pedata*, and *V. sororia* at 68°F (20°C) for a month, stratify them for a month more, then move them to 50°F (10°C) for germination. Sweet violet seeds will take 1 to 2 months to germinate. Sow seeds of *V. elatior*, *V. jooi*, *V. labradorica*, *V. obliqua*, *V. pumila*, *V. reichenbachiana*, *V. rupestris*, and *V. triloba* at 41°F (5°C) and let them germinate over several months.

TRANSPLANTING: Seedlings need about 10 weeks at 60 to 65°F (16 to 18°C) to reach transplanting size.

——————— FLOWER ALERT ———————

Johnny jump up (*V. tricolor*) can become an invasive weed in some areas, as can some species of perennial violets.

ZINNIA SEEDS

ZINNIA

Zinnia

> *Z. angustifolia* (**star zinnia**), *Z. elegans* (**common zinnia**), *Z. haageana* (**Mexican zinnia**), *Z. peruviana* (**Peruvian zinnia**)

FAMILY: Asteraceae; sunflower family
PLANT TYPE: Annual
SEED VIABILITY: 10 years
SPACING FOR SEED SAVING: Dwarf types, 8"; tall types, 12"

FLOWERING: Zinnia flower heads contain two types of flowers. The ray florets are pistillate, whereas the disk florets are perfect, small, and yellow. Both produce seeds. The plants are cross-pollinated by insects.

SEED COLLECTION: The fruits are achenes. Harvest the individual heads or the entire plants as soon as the flowers are spent and dried. Do not let the heads get wet just before harvest.

SEED CLEANING: Spread the plants or the individual heads on a screen to dry, then tumble or flail them to remove the seeds.

SEED TREATMENT: The seeds can carry powdery mildew, so do not harvest from infected plants. Soak seeds for 30 minutes in 125°F (52°C) water to control seedborne disease; if this is an issue in your garden, then sow.

GERMINATION: Seeds will germinate in 7 to 21 days at 70°F (21°C).

TRANSPLANTING: Seedlings need about 7 weeks to reach transplanting size. Sometimes transplanting zinnias can cause double-flowered types to revert to single-flowered types.

CHAPTER 11

ADVANCED SEED SAVING
NUTS, FRUITS &
WOODY ORNAMENTALS

Saving seeds from vegetables, herbs, and flowers is relatively easy, and it can be quite productive and practical in terms of generating lots of plants to fill your beds and borders. In this chapter, we step beyond the realm of the practical and easy and into the adventurous and artful world of saving seeds from nut trees; fruit crops, such as apples and cherries; and woody ornamental trees, shrubs, and vines. The conventional method of propagating many of these species is by cuttings, layering, or grafting, but it is possible to propagate them by seeds. Some gardeners find collecting seeds from a grapevine or serviceberry bush or pine tree and growing out the seedlings to be a fascinating challenge. When we give gardening talks, we field many questions from gardeners who are eager to try collecting and growing acorns, buckeyes, walnuts, and chestnuts. And planting seeds from a backyard apple tree or orange tree can be a great project to try with a school group or with your children or grandchildren.

One basic difference between saving seeds from woody plants and saving seeds from vegetables and flowers is that, with woodies, it's generally not possible to create duplicates of the parent plants from seed. Sometimes it's just a logistic impossibility — the

flowers on an oak tree may be 50 feet up in the air, so there's no way to control pollination. Or sometimes it's a matter of genetics. Many fruit crops are naturally heterozygous, and thus almost none of the popular fruit crops come true to seed (some crab apples do come true, and some peaches and cherries come pretty close). On top of that, some fruit crops are very difficult to save seed from (which is why you won't find instructions for saving seeds from strawberries or raspberries in this book). Another point to consider is that fruit trees, bushes, and vines grown from seed have a long juvenile period, and most will eventually produce inferior fruit.

If you do decide to venture into seed propagation of fruit and woody ornamentals, do it for the wonderment of producing the great variety of plants that nature provides. Keep your expectations realistic about producing a superior new variety. The chances are slim, but they're not nonexistent. After all, somewhere along the line, 'McIntosh' and 'Golden Delicious' apples, 'Montmorency' cherry, and 'Concord' grape all came into being as chance seedlings that some gardener carefully tended until a plant produced its first memorable and delicious fruit.

Here are some points to keep in mind as you consult the entries in this chapter.

POLLINATION. Unlike the entries on vegetables and herbs, we haven't included much information on the mechanism of pollination in this chapter. That's because, as we noted above, it's impractical to control pollination in most woody ornamentals and fruit crops, so how the plants are pollinated isn't relevant.

STRATIFICATION. Unless noted otherwise, you can assume that stratification temperatures should range between 34 and 41°F (1 and 5°C), and storage temperature for orthodox seeds should be between 32 and 41°F (0 and 5°C).

SEED STORAGE. As we explained on pages 49–54, orthodox seeds are always stored dry and in sealed containers, but recalcitrant seeds should not be dried excessively.

GERMINATING AND RAISING SEEDLINGS. Many times fresh seeds are sown directly outdoors in the fall, and stratified seeds are sown in spring. However, some folks like to start their seeds indoors, then prick out and transplant the emergents. Use the specifications we give under "Germination" if you want to start your seeds indoors. Unless stated otherwise, provide temperatures of 68°F at night and 86°F during the day (20° and 30°C). If light is called for, it's applied during the high-temperature period, the duration of which varies between 8 and 16 hours. The seeds of most tree fruits are not sensitive to light, so they can be planted at a depth of one to two times their larger diameter. Within a species, the largest seeds generally produce the most vigorous seedlings, so try to select the largest seeds for planting. Nuts such as hickories can simply be pressed into the soil with the heel of your boot. However, if nuts have begun to sprout, plant them so their hypocotyls just breach the surface of the soil and their radicles are pointed downward.

Throughout this chapter, we use "shorthand" terms such as 1+0 and 2+1 to describe the age and treatment of young plants. For example, 1+0 stock indicates that after a seed is planted, the seedling should be allowed to grow for 1 year in a nursery bed, then planted out in the garden or landscape. For more explanation of these terms, see page 78.

ABIES

Fir

A. alba (**silver fir**), *A. amabilis* (**Pacific silver fir**), *A. balsamea* (**balsam fir**), *A. bracteata* (**bristlecone fir**), *A. concolor* (**white fir**), *A. firma* (**Japanese fir**), *A. fraseri* (**Fraser fir**), *A. grandis* (**grand fir**), *A. guatemalensis* (**Guatemalan fir**), *A. homolepsis* (**Nikko fir**), *A. lasiocarpa* (**corkbark fir**), *A. magnifica* (**California red fir**), *A. mariesii* (**Mariesii fir**), *A. procera* (**noble fir**), *A. sachalinensis* (**Sakhalin fir**), *A. veitchei* (**Veitch fir**)

FAMILY: Pinaceae; pine family

FLOWERING: Fir flowers are unisexual. The small male strobili hang from the lower sides of the branches in the upper portions of the crown, whereas

SEED-SAVING FACTS FOR FIRS

TYPE OF FIR	FLOWERING SEASON	CONE RIPENING	SEED DISPERSAL	
Balsam	Spring	Summer to fall	Fall	
California	Spring	Summer	Fall	
Corkbark	Summer	Summer to fall	Fall	
Fraser	Spring	Fall	Fall to winter	
Grand	Spring	Summer	Fall	
Noble	Summer	Fall	Fall	
Pacific silver	Spring	Summer to fall	Summer to fall	
White	Spring	Fall	Fall	

the female strobili stand upright. This arrangement favors cross-pollination. Flowering time varies by species, as listed in the chart below.

SEED COLLECTION: The fruit is the erect female cone, which ripens in the fall. Scales, with two winged, mature seeds each, shed from the cone. The cones shatter easily, so hand-harvest them carefully before seed-shed. See the chart below for information on ripe cone color and seed ripening and dispersal times.

You can determine when cones of other species of fir are ripe, as follows: Japanese fir (*A. firma*) turns dull yellow-brown; Guatemalan fir (*A. guatemalensis*) turns dark green or purple, resin droplets form, wings assume an exterior yellow color; Nikko fir (*A. homolepsis*) turns dull yellow-brown; Mariesii fir (*A. mariesii*) turns dull brown; Sakhalin fir (*A. sachalinensis*) turns dull brown; and Veitch fir (*A. veitcheii*) turns dull brown.

SEED CLEANING AND STORAGE: Spread the cones to dry for a week or two at about 81°F (27°C), then rub them to separate the seeds. The seed coats are soft, so take care not to injure them during cleaning. Dry the seeds once more, then store them sealed in plastic bags at the temperatures indicated in the chart below. Seeds of fir will not maintain viability longer than about 4 years.

SEED TREATMENT: Stratification times and temperatures vary by species; refer to the chart below for details. Stratification for 1 to 3 months will be adequate for most species.

GERMINATION: Germination takes place under standard alternating temperatures of 68°F (20°C) for 16 hours and 86°F (30°C) for 8 hours, with 8 hours of light provided during the high temperature. Many species of fir have a germination percentage in the range of 22 to 28 percent; Fraser fir is 42 percent; white fir, 34 percent.

NURSERY PRACTICES: Sow fresh seeds in fall in moist humus or moist mineral soil, then provide a winter mulch and partial to heavy shade. Spring-sow stratified seeds, and provide them with conditions similar to the fall-sown seeds.

RIPE CONE COLOR OR SPECIFIC GRAVITY	RIPE SEED COLOR	AIR-DRY TIME (DAYS)	STORAGE TEMPERATURE	STRATIFICATION TIME (DAYS)	WHEN TO SOW SEEDS
Purple	Wing dark brown with purple edge	20–30	32.5°F–39°F (0.5°C–4°C)	30–60	Spring
SG <0.75	Wing brown with deep purple margin	8–21	41°F (5°C)	30–60	Spring
Green with yellow tinge, turning gray to purple	Seed coat tan; wing brown to pale purple	60–180	2°F (−17°C)	30–120	Fall
Blue-green turning brown	Translucent; light brown or tan	30–45	10°F (−12°C)	30–60	Spring
SG <0.90; light brown	Wing purple-brown	60–180	10°F–25°F (−12°C–−4°C)	28–120	Spring
SG <0.90; light brown	Wing brown	60–180	25°F (−4°C)	30–120	Spring
Green-yellow, then gray or purple	Seed coat tan; wing light purple with brown margin	60–180	1°F (−17°C)	30–120	Spring
SG 0.85–0.96	Wing brown with dark purple margin	7–14	0°F–32°F (−18°C–0°C)	28–60	Spring

A

ACER

Maple

A. buergerianum (**trident maple**), *A. campestre* (**hedge maple**), *A. carpinifolium* (**hornbeam maple**), *A. circinatum* (**Oregon vine maple**), *A. ginnala* (now *A. tataricum* ssp. *ginnala*; **Amur maple**), *A. japonicum* (**fullmoon maple**), *A. negundo* (**boxelder**), *A. palmatum* (**Japanese maple**), *A. pensylvanicum* (**striped maple**), *A. platanoides* (**Norway maple**), *A. pseudoplatanus* (**sycamore maple**), *A. rubrum* (**red maple**), *A. saccharinum* (**silver maple**), *A. saccharum* (**sugar maple**), *A. saccharum* ssp. *nigrum* (**black maple**), *A. tataricum* (**tatarian maple**), *A. truncatum* ssp. *mono* (**painted maple**)

FAMILY: Aceraceae; maple family

FLOWERING: Most species flower in mid- to late spring, though red maple (*A. rubrum*) and silver maple (*A. saccharinum*) flower beginning in February in some areas. The flowers are regularly polygamo-monoecious or polygamo-dioecious and rarely all monoecious or dioecious. Boxelder (*A. negundo*), striped maple (*A. pensylvanicum*), and silver maple are mostly dioecious, though individual trees may be monoecious.

SEED COLLECTION: Most species ripen their fruit from August to October, followed by fall seed dispersal. Silver maple and red maple ripen their fruit in spring and disperse their seeds in late spring and early summer. The fruit consists of two fused samaras with very hard pericarps. Collect the seeds of most species when the samaras' wings and pericarps have turned tan, yellow-brown, or brown. Seeds of some difficult-to-germinate species, such as striped maple, fullmoon maple (*A. japonicum*), Japanese maple (*A. palmatum*), and painted maple (*A. truncatum* ssp. *mono*) should be collected before they've dried completely. Collect seeds from the trees or from the ground after they are shed. There is no need to extract the seeds from the samaras.

MAPLE SEEDLING WITH SPENT SEED ATTACHED

SEED CLEANING AND STORAGE: Silver maple and sycamore maple (*A. pseudoplatanus*) have recalcitrant seeds. Most other maples have orthodox seeds. Clean the seeds, and be sure they are surface-dry before storage. Seeds of silver maple have a fairly high moisture content at shedding (about 60 percent) and must not be dried excessively; if they lose half their moisture, they will fail to germinate at all. Most maple seeds are sown right after being shed, but you can store them. Except for the two species named above, store seeds of the spring-ripening species for up to a year and those of fall-ripening species for up to 2 years, dried and sealed in containers held at about 34°F (1°C).

SEED TREATMENT AND GERMINATION: Silver maple and some red maple seeds have no dormancy. Those of red maple seeds may germinate without stratification, but others require it. Try germinating red maple seeds at 68°F (20°C) for 3 weeks. If none comes up, then use a general maple stratification protocol, as follows. Soak the seeds in 110°F (43.5°C) water for 1 to 2 days, then stratify them for 2 months.

Sugar maple (*A. saccharum*) and sycamore maple seeds require 2 to 3 months of stratification; hornbeam maple (*A. carpinifolium*) and black maple (*A. saccharum* ssp. *nigrum*) seeds 3 months; striped maple (*A. pensylvanicum*), Amur maple (*A. tataricum* ssp. *ginnala*), painted maple (*A. truncatum* ssp. *mono*), and tatarian maple (*A. tataricum*) seeds 4 months; Japanese maple seeds 5 months; hedge maple (*A. campestre*) and Oregon vine maple (*A. circinatum*) seeds 1 to 2 months of warm stratification followed by 4 to 6 months of stratification; full-moon maple seeds 3 months of warm stratification followed by 3 months of stratification.

Partial removal of the pericarp by nicking or filing is often helpful. If your maple seeds are particularly reluctant to germinate, try a warm stratification at 77°F (25°C) for 1 month, then stratify for 1 to 6 months before germinating at 68°F (20°C). Some maple seeds, especially those that are very dry, may not germinate until their second year after sowing; others will germinate during stratification.

To ensure that seeds have been stratified long enough, wait until the first seeds germinate before sowing them. Maple seeds will germinate in many media under almost any light conditions. Red maple seeds germinate best in an acid medium. Germination rates are about 46 percent for seeds of red maple, 76 percent for silver maple, and 39 percent for sugar maple.

NURSERY PRACTICES: Sow seeds of silver maple and red maple in the nursery bed after you collect them in late spring, and sow those of other maples when they mature in the fall. This reduces the chances that the seeds will enter dormancy or double dormancy, and those that do may enter only shallow phases of each. Fall sowing eliminates the need for artificial stratification as well. Sow the seeds ¼ to 1 inch deep. In most cases, seeds will have germinated in a few weeks, but seeds of some species, such as Amur maple, may not germinate until the following spring or even the second spring. After fall sowing, mulch beds carefully with a few inches of straw to protect seeds from heaving. Most maples do best in a moist, mineral soil rich in organic matter. Provide young maple seedlings with some light shade until they become established. Outplant as 1+0, 2+0, or 2+1 stock.

AESCULUS

Buckeye

A. californica (**California buckeye**), *A.* ×*carnea* (**red horsechestnut**), *A. flava* (**yellow buckeye**), *A. glabra* (**Ohio buckeye**), *A. hippocastanum* (**horsechestnut**), *A. indica* (**Himalayan buckeye**), *A. parviflora* (**bottlebrush buckeye**), *A. pavia* (**red buckeye**), *A. sylvatica* (**painted buckeye**), *A. turbinata* (**Japanese horsechestnut**)

FAMILY: Hippocastanaceae; horse chestnut family

FLOWERING: Plants are polygamo-monoecious, with flowers forming in spikes in spring.

SEED COLLECTION: The spiny capsules ripen in fall, turning yellow-brown and dehiscing, exposing up to three dark, chocolate brown seeds in each. Harvest the ripe capsules from the tree or from the ground.

SEED CLEANING AND STORAGE: The shiny brown seeds are full of lipids and fats. If you dry them too much, they'll begin to dull and shrivel, a sign that they have lost much of their viability. Spread them out to surface-dry only until their husks dehisce. The seeds are recalcitrant. Store the moist seeds in plastic bags at 34°F (1°C) for up to 10 months.

SEED TREATMENT: Presoak seeds in warm water for 2 days, then remove about one-third of each seed coat at the scar end. Stratify in moist sand for 3 to 4 months, 1 month for red buckeye (*A. pavia*). However, moist cold storage for more than about 3 months substitutes for the stratification period.

GERMINATION: Germinate the seeds in sand.

NURSERY PRACTICES: Fall-sow fresh seeds immediately after cleaning, and they will germinate in spring. Seeds of bottlebrush buckeye (*A. parviflora*) will germinate immediately after planting; those of red buckeye may do the same; California buckeye (*A. californica*) will germinate as soon as winter rains begin. Red horsechestnut (*A.* ×*carnea*), even though it's a hybrid, will come nearly true to seed.

ALNUS

Alder

A. cordata (**Italian alder**), *A. crispa* (now *A. viridis* ssp. *crispa*; **American green alder**), *A. firma* (**Japanese green alder**), *A. glutinosa* (**European alder**), *A. hirsuta* (**Manchurian alder**), *A. incana* (**gray alder**), *A. rhombifolia* (**Sierra alder**), *A. rubra* (**red alder**), *A. rugosa* (now *A. incana* ssp. *rugosa*; **speckled alder**)

FAMILY: Betulaceae; birch family

FLOWERING: The monoecious flowers are borne from early spring until June, depending upon species. The male catkins, or aments, become long and pendulous in spring; the female catkins develop into oval "conelets" after fertilization.

SEED COLLECTION: The small, slightly winged or wingless nutlets are borne in pairs at the base of each scale of the conelet. The fruits ripen from August to November, depending upon species, when the conelets become thick, woody, and dark brown. Hand-pick the fruits when they are ripe.

SEED CLEANING AND STORAGE: Spread the conelets on a canvas to dry in a warm room at about 90°F (32°C). During this process, many of the seeds will fall out of the conelets. Those that remain can be tumbled out. Winnow to separate the chaff. The seeds themselves are about ⅛ inch in diameter. Dry the seeds, then store them in sealed containers at 41°F (5°C) for up to a year and a half.

SEED TREATMENT AND GERMINATION: Seeds of some species show embryo dormancy, and those of others show no dormancy. Most seeds can be germinated easily in flats over a 3-week period. Sierra alder (*A. rhombifolia*), red alder (*A. rubra*), and speckled alder (*A. rugosa*) will germinate immediately if sown fresh. If you have an especially difficult-to-germinate seed lot, stratify the seeds for 2 to 3 months (6 months for speckled alder), then germinate them as above. About 27 percent of the seeds of red alder will germinate.

MALE (UPPER LEFT) AND FEMALE ALDER CATKINS

NURSERY PRACTICES: Sow seeds in the nursery bed right after harvest or sow stratified seeds in spring. Surface-sow seeds, then cover them lightly with coarse sand or a mixture of coarse sand and hardwood humus. Keep the beds moist and shaded until late summer of the sowing year. Germination is usually complete within a month of sowing. Outplant seedlings as 1+0 or 2+0 stock.

AMELANCHIER

Serviceberry

A. alnifolia (**Saskatoon serviceberry**), *A. arborea* (**downy serviceberry**), *A. asiatica* (**Asian serviceberry**), *A. canadensis* (**shadblow serviceberry**), *A. florida* (now *A. alnifolia* var. *semi-integrifolia*; **Pacific serviceberry**), *A. laevis* (**Allegheny serviceberry**), *A. sanguinea* (**roundleaf serviceberry**), *A. spicata* (**dwarf serviceberry**)

FAMILY: Rosaceae; rose family

FLOWERING: The perfect flowers open in spring just before or just after the leaves appear, depending upon species.

SEED COLLECTION: The fruits are berrylike pomes and ripen in summer. The color of ripe fruit varies by species. Saskatoon serviceberry (*A. alnifolia*) ripens to dark purple, downy serviceberry (*A. arborea*) to red-purple, and roundleaf serviceberry (*A. sanguinea*) to black. Pick the ripe fruits from the plants before the birds get them.

SEED CLEANING AND STORAGE: Macerate the fruits in a blender and dry the cleaned seeds. The seeds are orthodox and can be stored for a couple of years sealed in containers at 41°F (5°C). However, storage can make seed coats become quite resistant to imbibition, and you will compound your germination problems by not sowing fresh seeds.

SEED TREATMENT: Stratify the stored seeds for 3 to 4 months, or scarify the seeds mechanically, then stratify them for 1 to 2 months. Also, an acid soak for about 20 minutes (see page 59) followed by 3 months of stratification works well. Following treatments, germinate the seeds for up to 3 weeks. Seeds of Allegheny serviceberry (*A. laevis*) should be warm-stratified at 77°F (25°C) for 1 month, then stratified for 4 months before sowing. For dwarf serviceberry (*A. spicata*), 2 months of warm stratification followed by 2 to 3 months' stratification improves germination.

NURSERY PRACTICES: Fall-sow fresh seeds or spring-sow stratified seeds. Sometimes fall-sown seeds do not germinate until the second spring. Cover the seeds with ¼ inch of soil and provide partial shade for the first year. Be sure to supply a winter mulch for fall-sown seeds. Seedlings grow slowly and are transplanted after 1 year in the seedbed and outplanted as 1+2 or 1+3 plants.

AMORPHA

Indigobush

A. californica (**false indigo**), *A. canescens* (**leadplant**), *A. fruticosa* (**indigobush**), *A. microphylla* (**dwarf indigobush, fragrant false indigo**)

FAMILY: Fabaceae; legume family

FLOWERING: The perfect flowers open from May to July.

SEED COLLECTION: Fruit is a small, slightly curved, indehiscent pod containing one or two tiny, glossy black seeds. Pods ripen to light brown from July to September. Strip ripe pods from the plants by hand.

SEED CLEANING AND STORAGE: Place ripe pods on a screen to dry, then flail to release the seeds, though seed extraction is not necessary for germination. Seeds are orthodox and can be stored dried and sealed in containers at 41°F (5°C) for several years.

SEED TREATMENT: Seed dormancy varies among species. Fresh seeds of false indigo (*A. californica*) and leadplant (*A. canescens*) germinate without pretreatment, but stratification for 2 months or a 15-minute acid soak (see page 59) hastens the speed of germination in leadplant. Seeds of indigobush (*A. fruticosa*) and dwarf indigobush (*A. microphylla*) are soaked in acid for 5 to 10 minutes before sowing.

GERMINATION: Germinate in alternating temperatures with light during the high-temperature period. Germination takes from 1 week to 1 month.

NURSERY PRACTICES: Fall-sow fresh seeds or sow treated seeds in spring. Inoculating the medium with nitrogen-fixing bacteria may be beneficial, though is not required. Sow seeds ¼ inch deep and cover with a light sawdust mulch. The first leaves will be juvenile and round, so don't be fooled into wondering what species you've planted. Adult leaves will form beginning in midsummer. Seedlings are usually outplanted at 20 to 40 inches tall as 2+0 stock.

SEED CLEANING AND STORAGE: Dry the fruits, then flail or thresh to remove the seeds. Winnowing or screening will remove the debris and chaff.

SEED TREATMENT: No pretreatment needed.

GERMINATION: Germinate the seeds at standard temperatures.

NURSERY PRACTICES: Butterfly bushes are not fussy about soil, but make sure nursery soils are well drained.

BUXUS
Boxwood

> *B. microphylla* (**littleleaf boxwood**), *B. sempervirens* (**common boxwood**)

FAMILY: Buxaceae; boxwood family

FLOWERING: The flowers are imperfect and inconspicuous and open in April. They are highly attractive to bees.

SEED COLLECTION: The fruit is a three-celled capsule containing shiny black seeds. Harvest the capsules before they open.

SEED CLEANING AND STORAGE: Further dry the capsules on a screen, then flail or tumble to remove the seeds. Winnowing will further clean the seeds.

SEED TREATMENT: Seeds usually require no pretreatment. Stratifying littleleaf boxwood (*B. microphylla*) for 2 months and common boxwood (*B. sempervirens*) for 1 month often speeds germination.

GERMINATION: Germinate the seeds at standard temperatures.

NURSERY PRACTICES: Provide the seedlings with light shade and rich, well-drained soil. Mulching will keep the roots cool and protect the seedlings from windy conditions.

CARAGANA
Peashrub

> *C. arborescens* (**Siberian peashrub**)

FAMILY: Fabaceae; legume family

FLOWERING: The flowers, which are perfect, open from April to June.

SEED COLLECTION: The fruits are pods about 2 inches in length, each containing four to six reddish brown seeds, which are shed when the pods turn brown and dehisce in midsummer. There are only a couple of weeks between pod ripening and seed-shed, so take care to harvest in a timely manner.

SEED CLEANING AND STORAGE: Spread the pods on screens to dry until they open, then flail lightly to remove the seeds. Seeds are orthodox and will store for several years.

SEED TREATMENT: Although untreated seeds will germinate fairly well, germination will be slightly better if you soak the seeds in 185°F (85°C) water for 24 hours before planting.

GERMINATION: Germinate for a month or so at standard temperatures.

NURSERY PRACTICES: Fall-sow fresh seeds, or sow stored/treated seeds in spring ¼ to ½ inch deep. Provide a winter mulch for fall-sown seeds. Seedlings are outplanted as 1+0, 2+0, or 3+0 stock.

CARYA
Hickory, pecan

> *C. glabra* (**pignut hickory**), *C. illinoinensis* (**pecan**), *C. ovata* (**shagbark hickory**)

FAMILY: Juglandaceae; walnut family

FLOWERING: Hickory is monoecious but has imperfect flowers. Flowering occurs in spring.

SEED COLLECTION: The nuts are enclosed in a husk made of the floral involucres. The husks are green when immature and turn to a deep brown or black at maturity, then dry and open at their bases, releasing the nuts. Harvest the nuts from the tree or ground.

SEED CLEANING AND STORAGE: Remove the husks by hand. All hickory nuts show orthodox storage behavior. Clean and store the seeds at 32°F (0°C) in plastic freezer bags. Nuts of pecan (*C. illinoinensis*) can display vivipary in certain varieties.

SEED TREATMENT: Dormant seeds of pignut hickory (*C. glabra*) and pecan should be stratified for 3 to 4 months. Alternatively, presoak the seeds in water for 1 week, then stratify them for about 40 days. Shagbark hickory (*C. ovata*) seeds should be stratified for 3 to 5 months.

GERMINATION: Most species in this genus show little dormancy. Because pecan can display viviparous germination, a lack of dormancy may be worse than a true dormancy in this species. Germinate nondormant (fresh) seeds of pecans and hickories for 45 to 60 days. Germinate seeds at 86 to 95°F (30 to 35°C), with light provided for 15 hours per day. Germination at the high temperature can overcome growth restrictions induced by the seed coats.

NURSERY PRACTICES: Fall sowing of fresh seeds is preferred. Use mulch for winter protection, and cover the nuts with hardware cloth to prevent rodent predation. Sow the seeds about 1 inch deep in rows 10 inches apart.

Castanea

Chestnut

C. dentata (**American chestnut**), *C. mollissima* (**Chinese chestnut**)

FAMILY: Fagaceae; beech family

FLOWERING: The plants are for the most part monoecious but bear imperfect flowers. Male catkins are

CHESTNUT BURS

borne near the base of the flowering branch, whereas the female flowers are borne near the tips. Flowering occurs from April to June.

SEED COLLECTION: Chestnuts produce spiny round burs about 2 inches in diameter. Each bur contains from one to three seeds (nuts) that turn light brown to black when ripe. Harvest the fruits from the tree or from the ground as soon as the burs begin to split.

SEED CLEANING AND STORAGE: Chestnut seeds are recalcitrant. Seeds must not be allowed to dry out, and those harvested from the ground must be imbibed before planting or treatment. Seeds on the ground usually lose their viability within a week of shed.

There is no long-term storage for chestnuts because of their recalcitrance. Store seeds in partly opened plastic freezer bags at 32 to 37°F (0 to 3°C) for up to 1 year. This will serve as satisfactory stratification in many cases. Moisten them frequently, and don't let them dry out.

─────── CHESTNUT NOTE ───────

The beech family, Fagaceae, comprises more than 600 species of trees and shrubs in six genera, but only one genus, *Castanea*, produces nuts normally eaten by humans.

SEED TREATMENT: Fresh seeds have little dormancy. If seeds have been allowed to dry out even a bit, you will have to stratify them. Remove the husk, then presoak the seeds for 48 hours in cold water, then stratify them for 1 to 2 months.

GERMINATION: Sow the seeds, and they will produce seedlings within about a month. Sometimes weevils will burrow into the fruits and destroy the embryos. If you suspect an infestation, dip the seeds in 120°F (49°C) water for 30 minutes before stratification.

NURSERY PRACTICES: Sow fresh seeds in the fall or stratified seeds in spring. Sow them about 1 inch deep, 3 to 4 inches apart in rows spaced about 5 inches apart. Here's a tip: if you place the nuts on their sides when sowing, they may germinate more reliably. Outplant as 1+0 stock.

CATALPA
Catalpa

C. bignoniodes (**southern catalpa**), *C. bungei* (**bunge catalpa**), *C. ovata* (**Chinese catalpa**), *C. speciosa* (**northern catalpa**)

FAMILY: Bignoniaceae; trumpet creeper family

FLOWERING: The flowers are perfect and open from May to June.

SEED COLLECTION: The mature fruits are brown, beanlike, two-celled capsules 2 or more feet long. Each contains many thin-winged seeds. Collect dry, brown fruits before they split open and shed their seeds.

SEED CLEANING AND STORAGE: Dry the fruits further, then flail them to collect the fairly large seeds. Winter collection of pods of northern catalpa (*C. speciosa*) yields seeds with higher germination than that of pods collected in the fall. Seeds are orthodox and can be stored for a few years.

GERMINATION: Seeds germinate well over a 3-week period without pretreatment. About 63 percent of the seeds of northern catalpa will germinate at standard temperatures.

NURSERY PRACTICES: Sow seeds in late fall or early spring and cover with about ⅛ inch of soil. Southern catalpa (*C. bignoniodes*) germinates best when covered with a pine-needle mulch. Southern catalpa and northern catalpa are usually outplanted as 1+0 stock.

CEPHALOTAXUS
Plum-yew

C. harringtonii (**Harrington plum-yew**)

FAMILY: Cephalotaxaceae; plum yew family

FLOWERING: The plants are dioecious.

SEED COLLECTION: The fruits, fleshy arils containing a seed, are plum-shaped and take 2 years to mature. Hand-pick the mature fruits in October.

SEED CLEANING AND STORAGE: Clean the fruits by rubbing them on a screen to remove the fleshy material, or simply pick the seeds out of the fruits.

SEED TREATMENT: Stratify for 3 months.

GERMINATION: Germinate at standard protocol.

NURSERY PRACTICES: Fall-plant the seeds, or plant stratified seeds in spring.

CERCIS
Redbud

C. canadensis (**eastern redbud**), *C. canadensis* var. *mexicana* (**Mexican redbud**), *C. canadensis* var. *texensis* (**Texas redbud**), *C. chinensis* (**Chinese redbud**), *C. occidentalis* (now *C. canadensis* var. *occidentalis*; **western redbud**), *C. orbiculata* (**California redbud**), *C. siliquastrum* (**Judas tree**)

REDBUD SEED PODS

FAMILY: Fabaceae; legume family

FLOWERING: The perfect flowers open from February to May and are insect-pollinated.

SEED COLLECTION: Fruits are flat pods 2 to 4 inches long that ripen from July to September. Each contains from 4 to 10 hard brown seeds. The ripe fruits are brown and dehisce and shed their seeds in the fall. Hand-pick them from the trees.

SEED CLEANING AND STORAGE: Spread the pods to dry, then flail gently to remove the seeds. Seeds are orthodox and can be stored for many years at ⁻13°F (⁻25°C).

TREATMENTS FOR REDBUD SEED

TYPE OF REDBUD	ACID SOAK (MINUTES)	STRATIFICATION (MONTHS)
Eastern	30–45	3
Mexican	30–45	3
Western	30–45	3
Texas	60	1
California	60	3
Chinese	30	2

SEED TREATMENT: The seeds are hardseeded and may also have some embryo dormancy. Soak seeds in vinegar or in acid (see page 59) for the times indicated in the chart below, then stratify them immediately. Judas tree (*C. siliquastrum*) seeds should be soaked in hot water, then given a 3-month stratification. Sometimes nicking or filing the seed coats or a hot-water soak at 190°F (88°C) can substitute for the acid soak. Whatever scarification treatment is used, stratification must follow it immediately.

GERMINATION: Germination takes from 1 to 2 months.

NURSERY PRACTICES: Fall-sow fresh seeds or sow treated seeds in spring. Sow them about ½ inch deep, then cover them with 1 inch of sawdust mulch. Winter mulch is useful for fall-sown seeds. Seedlings are outplanted as 1+0 plants when they are about 12 inches tall.

CHAENOMELES

Flowering quince

C. japonica (**Japanese flowering quince**), *C. speciosa* (**common flowering quince**)

FAMILY: Rosaceae; rose family

FLOWERING: The flowers are perfect and bloom in early spring.

SEED COLLECTION: The fruit are large, fragrant pomes that develop a slight blush when they ripen in October. Harvest fruits once they ripen.

SEED CLEANING AND STORAGE: Cut open the fruits to extract the seeds, or soften them first by allowing them to freeze, then thaw. You can also partially ferment the fruits to ease seed extraction, but take care the fruits don't overheat.

SEED TREATMENT: Warm-stratify at 77°F (25°C) for 2 weeks, then stratify for 2 to 4 months.

C

GERMINATION: Germinate the seeds at standard temperatures.

NURSERY PRACTICES: Fall-sow fresh seeds or spring-sow stratified seeds. Cover the seeds with about ½ inch of soil and provide a winter mulch for fall-sown seeds.

CHAMAECYPARIS
Chamaecyparis

> *C. lawsoniana* (**Port Orford cedar**), *C. nootkat-ensis* (**Alaska cedar**), *C. obtusa* (**Hinoki false-cypress**), *C. pisifera* (**Japanese falsecypress**), *C. thyoides* (**Atlantic white cedar**)

FAMILY: Cupressaceae; cypress family

FLOWERING: The flowers are inconspicuous and imperfect and the plants monoecious. Male strobili are usually yellow, but those of Port Orford cedar (*C. lawsoniana*) are red. Female strobili produce cones. Bloom occurs in early spring.

SEED COLLECTION: The cones are ¼ to ½ inch in diameter and erect. They become red-brown and are covered with a powdery white wax. Atlantic white cedar (*C. thyoides*) cones are an exception, turning blue-purple. Most cones mature in the late summer and early fall of the first year, at which time they open to shed seeds. Some cones may remain on the plant into the second year. Each cone has 6 to 12 scales, and each scale has one to five tiny winged seeds. Collect the ripe cones from the trees.

SEED CLEANING AND STORAGE: Spread them to dry in a 100°F (38°C) room or oven. After drying, don't try to remove the wings, because the seeds are easily injured. Winnow to remove the chaff.

SEED TREATMENT: Stratify seeds of Atlantic white cedar for 3 months. Seeds of the other species may need stratification for only a month or two. Seeds of Port Orford cedar, Hinoki falsecypress (*C. obtusa*), and Japanese falsecypress (*C. pisifera*) germinate

without stratification, though germination is more rapid if the seeds are stratified.

GERMINATION: Germinate for 3 weeks. About 52 percent of the seeds of Port Orford cedar will germinate.

NURSERY PRACTICES: Broadcast fresh seeds in the fall or stratified seeds in spring, and cover with ¼ inch of soil. Provide the seedbed with half shade through midsummer of the first year. About half the seeds will germinate in the first 2 months after sowing.

CITRUS
Citrus

> *C. aurantiifolia* (**lime**), *C. limon* (**lemon**), *C. maxima* (**pumulo**), *C. medica* (**citron**), *C. ×paradisi* (**grapefruit**), *C. paradisi ×C. reticulata* (**tangelo**), *C. sinensis* (**sweet orange**)

FAMILY: Rutaceae; citrus family

FLOWERING AND FRUITING: Citrus have heavily scented flowers that are polygamo-dioecious: that is, the plants bear both perfect and imperfect flowers on separate plants.

SEED COLLECTION: Collect the fruits when they are fully ripe by harvesting from the tree. The only way to tell if citrus fruit is ripe is by tasting — if a fruit tastes good, it's ready for collection. Shed fruits may be bruised or rotted and seed germination poor.

SEED CLEANING AND STORAGE: Remove the rind and the flesh, and extract the relatively large seeds by hand. Surface-dry seeds slightly at room temperature if you plan to store them. Before storage, dip seeds for 10 minutes in 120°F (49°C) water to reduce seed-borne diseases, then store the seeds, moist, in poly bags at 40°F (4°C).

GERMINATION: No citrus seeds show true dormancy, only very slow germination, due probably to a semi-impermeable seed coat. They are also damaged if allowed to dry out. Germination may take up to a month at 86°F (30°C). Drying extends germination time to nearly twice the amount of time over fresh seeds.

Germinate lime (*C. aurantiifolia*) seeds at 86 to 95°F (30 to 35°C) for up to 42 days and lemon (*C. limon*) seeds at 86°F (30°C) for about a month.

NURSERY PRACTICES: Plant seeds in outdoor beds in spring after the soil has warmed to at least 60°F (16°C). Press the seeds into the soil, then cover them with about ¾ inch of clean sharp sand. Transplant the seedlings when they are 8 to 12 inches tall into a nursery row, then outplant in the fall or the following spring.

———————— CITRUS ALERT ————————

There is not much data about seed saving and seed starting for many species of citrus, but some of the information here may work for species other than the ones we've listed.

CLEMATIS

Clematis

C. columbiana (**rock clematis**), *C. drummondii* (**Drummond's clematis**), *C. flammula* (**plume clematis**), *C. ligusticifolia* (**western virgin's bower**), *C. pauciflora* (**rope-vine**), *C. virginiana*

CLEMATIS FRUITS

(now *C. ligusticifolia*; **eastern virgin's bower**), *C. vitalba* (**traveler's joy**), *C. viticella* (**Italian clematis**)

FAMILY: Ranunculaceae; buttercup family

FLOWERING: Some species are monoecious, some dioecious. Western virgin's bower (*C. ligusticifolia*) and eastern virgin's bower (*C. virginiana*) are dioecious, although the female flowers do have nonfunctional stamens. Traveler's joy (*C. vitalba*) is monoecious. Flower sizes and colors vary considerably among species.

SEED COLLECTION: The fruits are one-seeded achenes with feathery, persistent styles. Harvest the ripe brown fruits when the styles have become feathery, before they're dispersed by the wind.

SEED CLEANING AND STORAGE: Trample the ripe fruits lightly to separate them from their styles, then winnow lightly to clean. Store dry seeds in paper envelopes at room temperature for up to 2 years.

SEED TREATMENT: Seeds have dormant embryos, and procedures for pretreatment differ among species, varieties, and lots. You will need to work out suitable treatments for your own lots, but perhaps the best way to break seed dormancy is to provide warm stratification for about 3 months, followed by stratification for 2 to 5 months.

GERMINATION: Germinate the seeds for at least 2 months at standard temperatures. Good luck!

NURSERY PRACTICES: Plant fresh seeds in the fall or spring-sow treated seeds. Seeds may not germinate in a nursery bed for at least 1 year.

———————— CLEMATIS ALERT ————————

Propagation of clematis by seeds gives highly variable results. If you like a challenge, try growing clematis from seed. If you like a sure thing, multiply your clematis by cuttings instead.

Cornus

Dogwood

C. alba (**tatarian dogwood**), *C. alternifolia* (**blue dogwood**), *C. amomum* (**silky dogwood**), *C. asperifolia* (**roughleaf dogwood**), *C. canadensis* (**bunchberry**), *C. controversa* (**giant dogwood**), *C. florida* (**flowering dogwood**), *C. kousa* (**Kousa dogwood**), *C. mas* (**Cornelian cherry**), *C. nuttallii* (**Pacific dogwood**), *C. racemosa* (**gray dogwood**), *C. sanguinea* (**bloodtwig dogwood**), *C. stolonifera* (now *C. sericea;* **red osier dogwood**)

FAMILY: Cornaceae; dogwood family

FLOWERING: The flowers are perfect and open in spring. In some species, notably flowering dogwood (*C. florida*), the small flowers are surrounded by a colorful involucre of four to six white or pink-white scales.

SEED COLLECTION: The fruit is a drupe ⅛ to ¼ inch in diameter and ripens from July to October, depending on species. The drupe has mealy flesh containing a single stone with two bony seeds. Strip the fruits from the plants as soon as they are ripe, but avoid collecting from isolated plants, as they often have nonviable seeds.

SEED CLEANING AND STORAGE: Sow the fruits outdoors immediately, or clean them for storage by use of a blender. Do not allow the fruits to sit in a pile to ferment or to dry so the flesh becomes difficult to

KOUSA DOGWOOD FRUITS

remove. The seeds are orthodox and will store for up to 2 years, but viability is lost rapidly after 1 year.

SEED TREATMENT: Seeds of all dogwoods show embryo dormancy, and most also have a very hard pericarp, which delays imbibition. Pregermination protocol varies greatly among species (see box on facing page), but all can be germinated at standard temperatures.

NURSERY PRACTICES: Sow the stones or the fruits in the fall, or plant stratified seeds in late April or early May. Fresh, nontreated seeds may not germinate until the second spring. Cover seeds with ¼ to ½ inch of soil, and give them a winter mulch of leaves or straw. With luck you can expect about three-quarters of the seeds to germinate. Transplant 1-year-old seedlings to their permanent location, but be sure the species are adapted to the site. Most will do well in somewhat moist soils, but redosier dogwood does well on wet sites, and roughleaf dogwood should be planted only on dry sites.

DOGWOOD FRUIT COLOR	
DOGWOOD SPECIES	**COLOR OF RIPE FRUIT**
Bloodtwig, giant	Purple-black
Blue	Dark blue
Bunchberry, flowering	Scarlet
Cornelian cherry, Pacific	Red
Gray, roughleaf, tatarian,	White
Kousa	Pink to red
Red osier	Lead-colored
Silky	Pale blue

PREGERMINATING DOGWOODS

Tatarian dogwood (*C. alba*): stratify for 3 to 4 months, then germinate with 8 hours of light per day.

Blue dogwood (*C. alternifolia*): warm-stratify at 68 to 86°F (20 to 30°C) for 2 months, then stratify for 2 months. A 2-hour acid soak (see page 59) or mechanical scarification substitutes for warm stratification.

Silky dogwood (*C. amomum*): warm-stratify at 77°F (25°C) for 120 days, then stratify for 3 to 4 months.

Roughleaf dogwood (*C. asperifolia*): warm-stratify at 68 to 86°F (20 to 30°C) for 1 day, then stratify for at least 2 months.

Bunchberry (*C. canadensis*): warm-stratify at 77°F (25°C) for 1 to 2 months, then stratify at 33°F (1°C) for 4 to 5 months. A 2-hour acid soak (page 59) can substitute for warm stratification.

Giant dogwood (*C. controversa*): warm-stratify at 77°F (25°C) for 2 to 3 months, providing light for 8 hours per day, then stratify for 2 to 3 months.

Flowering dogwood (*C. florida*): stratify for 3 to 4 months.

Kousa dogwood (*C. kousa*): stratify for 3 to 4 months, then germinate with 8 hours of light per day.

Cornelian cherry (*C. mas*): warm-stratify at 77°F (25°C) for 4 months, providing light for 8 hours per day, then stratify for 1 to 4 months.

Pacific dogwood (*C. nuttallii*): soak the seeds for 2 hours in acid (page 59), then stratify for 4 months.

Gray dogwood (*C. racemosa*): warm-stratify at 68 to 86°F (20 to 30°C) for 2 months, then stratify for 4 months. A 2-hour acid soak (page 59) can substitute for warm stratification.

Bloodtwig dogwood (*C. sanguinea*): warm-stratify at 77°F (25°C) for 4 months, providing light 8 hours per day, then stratify for 2 to 3 months.

Redosier dogwood (*C. sericea*): scarify, then stratify for 3 to 4 months.

CORYLUS

Hazelnut

C. americana (**American hazelnut**), *C. avellana* (**European hazelnut**), *C. colurna* (**Turkish filbert**), *C. cornuta* (**beaked hazelnut**), *C. maxima* (**purple giant filbert**)

FAMILY: Betulaceae; birch family

FLOWERING: Male and female flowers are borne on separate twigs of the same plant and open in spring before the leaves unfurl. The male flowers form in clusters of two to five catkins, whereas the female flowers are budlike.

SEED COLLECTION: Fruits form in midsummer and ripen in late summer and early fall. The hard-shelled "nuts" contain a single embryo surrounded by a woody pericarp (shell). The nuts are enclosed in involucres (husks) made up of two or more hairy bracts. Harvest the fruits as soon as the edges of the husks begin to turn brown (about mid-August) to avoid predation.

SEED CLEANING AND STORAGE: Hazelnut seeds are suborthodox but should not be dried to a great degree. Remove the husks, then dry the nuts on a wire screen in a humid room for 1 month. Very dry conditions may induce mild to deep dormancy. Store the seeds for up to a year in a sealed container at about 35°F (2°C).

GERMINATION: Newly harvested nuts show no dormancy, but there may be some germination inhibitors present in the seed coat and pericarp. For that reason, germination is slow and erratic in nonstratified seeds. Most nurseries sow the fresh nuts in fall. These can be stored at 35°F (2°C) until sowing in October or November. Nuts that have been stored for longer periods should be stratified for 2 to 6 months before sowing. Give European hazelnut (*C. avellana*) warm stratification at 77°F (25°C) for 3 weeks, then stratification for another 3 weeks prior to sowing. Germinate the nuts over a 2-month period.

nuts, fruits, trees . . .

NURSERY PRACTICES: Sow the seeds 1 inch deep in well-drained soil, then cover with sawdust. Fall sowing of fresh seeds in outdoor nursery beds works well for all species. Allow about 5 square feet for each plant. After 1 year, the plants can be transplanted to their permanent location.

COTONEASTER

Cotoneaster

C. acutifolius (now *C. lucidus*; **Peking cotoneaster**), *C. adpressus* (**creeping cotoneaster**), *C. apiculatus* (**cranberry cotoneaster**), *C. horizontalis* (**rock cotoneaster**), *C. lucidus* (**hedge cotoneaster**)

FAMILY: Rosaceae; rose family

FLOWERING: The perfect flowers open in clusters in May or June.

SEED COLLECTION: The pomelike fruits ripen in September or October. Peking cotoneaster and hedge cotoneaster (both *C. lucidus*) ripen to black; others ripen to bright red. Those of some species may persist for a few months. Pick the ripe fruits from the plants before the birds eat them.

SEED CLEANING AND STORAGE: Each fruit contains one to five nutlets. Macerate the fruits in a blender, then clean the seeds, which are orthodox.

SEED TREATMENT: Seeds have a double dormancy. Warm-stratify the seeds for 6 months at 60 to 77°F (16 to 25°C), followed by stratification for 3 months. Alternatively, soak them for about 90 minutes in acid (see page 59). Soak hedge cotoneaster for 15 minutes only; creeping cotoneaster (*C. adpressus*), for 2 hours. After soaking, stratify seeds for 3 to 4 months.

GERMINATION: Germinate seeds in alternating temperatures with 8 hours of light during the high-temperature period.

NURSERY PRACTICES: Fall-sow fresh seeds, or sow stratified seeds the following spring. Most germination will take place the spring after sowing: that is, during the second spring.

CRATAEGUS

Hawthorn

C. arnoldiana (**Arnold hawthorn**), *C. crus-galli* (**cockspur hawthorn**), *C. ×lavalleei* (**Lavalle hawthorn**), *C. mollis* (**downy hawthorn**), *C. monogyna* (**single-seed hawthorn**), *C. ×nitida* (**glossy hawthorn**), *C. phaenopyrum* (**Washington hawthorn**)

FAMILY: Rosaceae; rose family

FLOWERING: The perfect flowers open from April to early summer.

SEED COLLECTION: The fruits are pomes, which turn bright scarlet when they ripen in early September. They fall quickly when ripe and are a favored food of birds, so hand-pick them as soon as they ripen.

SEED CLEANING AND STORAGE: Each fruit contains four or five bony nutlets, each with a single seed. Macerate the fruits in a blender and dry the seeds. The orthodox seeds will store for a couple of years.

SEED TREATMENT: Warm-stratify the seeds at 77°F (25°C) for 1 to 4 months, then stratify them for 3 to 4 months. You may also mechanically scarify the seeds, then stratify them for 1 to 2 months. Seeds of cockspur hawthorn (*C. crus-galli*) and downy hawthorn (*C. mollis*) are particularly difficult to germinate and show best germination if soaked in acid for 2 hours (see page 59), then warm-stratified at 68°F (20°C) for 4 months, followed by stratification for 5 months. Seeds of Arnold hawthorn (*C. arnoldiana*) should be acid-soaked for 4 to 5 hours, then stratified for 6 months.

GERMINATION: Germinate seeds at standard temperatures.

NURSERY PRACTICES: Fall-sown seeds will germinate either the following spring or after two springs, but be sure to give them a winter mulch. Try to sow seeds early in the fall to provide a few weeks of warm weather before cold sets in. Spring-sow stratified seeds in rich soil and mulch them until the seedlings emerge. Cover seeds with about ¼ inch of soil.

DAPHNE

Daphne

D. giraldii (**Giraldi daphne**), *D. mezereum* (**February daphne**)

FAMILY: Thymelaeaceae; mezereum family

FLOWERING: The fragrant flowers open in early spring.

SEED COLLECTION: The fruit is a fleshy or leathery drupe, either red or yellow-brown, depending on species. Harvest the fruits when they have softened and are "dead" ripe.

SEED CLEANING AND STORAGE: Macerate the fruits in a blender, then rinse and dry the seeds.

SEED TREATMENT: Warm stratification for 1 month followed by stratification for 2 months gives good germination in seeds of some species. Giraldi daphne (*D. giraldii*) seeds germinate well after 3 months of stratification, whereas seeds of February daphne (*D. mezereum*) should be given a simple water soak before germination.

GERMINATION: Germinate at 50°F (10°C).

NURSERY PRACTICES: Fall-sow fresh seeds or sow stratified seeds in spring. Nursery soil should be light and well drained.

JAPANESE PERSIMMON FRUITS

DIOSPYROS

Persimmon

D. kaki (**Japanese persimmon**), *D. texana* (**Texas persimmon**), *D. virginiana* (**American persimmon**)

FAMILY: Ebenaceae; ebony family

FLOWERING: Persimmons bear small, inconspicuous, axillary flowers from March to mid-June on dioecious plants.

SEED COLLECTION: Fruit are large berries that turn to orange/yellow or sometimes black as they ripen, and are shed from September to November. Harvest ripe fruits from the ground soon after they fall.

SEED CLEANING AND STORAGE: Each fruit contains three to eight seeds, each about ½ inch long, with a thick, brown, rough seed coat. Clean them in a blender, then dry the seeds, which are orthodox in storage behavior.

SEED TREATMENT: If dried and stored, persimmon seeds will develop considerable dormancy due to a fairly impermeable seed coat. To overcome dormancy in Texas persimmon (*D. texana*), stratify the seeds for 1 to 6 months before sowing. Mechanical scarification can in part substitute for stratification. Dormancy in seeds of Japanese persimmon (*D. kaki*) and American persimmon (*D. virginiana*) can be overcome by stratification for 2 to 3 months at 50°F

(10°C) before sowing. If this does not work well for you, use a file to remove part of the seed coat over the radicle, then subject the seeds to germination temperatures.

GERMINATION: Light does not substantially influence germination, but this genus is particularly sensitive to temperature. Seeds of Texas persimmon show no dormancy when shed and can be germinated readily at 86°F (30°C). However, if they are chilled for a long period, they will enter dormancy. Those of American persimmon, if sown fresh, will germinate well at standard temperatures for 125 days. If held at temperatures that fluctuate daily from 68°F (20°C) during a 16-hour night to 86°F (30°C) for an 8-hour day under fluorescent light (160 foot-candles), they may germinate far more quickly.

NURSERY PRACTICES: Seeds can be fall-sown outdoors or stratified and planted in late winter. Plant seedlings in moist, well-drained, rich soil, placing seeds ½ to ¾ inch deep and covering with a loose straw mulch, which should be removed after emergence. Give all young persimmon seedlings light shade in the nursery bed.

ENKIANTHUS
Enkianthus

E. campanulatus (**redvein enkianthus**), *E. cernuus*, *E. deflexus* (**bent enkianthus**), *E. perulatus* (**white enkianthus**)

FAMILY: Ericaceae; heath family

FLOWERING: The perfect flowers open in spring.

SEED COLLECTION: The fruit is a five-valved dehiscent capsule that you must collect when it is yellow-brown, before seed-shed.

SEED CLEANING AND STORAGE: Dry the capsules on a screen, then flail to extract the seeds. The tiny seeds are orthodox and will store well in a sealed container at about 37°F (3°C).

SEED TREATMENT: Seeds need no pretreatment.

GERMINATION: Germinate the seeds at standard temperatures.

NURSERY PRACTICES: Sow the seeds on finely milled sphagnum under mist. Transplant seedlings to the nursery bed when they are about 2 months old.

EUONYMUS
Euonymus

E. alatus (**winged euonymus**), *E. americanus* (**strawberry-bush**), *E. atropurpureus* (**eastern wahoo**), *E. bungeanus* (**winterberry euonymus**), *E. europaeus* (**European euonymus**), *E. fortunei* (**wintercreeper euonymus**), *E. verrucosus* (**wartybark euonymus**)

FAMILY: Celastraceae; staff tree family

FLOWERING: The flowers are usually perfect and inconspicuous. They open in spring.

SEED COLLECTION: The fruits, four- or five-celled capsules, ripen in late summer or early fall. Each cell contains one or two seeds enclosed in a fleshy aril. Harvest the ripe fruits by picking from the plants or by shaking them from the plants onto sheets spread on the ground. Colors of the ripe fruits vary according to species. The eastern wahoo (*E. atropurpureus*) has pink to crimson fruit, scarlet arils, and brown seeds; the European euonymus (*E. europaeus*) red to pink fruits, orange arils, and white seeds; the wartybark euonymus (*E. verrucosus*) has yellow-red fruits, orange arils, and black seeds. The other species all

EUONYMOUS FRUITS

bear conspicuous red or orange fruit. As long as the wintercreeper euonymus (*E. fortunei*) trails along the ground, it remains juvenile. Allow it to climb a wall, and it will enter its adult phase and set fruit.

SEED CLEANING AND STORAGE: Spread the capsules in a thin layer on screens to dry for a couple of days, then rub or flail the capsules to remove the fruits. Winnow to separate the fruits from the chaff. Once winnowed, rub the fruits on a fine-mesh screen to remove the pulp from the seeds, but take care not to damage the thin seed coats. The seeds are orthodox and will keep for about 2 years.

SEED TREATMENT: The seeds have dormant embryos and are difficult to germinate. Stratify the cleaned seeds for 3 to 4 months, then germinate them for 1 month. If this doesn't work, give the seeds a warm stratification at 77°F (25°C) for 2 to 3 months, then stratify them for 2 to 4 months before placing them in germination flats. The alternating stratification is preferred for strawberry-bush (*E. americanus*) and eastern wahoo.

GERMINATION: Sowing seeds directly into a nursery bed in spring following stratification is more effective than sowing them in flats under lights. Germinate under standard temperatures.

NURSERY PRACTICES: Spring-sow stratified seeds in outdoor nursery beds. They require fertile soil. Expect between 30 percent and 70 percent of the seeds to germinate.

Fagus

Beech

F. grandifolia (**American beech**), F. sylvatica (**European beech**)

FAMILY: Fagaceae; beech family

FLOWERING: The monoecious plants flower from March to May, about the time the leaves are roughly one-quarter expanded. Flowers are imperfect.

SEED COLLECTION: The fruits are prickly burs that open when brown and shed their seeds from September to November. Each fruit contains two or three yellow or brown nuts. Rake the ripe fruits from the ground soon after they are shed, or harvest them from the tree.

SEED CLEANING AND STORAGE: Spread ripe fruits on a screen to dry, then shake the seeds from them. Screen to separate seeds from the chaff. Seeds are suborthodox but can be dried and stored in sealed containers at 5 to 23°F (−15 to −5°C) for several years. However, given their high lipid content, it's a good idea to germinate the seeds as soon as you can.

SEED TREATMENT: Stratify American beech (*F. grandifolia*) seeds for 3 months and European beech (*F. sylvatica*) seeds for 4 to 5 months.

GERMINATION: Germinate the stratified seeds for up to 4 months. Some seeds may germinate without stratification.

NURSERY PRACTICES: Sow fresh seeds in fall or stratified seed in spring, covering them with about ½ inch of soil. Protect fall-sown seeds from rodents with a screen, and give them winter protection. Seedbeds with 50 percent shade provided until after midsummer produce stocky seedlings.

Fraxinus

Ash

F. americana (**white ash**), F. dipetala (**two-petal ash**), F. excelsior (**European ash**), F. nigra (**black ash**), F. oregona (now F. latifolia; **Oregon ash**), F. pennsylvanica (**green or red ash**), F. quadrangulata (**blue ash**), F. velutina (**velvet ash**)

FAMILY: Oleaceae; olive family

FLOWERING: The small, inconspicuous flowers appear in early spring and are sometimes perfect, but the trees are usually dioecious or polygamodioecious.

SEED COLLECTION: The single-seeded samaras are borne in clusters and ripen in late summer or fall, when they assume a light yellowish brown color and become crisp. Seeds are naturally dispersed by wind in fall or early winter in most species. Harvest them from the tree or ground before they blow away.

SEED CLEANING AND STORAGE: Spread the seeds on a screen to dry. This is especially important if you harvested early in the season, before complete natural drying has taken place. Separate the seeds from debris. Storage life is variable and species dependent. In general, treat the seeds as orthodox. Seeds of European ash (*F. excelsior*) will keep for several years, those of black ash (*F. nigra*) and some other species for about 1 year. The key is to keep them very dry.

SEED TREATMENT: Fresh seeds are often less dormant than older seeds. Most species show embryo dormancy (immature embryo), though germination is also inhibited by the seed coat in black ash, blue ash (*F. quandrangulata*), and European ash. Water soaking at room temperature for 10 days to 2 weeks prior to stratification treatments may also improve germination. Seeds of European ash, white ash (*F. americana*), black ash, and blue ash are warm-stratified at 77°F (25°C) for 3 months, then stratified for 4 months. Seeds of velvet ash (*F. velutina*), Oregon ash (*F. latifolia*), and green ash (*F. pennsylvanica*) are given only the stratification treatment. Stratify seeds of two-petal ash (*F. dipetala*) for 4 to 5 months.

GERMINATION: Germination time of stratified ash seeds varies. Here are some estimated germination times.

- Oregon ash: 2 weeks
- Two-petal ash, black ash, and blue ash: 1 month
- White ash, green ash, and velvet ash: 2 months

About 42 percent of the seeds of green ash and 38 percent of those of white ash will germinate.

NURSERY PRACTICES: Fall-sow the seeds; most species will germinate the following spring. Seeds of European ash, black ash, and blue ash will germinate during the second spring after sowing. Moist, well-drained, fertile loam makes the best soil for germination of most ash species, though blue ash does better on limestone-rich soils partially covered with ash leaf litter. Sow seeds ¼ to ½ inch deep, and provide a winter mulch of straw to reduce heaving. Red ash and green ash seedlings may take only a year or two in the nursery before they are planted to their permanent location as 1+0 or 2+0 stock, while seedlings of other species may require up to 4 or 5 years before planting as 2+2 or 2+3 stock. Untreated seeds of European ash, black ash, and blue ash are better sown in spring and the beds mulched until the following spring. If seeds of these species are stratified first, sow them in the spring and they will germinate immediately. Provide them with partial shade during the first summer.

GINKGO

Ginkgo

G. biloba (**ginkgo**)

FAMILY: Ginkgoaceae; ginkgo family

FLOWERING: This species is dioecious. The female flowers appear in late April before the leaves emerge, and the male flowers appear in early April.

SEED COLLECTION: The fruits, really the seeds, are drupelike and smell like rancid butter. Collect them from the ground 6 to 8 weeks after they drop from the trees, after they have been exposed to some cold.

GINKGO SEEDS AND LEAVES

SEED CLEANING AND STORAGE: Soak the "fruits" in water for several days until the flesh softens, then macerate in a blender. Seeds are orthodox and can be cleaned and stored in a sealed container at 50°F (10°C).

SEED TREATMENT: Warm-stratify the seeds for 1 to 2 months, then stratify them for a similar period of time before germination.

GINKGO ALERT

Ginkgo, the only genus in the ginkgo family, is native to China. The odor of the ginkgo fruit is quite offensive, and therefore male clones are usually planted, which will be frustrating to you if you want to save ginkgo seeds. You'll need to search your neighborhood to find a tree that is producing fruit, and collect the drupes after they drop.

GERMINATION: Germinate at standard temperature.

NURSERY PRACTICES: Sow seeds in late fall and cover them with 2 to 3 inches of soil; then mulch them lightly with sawdust. You may also sow stratified seeds in spring. Seedlings are outplanted as 2+0 stock.

GLEDITSIA

Locust, honeylocust

G. aquatica (**waterlocust**), *G. ×texana* (**Texas honeylocust**), *G. triacanthos* (**honeylocust**)

FAMILY: Fabaceae; legume family

FLOWERING: Trees are polygamo-dioecious, with the greenish-white flowers of the two species and the orange-yellow flowers of their hybrid opening from April to June.

SEED COLLECTION: The fruits are large (7 to 20 inches long), flat, twisted, indehiscent pods that turn dark brown when ripe. Waterlocust (*G. aquatica*) fruits contain one to three seeds each; honeylocust

(*G. triacanthos*) fruits contain up to a dozen seeds. The fruits ripen from August to October, and the seeds are shed during the fall. Collect the fruits from the trees or the ground before they disintegrate and shed their seeds.

SEED CLEANING AND STORAGE: Spread the fruits on screens to dry, then flail to remove the seeds. The seeds are orthodox.

SEED TREATMENT: This species is hardseeded. Soak the seeds in acid (see page 59) for 1 to 2 hours (precise time depends upon the seed lot). Seeds collected when fully mature have a thicker seed coat and need a longer soak time. Nicking, filing, and 194°F (90°C) water soaks for about 1 day are also effective with some lots.

GERMINATION: Germinate treated seeds in alternating temperatures (68°F [20°C] night/86°F [30°C] day), with light applied for 8 hours during the warm period.

NURSERY PRACTICES: Sow seeds ½ to ¾ inch deep; outplant the seedlings as 1+0 stock.

ILEX

Holly

I. aquifolium (**English holly**), *I. crenata* (**Japanese holly**), *I. decidua* (**possumhaw**), *I. glabra* (**inkberry**), *I. opaca* (**American holly**), *I. pedunculosa* (**longstalk holly**), *I. pernyi* (**Perny holly**), *I. serrata* (**Japanese winterberry**), *I. verticillata* (**common winterberry**)

FAMILY: Aquifoliaceae; holly family

FLOWERING: Most hollies flower in spring, though common winterberry (*I. verticillata*) flowers in early to midsummer. The plants are normally dioecious, but occasionally you'll find some monoecious specimens. Fertile flowers are usually solitary and the infertile ones are clustered.

SEED COLLECTION: The fruits, which are drupes containing up to nine bony nutlets (pyrenes), ripen in early to mid-fall and are shed from winter to spring in most species. Common winterberry sheds its fruit from fall to winter. Ripe fruits of English holly (*I. aquifolium*) are bright red, those of inkberry (*I. glabra*) jet black, those of American holly (*I. opaca*) red or yellow, and those of common winterberry bright red or bright yellow. Harvest the ripe fruits from the plants.

SEED CLEANING AND STORAGE: Immediately after harvest, macerate the fruits in a blender, then spread the seeds on a screen to dry. The seeds are orthodox. Store them dry in paper envelopes at 41°F (5°C). Holly seeds are not usually stored for longer than over the winter.

SEED TREATMENT: Germinating seeds of this genus is a real challenge. Dormancy is difficult to overcome, but you can do it if you're persistent. In nature, holly seeds often require 1 to 3 years to germinate because of their very hard seed coats and dormant or immature embryos. Many times boiling water, stratification, and scarification all fail to speed germination. Use a combination of warm and cold stratification: First soak the seeds for 2 days at 80°F (27°C), then warm-stratify them for 2 months, with light applied for 12 hours per day. Next, place the moist seeds in a plastic bag and hold them at 35°F (2°C) for 3 months, then germinate them in flats, with light applied for 12 hours. Germination may take up to 2 years. If this doesn't work, increase or decrease the period of warm stratification.

Here are some protocols for particular species.

- English and American hollies: warm-stratify at 77°F (25°C) for 10 months, then cold stratify for 6 months.
- Common winterberry: Warm-stratify for 5 months, then stratify for 3 months.
- Japanese winterberry (*I. crenata*): Stratify for 3 months.

GERMINATION: Germinate at standard temperature.

NURSERY PRACTICES: Probably the simplest way to germinate holly seeds is to sow them about ⅛ inch deep in nursery beds in fall or in spring and then be patient for about 3 years. Mulch fall-sown seeds, and leave the mulch on until the second spring. Provide the young seedlings, especially the evergreen species, with light shade. Most hollies do well in well-drained, rich soil. Common winterberry does a bit better in moist soils. Transplant the seedlings to their permanent location when they are dormant.

JUGLANS
Walnut

J. cinerea (**butternut**), *J. hindsii* (**California black walnut**), *J. nigra* (**black walnut**), *J. regia* (**English or Persian walnut**)

FAMILY: Juglandaceae; walnut family

FLOWERING: Walnut is monoecious with imperfect flowers. The greenish male catkins and the female flowers appear right after the leaves emerge in spring.

SEED COLLECTION: The nut is enclosed in a thick dehiscent husk that darkens at maturity. Harvest the nuts from the ground or the tree.

SEED CLEANING AND STORAGE: Remove the husks before they dry around the nuts. Walnut husks will stain your hands a deep black, so be sure to wear gloves during this procedure. Placing the nuts beneath a tree or other shade to rot sometimes helps, but if the husks dry too much, they will be very difficult to

BLACK WALNUTS WITH HUSKS
AND WITHOUT (LOWER RIGHT)

remove. Float off unfilled nuts, then surface-dry the remainder. The seeds are suborthodox and may be kept at 37°F (3°C) in plastic freezer bags with 80 to 90 percent humidity for up to 1 year. Scant data is available on longer storage.

SEED TREATMENT: Dormant seeds of most species are stratified for several months before germination. Although only a couple of months of stratification are necessary, additional time (up to 6 months) can increase the rate of germination. Seeds of black walnut (*J. nigra*) are either stratified for 4 months or warm-stratified for 2 months at 70°F (21°C), then stratified for 4 months.

GERMINATION: Fresh seeds are nondormant and can be sown indoors or directly outdoors in fall. Indoors, sow butternut (*J. cinerea*) and black walnut in sand, peat, or soil, with light provided for 12 hours. Do the same for California black walnut (*J. hindsii*) and English walnut (*J. regia*), but keep them in darkness. Pretreated dormant seeds are germinated at 86°F (30°C). Removal of the seed coat over the radicle will improve germination, but be careful you don't damage the embryo.

NURSERY PRACTICES: Sow fresh nuts in the fall with the husks removed. A hot-water soak of 1½ to 2 minutes preceding fall sowing of California black walnut is helpful. Sow stratified nuts in spring. Cover the nuts with 1 to 2 inches of soil. Thin the seedlings to 8 inches between plants in the row. A black walnut seedling should be about 18 inches tall and have a stem diameter of ⅓ inch before you plant it out into the landscape. *Juglans* seedlings are very sensitive to waterlogged soils.

JUNIPERUS

Juniper

J. chinensis (**Chinese juniper**), *J. communis* (**common juniper**), *J. horizontalis* (**creeping juniper**), *J. monosperma* (**one-seed juniper**), *J. occidentalis* (**Sierra juniper, western juniper**), *J. scopulorum* (**Rocky Mountain juniper**), *J. utahensis* (**Utah juniper**), *J. virginiana* (**eastern redcedar**)

FAMILY: Cupressaceae; cypress family

FLOWERING: Most junipers flower in spring. The small, inconspicuous flowers are imperfect, and plants of most species are dioecious. Occasionally, plants are monoecious in some species. The male flowers are yellow and form in what looks like a small catkin; greenish female flowers are enclosed in scales that eventually become fleshy strobiles. The "berries" ripen in the first, second, or sometimes third fall.

SEED COLLECTION: Fruits ripen from August to December, depending on species. Ripe fruit color is deep blue to black in common juniper (*J. communis*), one-seed juniper (*J. monosperma*), Sierra juniper *J. occidentalis*), and eastern redcedar (*J. virginiana*); dark red to brown in Utah juniper (*J. utahensis*); and

PREGERMINATION TREATMENTS FOR JUNIPERS

SPECIES	TYPE OF DORMANCY	MEDIUM	TREATMENT
Common Sierra	Embryo and impermeable seed coat	Sand or peat	Warm-stratify for 2 to 3 months at 68 to 86°F (20 to 30°C), then stratify for 4 months.
Chinese Creeping	Embryo and impermeable seed coat	Sand and peat	Soak in acid for 60 minutes (see page 59), then stratify for 3 months.
One-seed	Embryo and impermeable seed coat	Sand and peat	Stratify for 2 months.
Rocky Mountain Utah	Embryo and impermeable seed coat	Sand and peat	Warm-stratify for 4 months at 68 to 86°F (20 to 30°C), then stratify for 4 months.
Eastern redcedar	Embryo dormancy	Sand and peat	Stratify 3 to 4 months.

bright blue in Rocky Mountain juniper (*J. scopulorum*). Each "berry" contains one to six brown seeds. Seeds are dispersed in February and March in eastern redcedar but persist on the plants for a few years in the other junipers. Hand-pick fruits only when they show their ripe color. Green or blue-green fruits contain unripe seeds.

SEED CLEANING AND STORAGE: Run fruits through a blender to remove the seeds. Alternatively, dry the fruits and rub them on a screen to separate the seeds. The orthodox seeds will store for several years.

SEED TREATMENT: Most of the junipers show delayed germination because of embryo dormancy and sometimes due to an impermeable seed coat. Some may germinate the first year after sowing, some the second or third year. See the chart on page 247 for specific protocols.

GERMINATION: After treatment, the seeds can be germinated at 59 to 68°F (15 to 20°C) for 1 month. Germination of Rocky Mountain juniper seeds is about 22 percent.

NURSERY PRACTICES: Fall-sow fresh seeds or spring-sow stratified seeds. It is especially important to sow stratified seed as early in the spring as possible so that germination will be complete before temperatures rise above 70°F (21°C). Sow seeds about 8 inches apart, then cover them with about ¼ inch of firmed soil. Keep the soil surface moist, and mulch the seeds lightly. Germination should be complete in 4 to 5 weeks after sowing. Provide the bed with light shade during the first season.

KALMIA

Mountain laurel

K. angustifolia (**lamb-kill**), *K. latifolia* (**mountain laurel**), *K. polifolia* (**bog kalmia**)

FAMILY: Ericaceae; heath family

FLOWERING: The flowers are perfect and pollinated by insects; they open from March to July. The plants are considered naturally cross-pollinated.

SEED COLLECTION: The fruit is a globe-shape, dry, dehiscent, five-valved capsule and contains many very tiny brown seeds. Seeds are dispersed when the capsule splits, usually around September or later. Pick the capsules from the plants as soon as they turn deep brown and before they split.

SEED CLEANING AND STORAGE: Dry the fruits in a paper bag for a month after harvest, then rub them between your hands or across a fine screen or flail them to separate the seeds. Separate the seeds from the chaff by shaking them down a creased paper. The seeds, which move faster than the chaff, are very small; those of mountain laurel require about 1.4 million to weigh 1 ounce. The seeds are orthodox and will store for several years.

SEED TREATMENT AND GERMINATION: Seeds of this genus show little dormancy, but germination can be improved if they are stratified for 2 months. They require light for germination. Sow them in peat and germinate them at 77°F (25°C) under cool white fluorescent lights for 12 to 18 hours per day. Seedlings grow very slowly, but maintain them at 77°F (25°C) in 9 hours of light and 15 hours of darkness. When they are 2 to 6 months of age, transplant them to pots and place them in the pot bed. The potting medium should be made of 2 parts peat, 1 part perlite, and ½ part coarse sand.

NURSERY PRACTICES: Sow the seeds in flats filled with sandy or peaty soil, then expose them to winter conditions for 2 to 3 months. As soon as seedlings are large enough to handle, transplant them to pots filled with the medium described above. Or sow seeds indoors according to the directions above; raise the plants in a greenhouse, then move pots to the pot bed. Give the young seedlings a complete liquid fertilizer every 3 weeks. The following year, transplant them to a nursery bed for 1 to 2 years before setting them in their permanent location.

KOELREUTERIA

Goldenraintree

> *K. bipinnata* (**bipinnate goldenraintree**),
> *K. paniculata* (**panicled goldenraintree**)

FAMILY: Sapindaceae; soapberry family

FLOWERING: The perfect flowers bloom in large clusters in midsummer.

SEED COLLECTION: The inflated three-valved capsules bear one or two black seeds each. Harvest the fruits when they turn brown but before they dehisce and shed their seeds.

SEED CLEANING AND STORAGE: Dry the capsules on screens, then flail or thresh to remove the seeds. The seeds are orthodox and will store sealed at 41°F (5°C) for a long time.

SEED TREATMENT: Scarify the seeds by filing or with an acid soak for 30 minutes (see page 59), followed by stratification for 3 months.

GERMINATION: Seed germination is variable, taking anywhere from 10 to 100 days at standard temperatures.

NURSERY PRACTICES: Fall-sow or plant treated seeds in spring in rich, well-drained soil. When transplanting, try to include as much soil and root-ball as possible, because bare-root plants don't transplant well.

LABURNUM

Goldenchaintree

> *L. alpinum* (**Scotch laburnum**), *L. anagyroides* (**common laburnum**), *L. ×watereri* (**goldenchaintree, waterer laburnum**)

FAMILY: Fabaceae; legume family

FLOWERING: The hanging clusters of perfect flowers open in May.

SEED COLLECTION: Collect the pods by hand in September when they turn brown but before they dehisce.

SEED CLEANING AND STORAGE: Place the pods on a screen to dry further, then flail to extract the seeds. Dry and store the seeds in a sealed container under refrigeration for up to 2 years.

SEED TREATMENT: The seeds are hardseeded. For Scotch laburnum (*L. alpinum*) and goldenchain tree (*L. ×watereri*), soak the seeds in acid for 2 hours (see page 59). Soak seeds of common laburnum (*L. anagyroides*) in acid for 1 hour.

GERMINATION: Germinate the seeds at the standard temperatures.

NURSERY PRACTICES: Move seedlings carefully with a full root-ball. Soil should be well drained. Give seedlings some light shade for part of the day.

LAGERSTROEMIA

Crapemyrtle

> *L. indica* (**common crapemyrtle**)

FAMILY: Lythraceae; loosestrife family

FLOWERING: The flowers, which are perfect, bloom all summer.

SEED COLLECTION: The fruit is a six-valved, dehiscent capsule that ripens in mid-fall. Collect the fruits when they turn brown but before they dehisce.

SEED CLEANING AND STORAGE: Spread capsules on screens to dry further, then flail or tumble to extract the seeds. Seeds can be dried and stored under refrigeration for a short period.

SEED TREATMENT AND GERMINATION: No pretreatment is required but a short stratification improves germination. Germinate at standard temperatures.

NURSERY PRACTICES: Sow seeds in early spring in moist, well-drained soil and full sun.

LAVANDULA

Lavender

L. angustifolia (**common lavender**)

FAMILY: Lamiaceae; mint family

FLOWERING: The flowers are perfect, forming in long spikes and opening during the summer.

SEED COLLECTION: The fruit is a nutlet. Harvest it when it is brown and dry.

SEED CLEANING AND STORAGE: Dry the fruits on screens, then store sealed in containers under refrigeration.

SEED TREATMENT: No pretreatment is needed, but a short stratification may improve germination. Stored seeds should be stratified for about 5 weeks before sowing.

GERMINATION: Germination takes from 1 to 3 months at standard temperatures.

NURSERY PRACTICES: Sow seeds directly after harvest in well-drained soil and full sun.

LIGUSTRUM

Privet

L. japonicum (**Japanese privet**), *L. lucidum* (**glossy privet**), *L. obtusifolium* (**border privet**), *L. ovalifolium* (**California privet**), *L. sinense* (**Chinese privet**), *L. vulgare* (**European privet**)

FAMILY: Oleaceae; olive family

FLOWERING: The flowers, which are perfect, are borne in clusters in early summer.

SEED COLLECTION: The fruits are one- to four-seeded, berrylike drupes that ripen in early fall. The color of the ripe fruit is generally a deep blue-black, though some varieties of European privet (*L. vulgare*) may have yellow, green, or white fruit. Hand-pick the ripe fruits.

SEED CLEANING AND STORAGE: Macerate the fruits in a blender, then spread the clean seeds to dry. Some species have a soft seed coat, so take care not to damage it in the cleaning process. Store the orthodox seeds in plastic bags at 0°F (−18°C).

SEED TREATMENT: Stored seeds develop various stages of hardseededness and should be stratified for 2 to 3 months. Germination may take about 2 months. Hold the germination temperature at 50°F (10°C) for 16 hours per day, then increase it to 86°F (30°C) for the remaining 8 hours.

GERMINATION: Fresh seeds will germinate immediately.

NURSERY PRACTICES: Fall-sow fresh seeds or spring-sow stratified seeds. Outplant the seedlings when they are a year or two old.

LIQUIDAMBAR

Sweetgum

L. formosana (**Formosan sweetgum**), *L. styraciflua* (**sweetgum**)

FAMILY: Hamamelidaceae; witch hazel family

FLOWERING: The flowers are imperfect but the plants monoecious. The flowers open in March to May. Formosan sweetgum (*L. formosana*) is self-incompatible and must have another tree of the same species in range for fertilization.

SEED COLLECTION: The seed heads turn from a greenish to yellow as they ripen. Hand-pick the ripe seed heads.

SEED CLEANING AND STORAGE: You may need to dry the heads for a month more until they open, shedding the beaklike capsules, each of which contains one or two seeds. Clean the orthodox seeds, then store them for up to 5 years.

SEED TREATMENT: Stratify the seeds for 1 to 3 months to break shallow dormancy.

GERMINATION: Germinate seeds for 1 month. About 70 percent of the seeds of sweetgums germinate.

NURSERY PRACTICES: Sow stratified seeds in spring by spreading them on the surface of the medium and pressing them in with a board. Cover the seeds with ¼ to ½ inch of sawdust, and keep the surface moist at all times.

LIRIODENDRON

Tulip poplar

L. tulipifera (**tulip poplar, tulip tree**)

FAMILY: Magnoliaceae; magnolia family

FLOWERING: The perfect flowers open from April to June and are highly attractive to bees.

SEED COLLECTION: The fruits turn tan or light brown as they ripen from September to November, and the seeds disperse from October into the winter. Each fruit, or "cone," contains from 100 to about 120 seeds. Hand-pick the cones before they fall if you can, or gather them off the ground right after they drop.

SEED CLEANING AND STORAGE: Spread the cones to dry; then shake, flail, or tumble the seeds from them. Cones from the tops of the crowns usually produce the best seeds. The seeds are orthodox, but viability deteriorates rapidly.

SEED TREATMENT: Seeds possess seed-coat dormancy and some shallow embryo dormancy. Stratify the seeds for 10 weeks, with weekly temperature fluctuations from 32 to 50°F (0 to 10°C).

GERMINATION: Germinate seeds at standard temperatures. Expect a germination of about 5 percent.

NURSERY PRACTICES: It is easiest to fall-sow seeds, covering them with about ¼ inch of soil and supplying them with a winter mulch. When germination begins, remove the mulch and provide the seedlings with light shade for their first month or two.

MAGNOLIA

Magnolia

M. acuminata (**cucumber magnolia**), *M. ashei* (now *M. macrophylla* ssp. *ashei*; **Ashe magnolia**), *M. cylindrica, M. denudata* (**Yulan magnolia**), *M. fraseri* (**Fraser's magnolia**), *M. grandiflora* (**southern magnolia**), *M. hypoleuca* (**whiteleaf Japanese magnolia**), *M. kobus* (**Kobus magnolia**), *M. macrophylla* (**bigleaf magnolia**), *M. officinalis, M. pyramidata* (**pyramid magnolia**), *M. splendens* (**shining magnolia**), *M. stellata* (**star magnolia**), *M. tripetala* (**umbrella magnolia**), *M. virginiana* (**sweetbay magnolia**)

FAMILY: Magnoliaceae; magnolia family

SOUTHERN MAGNOLIA FRUITS

FLOWERING: The perfect flowers open from early spring through June. The flowers are protandrous and pollinated by beetles.

SEED COLLECTION: The fruits are follicles containing one or two seeds each. The follicles are clustered into a larger arrangement called a *follicetum*, which contains between 2 and 60 seeds. Seeds are shed when the follicles are ripe. Harvest them from the trees in August or September, before the follicles open.

SEED CLEANING AND STORAGE: Spread folliceta on screens to dry. After they are dry, remove oily residue on seeds by rinsing in soapy then clean water, then dry them again. The seeds appear to be recalcitrant or suborthodox and can be stored moist packed in plastic bags and moist moss at 32°F (0°C).

SEED TREATMENT: *Magnolia* seeds have a double dormancy. Stratify them for 3 to 4 months, then warm-stratify them for 2 months at about 68°F (20°C). Ashe magnolia (*M. macrophylla* ssp. *ashei*), *M. cylindrica*, Yulan magnolia (*M. denudata*), Fraser's magnolia (*M. fraseri*), southern magnolia (*M. grandiflora*), whiteleaf Japanese magnolia (*M. hypoleuca*), Kobus magnolia (*M. kobus*), *M. officinalis*, and star magnolia (*M. stellata*) need stratification for only 3 to 4 months.

GERMINATION: Germinate seeds for up to 2 months at alternating temperatures, with about 18 hours of light provided during the high-temperature period. Species differ on optimum treatment, so you'll have to experiment to find the one that's right for you.

─────── MAGNOLIA ALERT ───────

Most magnolia trees raised from seeds have a juvenile period of 5 to 10 years.

NURSERY PRACTICES: Sow fresh seeds in October or stratified seeds in spring, and provide protection from rodents. Mulch the spring-sown seeds with about ½ inch of mulch, then with the same amount of pine bark or pine straw for winter protection. Seedlings are outplanted as 2+0 stock in northern locations or as 1+0 in the South. Fertilize the seedlings as needed.

MAHONIA
Mahonia, Oregon grape

M. aquifolium (**hollyleaf mahonia**), *M. bealei* (now *M. japonica*; **Beale Oregon grape**), *M. fortunei* (**Chinese mahonia**), *M. fremontii* (**Fremont's mahonia**), *M. haematocarpa* (**red barberry**), *M. japonica* (**Japanese mahonia**), *M. nervosa* (**Cascades Oregon grape**), *M. nevinii* (**Nevin's barberry**), *M. ×wagneri* 'Pinnacle' (**cluster mahonia**), *M. repens* (**Oregon grape**)

FAMILY: Berberidaceae; barberry family

FLOWERING: The perfect flowers open in clusters in spring.

SEED COLLECTION: The fruits are berries containing from one to several seeds each. The fruits ripen in late summer or fall. Hand-pick them as soon as they are ripe.

SEED CLEANING AND STORAGE: Macerate the fruits in a blender, then spread the cleaned seeds to dry. Store seeds in sealed containers at 41°F (5°C).

SEED TREATMENT: Seeds of Fremont's mahonia (*M. fremontii*) and red barberry (*M. haematocarpa*) need no pretreatment, but short stratification for 2 months will improve germination. Stratify seeds of Beale Oregon grape (*M. japonica*), Oregon grape (*M. repens*), Japanese mahonia (*M. japonica*), and most other mahonias for 2 months. Difficult lots of Oregon grape may require a cold + warm + cold stratification for 4 months each, and difficult lots of Beale Oregon

MAHONIA FRUIT COLOR	
SPECIES OF MAHONIA	**COLOR OF RIPE FRUIT**
Beale Oregon grape	Light blue
Chinese, cluster, Oregon grape	Purple-black
Hollyleaf, Fremont's	Blue-black
Japanese, Cascades Oregon grape	Blue
Nevin's barberry	Yellow-red to deep red
Red barberry	Blood red

grape may require 5 months' stratification. Seeds of hollyleaf mahonia (*M. aquifolium*) require 4 months of warm stratification followed by 4 months of cold stratification.

GERMINATION: Germinate for up to 3 months at standard alternating temperatures.

──────── MAHONIA ALERT ────────

Some researchers include this genus with the true barberries (*Berberis*). Like certain barberries, some mahonias are alternate hosts for the black stem rust of some small grains.

──────────────────────────────

NURSERY PRACTICES: Sow cleaned fresh seeds in fall or stratified seeds in spring. Sow the seeds about ¼ inch deep in soil, then cover them with another ¼ inch of sand. Mulch fall-sown seeds with a light layer of straw until the seedlings emerge.

MALUS

Apple

M. angustifolia (**southern crab apple**), *M. baccata* (**Siberian crab apple**), *M. coronaria* (**American crab apple**), *M. fusca* (**Oregon crab apple**), *M. glabrata* (**Biltmore crab apple**), *M. ioensis* (**prairie crab apple**), *M. pumila* (**common apple**), *M. ×robusta* (**Siberian crab apple**), *M. sargentii* (**Sargent crab apple**), *M. sylvestris* (**European crab apple**)

FAMILY: Rosaceae; rose family

FLOWERING: The perfect flowers appear in the spring with or before the leaves. Flowering time varies from March to June.

SEED COLLECTION: The fruits are fleshy pomes with usually five carpels, each containing two seeds. Seeds have a thin lining of endosperm, except in the common eating apple; its seeds have almost no endosperm. Depending on species, fruits ripen from August to November. Color of ripe fruits varies among species, but in general a fruit's undercolor will turn yellow when the fruit is ripe. Collect very ripe apples from the tree or ground. Seeds from cider mills are often damaged, so do not collect them.

SEED CLEANING AND STORAGE: Accepted methods for seed extraction from the ripe fruits can be cumbersome. First, there is an afterripening period involving partial fermentation of the fruit. Place the fruits in a container maintained at 50 to 64°F (10 to 18°C) for 2 to 4 weeks to soften, then cover softened fruits with water and mash. Avoid high temperatures and excessive fermentation. Let the seeds settle out while the pulp is floated over the top with running water. You can also extract the seeds by putting the fruits through a blender, if you have only a few to do. Even easier, pick out the seeds.

Apple seeds are orthodox and can be stored for over 2 years at 41°F (5°C), for 35 to 40 years at 23°F (-5°C), or for 100 years at 0°F (-18°C). Two-year storage is probably satisfactory for most of us.

SEED TREATMENT: Fresh seeds don't need treatment if sown outdoors in fall. If there is a need for overwintering, air-dry the seeds at room temperature for 3 months, then soak them in water for several days, and place them in the refrigerator to stratify in a 50:50 sand-to-peat mixture for an additional 3 to 4 months to break embryo dormancy. Apple seeds may show vivipary and germinate in cold storage, resulting in difficult sowing.

Here's an alternative method of pregermination treatment: About 2 months before sowing, remove air-dried seeds from storage, mix them with about one-third of their weight with crushed charcoal, place the mixture in a cloth or nylon bag (panty hose works well), and plunge the bag into 165°F (74°C) water, moving it around for 10 seconds. Remove the bag with its seeds, lay it on a screen to cool for about half an hour, and repeat the dip. After removing it the second time, plunge it into cold water, then move it into storage at about 36°F (2°C). Keep the bag wet but aerated. Water the bag every day if needed, but also turn and shake it daily. Seeds will germinate in the bag in 6 to 7 weeks and then should be picked out and sown.

GERMINATION: After stratification, allow 2 to 3 months for germination at fluctuating temperatures of 68°F (20°C) night and 86°F (30°C) day. Seeds of American crab (*M. coronaria*) germinate better when held at a constant temperature of 50°F (10°C).

——————— APPLE ALERT ———————

There are more than two dozen species of edible and ornamental apples in the genus *Malus*. Some of the crab apples come true to seed, but most crabs and all eating apples do not. If you save *Malus* seeds and grow out some seedlings, the trees that result may or may not produce tasty fruits, but you will be able to enjoy their beautiful blossoms.

NURSERY PRACTICES: Untreated seeds sown in late fall and stratified seeds sown in the spring produce vigorous seedlings. A thin sawdust mulch aids seedling emergence on soils that crust. Apple seeds are among the first to germinate in the spring, often while soil temperatures are lower than 40°F (4°C), and the seedlings are generally hardy to spring frosts. By the end of the growing season, most of the seedling stems should be pencil-thick, about 15 inches high, and ready for outplanting as 1+0 stock or for grafting.

METASEQUOIA
Dawn redwood

M. glyptostroboides (**dawn redwood**)

FAMILY: Cupressaceae; cypress family

FLOWERING: This monoecious plant has imperfect flowers. Pollination is in March before needles emerge.

SEED COLLECTION: Mature female cones are light brown and should be collected just before they begin to open late in the year.

SEED CLEANING AND STORAGE: Dry the cones spread out on a screen for about 2 weeks, then tumble or flail them lightly to separate the seeds, taking care not to damage the fragile seed coats. The orthodox seeds will store well for several years.

SEED TREATMENT AND GERMINATION: No pretreatments are necessary.

NURSERY PRACTICES: Sow the seeds in good soil, and maintain relatively high humidity in the seedbed. Germination occurs about a week after sowing.

——————— A FOSSIL TREE ———————

The dawn redwood was known only from its fossil record before a living specimen was discovered in China in 1946. It's the only species in its genus.

OXYDENDRUM
Sourwood

O. arboreum (**sourwood**)

FAMILY: Ericaceae; heath family

FLOWERING: The urn-shaped flowers are perfect and open in midsummer.

SEED COLLECTION: The fruits are five-chambered, dehiscent capsules borne in clusters. The fruits and seeds ripen in September and October, at which time they should be picked from the trees.

SEED CLEANING AND STORAGE: Dry the fruits, then rub them on a screen to separate the seeds. The seeds are orthodox and can be stored for several years.

GERMINATION: The fresh, tiny seeds will germinate readily but require light for germination. Sow them with light provided for up to 4 hours during the high-temperature period of an alternating-temperature regimen.

NURSERY PRACTICES: Sprinkle the seeds on the soil surface, then cover them with sifted milled sphagnum and vermiculite (in a 1:1 ratio). Provide them with continuous light and moist conditions. The seedlings grow rapidly and can be transplanted to peat pots at the three-leaf stage. Use an acidic, highly organic medium in the pots.

PERSEA

Avocado

P. americana (**avocado**)

FAMILY: Lauraceae; laurel family

FLOWERING: Flowering in avocado is unusual. The flowers are partially self-pollinating, but there is difficulty because of dichogamy. Cross-pollinated flowers produce higher-quality seeds. A-type flowers open as females in the morning of the first day and close in the afternoon. On the second day, they open as male flowers in the afternoon. B-type flowers open as female on the afternoon of the first day, close in late afternoon, then open as male flowers the next morning.

SEED COLLECTION: Collect the fruit, a large berry containing a single seed, when it is fully ripe by picking it from the tree. Don't use shed fruits.

SEED CLEANING AND STORAGE: Cut open a fruit, extract the very large seed, and surface-dry it. Seeds can be stored for 6 to 8 months if placed in dry peat moss and held at 41°F (5°C) with about 90 percent relative humidity.

SEED TREATMENT: Reduce the incidence of pathogens by dipping the seeds in hot water at 120 to 125°F (49 to 52°C) for 30 minutes before planting.

AVOCADO SEED ROOTING

GERMINATION: The seeds will germinate readily if planted in soil immediately, or they can be germinated partially submerged in water. Germination may take 4 to 6 weeks.

NURSERY PRACTICES: Plant the young seedlings in fertile soil where they are to be permanently located, or plant them in a nursery bed, then transplant them to a permanent location in a year. Avocado has a long period of juvenility, and seedlings may take 5 to 6 years to produce their first crop.

PHELLODENDRON

Corktree

P. amurense (**Amur corktree**), *P. chinense* (**Chinese corktree**), *P. japonicum* (**Japanese corktree**)

FAMILY: Rutaceae; rue family

FLOWERING: Corktree plants are dioecious. The yellow flowers appear in May and June and are held in terminal panicles.

SEED COLLECTION: The fruits are persistent black drupes. Hand-pick them when they have turned fully black, after the leaves have fallen.

SEED CLEANING AND STORAGE: Macerate to clean. Dry the seeds, and then store them sealed in containers at 41°F (5°C).

GERMINATION: No pretreatment is needed for seeds of Amur corktree (*P. amurense*) and Japanese corktree (*P. japonicum*). Seeds of Chinese corktree (*P. chinense*) should be stratified for 3 months. Germination of fresh seeds is about 17 percent and of stratified seeds, about 93 percent.

NURSERY PRACTICES: Sow seed in fall or spring outdoors.

SEED COLLECTION: One group of willows disperses seeds in late spring and early summer. This group includes feltleaf willow (*S. alaxensis*), Bebb willow (*S. bebbiana*), pussy willow (*S. discolor*), cordate willow (*S. eriocephala*), black willow (*S. nigra*), meadow willow (*S. petiolaris*), and Scouler willow (*S. scouleriana*). Another group, which includes Arctic willow (*S. arctica*) and white willow (*S. glauca*), disperses seeds in fall after leaves drop. Collect the catkins as soon as seeds begin to disperse and when they (catkins) turn from green to yellow-brown and the capsules just begin to open.

SEED CLEANING AND STORAGE: Place loosely packed catkins in a paper bag and allow them to dry further. Do not expose them to direct sunlight. Tumble the dried catkins and rub the seeds in your hands to remove the cotton; then winnow carefully. Immediately after cleaning, store the seeds in sealed containers at ‾40 to 23°F (‾40 to ‾5°C) for long storage or at 34 to 41°F (1 to 5°C) for storage up to 6 months.

——— SEEDS VS. STUMP SPROUTS ———

Salix is a huge genus, with about 400 species of mostly Northern Hemisphere trees and shrubs. Plants in this genus reproduce both by seed and vegetatively by stump sprouts. Saving and sowing *Salix* seeds is an exciting and fun project, but if you want results fast, train a stump sprout into a tree instead.

GERMINATION: Willow seeds are tiny and the seed coats are transparent; the green cotyledons are easily seen before germination. Fresh seeds are viable only for a couple of days. Seeds of summer-dispersed species (most willows) germinate within a day of shedding at temperatures from 41 to 86°F (5 to 30°C). Provide them with light for 8 hours each day. Fall-dispersed seeds exhibit shallow dormancy; stratify them for 1 month, then germinate at the temperature range given above.

NURSERY PRACTICES: Sprinkle the seeds on the soil surface and press them in lightly with a board; then shade the bed to conserve moisture and provide protection from the sun. Outplant as 1+0 stock.

SHEPHERDIA

Buffaloberry

S. argentea (**silver buffaloberry**), *S. canadensis* (**russet buffaloberry**)

FAMILY: Elaeagnaceae; oleaster family

FLOWERING: The plants are dioecious and bear flowers in spring.

SEED COLLECTION: Fruit is drupelike and ripens in midsummer; the seeds are dispersed in fall.

SEED CLEANING AND STORAGE: Hand-pick the fruits from the trees, then macerate them in a blender. The orthodox seeds will keep for up to 5 years. Do not allow seeds to dry out during handling.

SEED TREATMENT: Seeds have both an impermeable seed coat and embryo dormancy. Mechanically scarify the seeds or soak them in acid for 20 minutes (see page 59), then stratify them for 3 months.

GERMINATION: Pretreated seeds should germinate in about 2 months.

NURSERY PRACTICES: Sow seeds at a density of about 30 per foot of row and cover with ¼ inch of soil. Transplant the seedlings to their permanent spot when they are 2 years old.

SORBUS

Mountain ash

S. alnifolia (**Korean mountain ash**), *S. americana* (**American mountain ash**), *S. aucuparia* (**European mountain ash**), *S. cashmiriana* (**Kashmir mountain ash**), *S. decora* (**showy mountain ash**), *S. intermedia* (**Swedish mountain ash**)

SORBUS FRUITS

FAMILY: Rosaceae; rose family

FLOWERING: Perfect flowers, sometimes offensive smelling, open in spring in clusters. Mountain ashes outcross easily, so your seedlings may have characteristics far different from those of the parent. Flowers are insect-pollinated.

SEED COLLECTION: The fruit is a round or oblong, two- to five-celled pome that may be orange, red, scarlet, or pink when ripe, depending on variety and species. Harvest the ripe fruits from the trees from August to October before the birds do.

SEED CLEANING AND STORAGE: Macerate the fruits in a blender, then spread the clean seeds to dry. The seeds are orthodox and can be stored for from 2 to several years.

SEED TREATMENT: Seeds of most species have dormant embryos. Stratify the seeds for 6 months. If your seed lot is particularly difficult to germinate, warm-stratify at 77°F (25°C) for 2 weeks, then stratify for 4 months before germination. Stratified seeds exposed to a high temperature will enter secondary dormancy.

GERMINATION: Germinate seeds at the standard temperatures.

NURSERY PRACTICES: Fall-sow fresh seeds or spring-sow stratified seeds, covering them with about ¹⁄₁₆ inch of soil. They may take 2 years to germinate. The hardy seedlings are outplanted as 1+1 or 2+0 stock.

SPIRAEA

Spirea

S. alba (**meadowsweet**), *S. betulifolia* (**birchleaf spirea**), *S. douglasii* (**Douglas spirea**), *S. japonica* (**Japanese spirea**), *S. stevenii* (**Beauverd spirea**), *S. virginiana* (**Virginia spirea**)

FAMILY: Rosaceae; rose family

FLOWERING: Flowering is highly dependent on species, photoperiod, and microclimate, but in general extends from May to July. The very small, perfect flowers usually form in clusters.

SEED COLLECTION: The fruits are dehiscent follicles that, when they become straw-colored, begin to shed their extremely small seeds. Hand-pick the fruit before this happens.

SEED CLEANING AND STORAGE: Spread the fruits to dry, then gently flail them to separate the seeds. Winnow the seeds to clean them of chaff. The seeds are probably orthodox and could be stored sealed in containers at 41°F (5°C).

GERMINATION: Fresh seeds germinate readily, and germination is greater than that of stored seeds. If your seeds have been stored for a while, you will find they might show better germination if stratified at 35°F (2°C) for 1 month before sowing. Germinate at standard temperatures.

S: nuts, fruits, trees...

NURSERY PRACTICES: Sow fresh seeds in fall. Seeds of some species, such as birchleaf spirea (*S. betulifolia*), will germinate beneath the snowmelt at temperatures of 32 to 35°F (0 to 2°C), while those of other species, such as Beauverd spirea (*S. stevenii*), will germinate only at about 77°F (25°C).

——————— SPIREA ALERT ———————

Some spireas, in particular *S. japonica* (syn. *S. ×bumalda*) have become invasive in parts of the eastern United States.

SYMPHORICARPOS

Snowberry

S. albus (**common snowberry**), *S. ×chenaultii* (**chenault snowberry**), *S. orbiculatus* (**Indian-currant coralberry**)

FAMILY: Caprifoliaceae; honeysuckle family

FLOWERING: Perfect flowers are borne on long spikes that open in June.

SEED COLLECTION: The fruits are white, berry-like drupes that ripen from September through November.

SEED CLEANING AND STORAGE: Macerate the fruits in a blender, then dry the seeds on a screen.

SEED TREATMENT: Warm stratification at 77°F (25°C) for 3 to 4 months followed by stratification for 4½ to 6½ months will improve germination.

GERMINATION: Germinate in alternating temperatures, with 8 hours of light per day supplied during the high-temperature period.

NURSERY PRACTICES: Fall-sow the seeds right after harvest, or sow stratified seeds in spring in fertile, well-drained soil. Provide a winter mulch for fall-sown seeds, and keep the soil surface moist until germination is complete.

TAXUS

Yew

T. baccata (**English yew**), *T. brevifolia* (**Pacific yew**), *T. canadensis* (**Canada yew**), *T. chinensis* (**Chinese yew**), *T. cuspidata* (**Japanese yew**), *T. floridana* (**Florida yew**), *T. globosa* (**Honduran yew**), *T. ×media* (**Anglojap yew**), *T. wallichiana* (**Himalayan yew**)

FAMILY: Taxaceae; yew family

FLOWERING: Canada yew (*T. canadensis*) is monoecious, but all other species are dioecious. The small, relatively inconspicuous, imperfect flowers open from March to June, depending on species and location. Those of Florida yew (*T. florida*) bloom from January to March.

SEED COLLECTION: The berrylike ripe fruit is scarlet in color and composed of a fleshy aril bearing a single greenish brown to black seed. Hand-pick the ripe fruits from August to October. As ripening takes place over a long periods, the birds and squirrels may get to the fruits before you do. One trick is to bag some ripening fruits to keep the birds from getting to them.

SEED CLEANING AND STORAGE: Macerate the fruits in a blender, then extract the seeds; then dry and store the seeds. Yew seeds are orthodox and should be stored in sealed containers at 34 to 36°F (1 to 2°C), where they should keep for several years.

SEED TREATMENT: Many protocols have been devised for different species of yew; we can't discuss all of them here. In general, warm-stratify the seeds at 60 to 65°F (16 to 18°C) for 5 to 7 months, then stratify them for 1 year. Seeds of Canada yew can be warm-stratified for 3 months, then stratified for 3 months; English yew (*T. baccata*) seeds may be warm-stratified for 4 months, then stratified for another 4 months.

GERMINATION: Germinate the seeds of Pacific yew (*T. brevifolia*) at 68°F (20°C) night/86°F (30°C) day and those of English yew (*T. baccata*) and Japanese yew (*T. cuspidata*) at 50°F (10°C) night/60°F (16°C) day. Temperatures for other species have not been determined, but the standard 68°F night/86°F day (20°C/30°C) is a good place to begin. Because of deep dormancy, the seeds may take several years to germinate, so you would be wise to test some of your seeds first by placing them in moist conditions at room temperature for 2 months, then cutting them open. Sound, viable seeds will be white and firm inside; nonviable seeds will be discolored or decayed.

NURSERY PRACTICES: Fall-sow fresh seeds or spring-sow treated seeds, and cover them with ½ to ¾ inch of soil. Apply a light winter mulch, and shade the seedbeds and the seedlings through the first summer after emergence. You might place a wire mesh over the seeds until the seedlings emerge to prevent rabbits and birds from eating the seeds.

--- YEW ALERT ---

The leaves and fruits of some species are toxic to humans. If you collect the fruits, take care to keep them away from small children who might try to eat them.

NORTHERN WHITE CEDAR

THUJA
Cedar

T. koraiensis (**Korean arborvitae**), *T. occidentalis* (**northern white cedar**), *T. orientalis* (now *Platycladus orientalis*; **oriental arborvitae**), *T. plicata* (**western red cedar**), *T. standishii* (**Japanese arborvitae**)

FAMILY: Cupressaceae; cypress family

FLOWERING: The plants are monoecious, the flowers imperfect. Reddish male flowers are borne on branches near the base of a shoot. The very small green- or purple-tinted female flowers develop into solitary cones about ½ inch long, with 3 to 10 pairs of scales. Usually, only the middle two or three pairs are fertile, and each bears two to five chestnut brown, winged seeds.

SEED COLLECTION: Harvest the fruits from the tree as soon as they have turned brown and firm. In northern white cedar (*T. occidentalis*), the time between cone ripening and seed-shed is only about a week. It's somewhat longer in western red cedar (*T. plicata*).

SEED CLEANING AND STORAGE: Spread the cones on a screen to dry for several weeks, being sure to place a sheet beneath the screen to catch any seeds. Dry and winnow the orthodox seeds, then store them in cold storage for as long as several years.

SEED TREATMENT AND GERMINATION: Some seed lots germinate easily, some require stratification. Stratify in moist sand or peat for 2 months, then germinate for a month. Germination protocols for Korean arborvitae (*T. koraiensis*), northern white cedar, and western red cedar are alternating temperatures of 68°F (20°C) day/86°F (30°C) night, with light supplied during the high-temperature period. The other species germinate well at a constant 68°F (20°C) without light. About 51 percent of the seeds of western red cedar will germinate; of northern white cedar, about 46 percent.

T

NURSERY PRACTICES: Fall-sown seeds of western red cedar will germinate the same autumn; those of northern white cedar the following fall. Sow stratified seeds in spring. Cover seeds with ⅛ to ¼ inch of soil and lightly mulch them. Seedbeds of moist rotten wood, decayed leaf litter, or peat moss are excellent, providing slightly acid to close to neutral soil pH. Provide partial shade for the first season, and keep seedlings mulched the first year and through their first winter.

TILIA

Linden

T. americana (**American basswood, American linden**), *T. americana* var. *caroliniana* (**Carolina basswood**), *T. americana* var. *heterophylla* (**white basswood**), *T. cordata* (**littleleaf linden**), *T. ×europea* (**European linden**), *T. mongolica* (**Mongolian linden**), *T. platyphyllos* (**bigleaf linden**), *T. tomentosa* (**silver linden**)

FAMILY: Tiliaceae; linden family

FLOWERING: The flowers are perfect and open in June and July.

SEED COLLECTION: The fruits are gray, nutlike, round capsules that mature in fall but may persist into winter. Each fruit contains a single seed. Harvest the fruits by hand when the pericarps are turning from green to gray-brown and before they become tough and leathery.

SEED CLEANING AND STORAGE: Rub the seeds to de-wing them and to break the pericarp along the suture; then winnow to remove chaff. The very tough seeds of American linden (*T. americana*) might need to be run through a coffee grinder to break the pericarp, or soaked in acid (see page 59). Seeds are orthodox and can be stored for several years.

SEED TREATMENT: Linden seeds have not only tough pericarps but also double dormancy, and seeds of different species and lots may require different

LINDEN CAPSULES

treatments. The simplest method works for littleleaf linden (*T. cordata*): stratification for 8 months. Bigleaf linden (*T. platyphyllos*) requires 3 to 5 months of warm stratification prior to the 8-month stratification. Unfortunately, most species require far more work. In most cases, soak the dried seeds in acid for about 25 minutes (see page 59), rinse them well, then soak in water for 1 to 2 days. Follow this with stratification for about 4 months. Germination will average 34 percent. A note to the determined seed saver: No

LOVELY LINDEN

THERE ARE ABOUT 40 species of linden, also known as basswood and lime, native to the temperate zones of the Northern Hemisphere. Most are characterized by their virtues of fragrance; being sources of excellent honey; and having light, easily worked wood. From a seed saver's perspective, though, these trees do have a downside: hardseededness.

one has yet found a series of pregermination treatments that works well and consistently for all species of lindens. If you come up with one, let us know.

GERMINATION: Germinate the seeds at about 86°F (30°C), with light provided for 12 hours per day.

NURSERY PRACTICES: Fall-sow fresh seeds, which may take several years to germinate, or spring-sow treated seeds.

ULMUS

Elm

U. alata (**winged elm**), *U. americana* (**American elm**), *U. crassifolia* (**cedar elm**), *U. glabra* (**Scots elm**), *U. japonica* (**Japanese elm**), *U. laevis* (**Russian elm**), *U. minor* (**smoothleaf elm**), *U. minor* var. *vulgaris* (**English elm**), *U. parvifolia* (**Chinese elm**), *U. pumila* (**Siberian elm**), *U. rubra* (**slippery elm**), *U. serotina* (**September elm**), *U. thomasii* (**rock elm**)

FAMILY: Ulmaceae; elm family

FLOWERING: Although elm flowers are inconspicuous and perfect, there is a high degree of self-incompatibility, thus requiring cross-pollination. Most species have red anthers, making otherwise dull-looking flowers interesting.

SEED COLLECTION: The fruit is a one-celled samara that ripens a few weeks after bloom. Each fruit contains one nutlet. Collect the fruits from the ground as soon as they are shed.

SEED CLEANING AND STORAGE: De-winging can damage the seeds, so simply dry and rub them gently across a screen; then winnow to remove chaff. The orthodox seeds are dried and stored in sealed containers. Store seeds of winged elm (*U. alata*), American elm (*U. americana*), cedar elm (*U. crassifolia*), Scots elm (*U. glabra*), and Siberian elm (*U. pumila*) at 39°F (4°C), those of Russian elm (*U. laevis*) at 70°F (21°C),

and those of Chinese elm (*U. parvifolia*) at 32°F (0°C) for from 1 to several years.

SEED TREATMENT AND GERMINATION: Seeds that ripen in spring germinate the same season; those that ripen in fall germinate the following spring. Seeds of winged elm, Scots elm, Russian elm, and Siberian elm can be sown in summer, while those of American elm, cedar elm, Chinese elm, slippery elm (*U. rubra*), September elm (*U. serotina*), and rock elm (*U. thomasii*) are sown in fall. Most require no pretreatment; however, some stratification for 2 to 3 months improves germination in most species. Germinate at 68°F (20°C) night/86°F (30°C) day, with light applied for 8 hours during the high-temperature period, for 2 to 3 weeks. Sow seeds about ¼ inch deep.

NURSERY PRACTICES: After sowing, keep the seedbeds moist until the seeds germinate. Most seedlings are ready for outplanting as 1+0 to 2+0 stock.

———————— ELM ALERT ————————

Siberian elm can become invasive.

VIBURNUM

Viburnum

V. acerifolium (**mapleleaf viburnum**), *V. betulifolium* (**birchleaf viburnum**), *V. bracteatum* (**bracted viburnum**), *V. ×burkwoodii* (**burkwood viburnum**), *V. carlesii* (**Korean spice viburnum**), *V. carlesii* var. *bitchiuense* (**bitchiu viburnum**), *V. carlesii* 'Juddii' (**Judd viburnum**), *V. cassinoides* (**withe-rod viburnum**), *V. cylindricum* (**tubeflower viburnum**), *V. dentatum* (**arrowwood**), *V. dilatatum* (**linden viburnum**), *V. erosum* (**beech viburnum**), *V. hupehense* (**Hupeh viburnum**), *V. lantana* (**wayfaringtree**), *V. lantanoides* (**hobblebush**), *V. lentago* (**nannyberry**), *V. lobophyllum, V. nudum* (**smooth withe-rod**), *V. opulus* (**European cranberrybush**), *V. plicatum* var. *tomentosum* (**doublefile viburnum**), *V. prunifolium* (**blackhaw**), *V. pubescens* (**downy viburnum**), *V. rhytidiphylloides* (**leatherleaf viburnum**), *V. rufidulum* (rusty blackhaw), *V. sargentii* (**Sargent viburnum**), *V. setigerum* (**tea viburnum**), *V. seiboldii* (**Siebold viburnum**), *V. trilobum* (**American cranberrybush**), *V. wrightii* (**Wright viburnum**)

FAMILY: Caprifoliaceae; honeysuckle family

FLOWERING: Viburnum flowers are perfect and borne in small clusters. They flower in spring and early summer.

SEED COLLECTION: The fruit is a drupe containing a single stone and usually ripens in late summer or fall. Pick the ripe fruits from the bush. Ripe fruit of most species is blue-black, but the fruits of both types of cranberrybush ripen to bright red.

SEED CLEANING AND STORAGE: Run the fruits through a blender, then dry the seeds before use or storage. The seeds of many viburnums can be dried and then stored in sealed containers or envelopes at 33 to 41°F (1 to 5°C) for up to 2 years.

VIBURNUM FRUITS

SEED TREATMENT: The seeds germinate slowly. Most species have embryo dormancy, and some also have an impermeable seedcoat. If planted outdoors in the nursery bed, the seeds may not germinate until the second spring after sowing. If you want to try to handle the seeds indoors, stratify them as explained below. We remind you that warm-stratification temperatures range from 68 to 86°F (20 to 30°C) and stratification temperature is about 41°F (7°C).

- Smooth withe-rod (*V. nudum*): No pretreatment is necessary.
- Doublefile viburnum (*V. plicatum* var. *tomentosum*): Fall-sow fresh seeds.
- Rusty blackhaw (*V. rufidulum*): Warm-stratify for 6 to 17 months, then stratify for 3 to 4 months.
- Bracted viburnum (*V. bracteatum*) and arrowwood (*V. dentatum*): Warm-stratify for 6 to 18 months, then stratify for 15 to 60 days.
- Birchleaf viburnum (*V. betulifolium*), linden viburnum (*V. dilatatum*), beech viburnum (*V. erosum*), Hupeh viburnum (*V. hupehense*), hobblebush (*V. lantanoides*), nannyberry (*V. lentago*), *V. lobophyllum*, and Wright viburnum (*V. wrightii*): Warm-stratify for 5 to 9 months, then stratify for 2 to 4 months.
- Bitchiu viburnum (*V. carlessi* var. *bitchiuensis*), Burkwood viburnum (*V. ×burkwoodii*), Korean spice viburnum (*V. carlesii*), withe-rod viburnum (*V. cassinoides*), Judd viburnum (*V. carlessi* 'Juddii'), European cranberrybush (*V. opulus*), Sargent viburnum (*V. sargentii*), Siebold viburnum (*V. seiboldii*): Warm-stratify for 2 to 3 months, then stratify for 1 to 2 months.
- Blackhaw (*V. prunifolium*): Warm-stratify for 5 to 7 months, then stratify for 1 to 2 months.
- American cranberrybush (*V. trilobum*), downy viburnum (*V. pubescens*), and tea viburnum (*V. setigerum*): Warm-stratify for 3 to 5 months, then stratify for 2 months.
- Tubeflower viburnum (*V. cylindricum*) and wayfaringtree (*V. lantana*): Stratify for 3 months.

GERMINATION: Germinate all species at about 77°F (25°C), with about 8 hours of light per day, but expect to wait at least 2 to 4 months to see results.

NURSERY PRACTICES: Sow the seeds in fertile soil in rows about 10 inches apart in spring or in mid-summer, just early enough so they will be exposed to about 2 months of warm days prior to winter. Cover the seeds with ½ inch of soil, and mulch them well for the winter.

VITEX

Chaste tree

V. agnus-castus (**lilac chaste tree**), *V. negundo* (**chaste tree**)

FAMILY: Verbenaceae; verbena family

FLOWERING: The perfect flowers are borne in clusters and open in mid- to late summer.

SEED COLLECTION: The fruits are small drupes. Hand-pick entire clusters of fruits when the drupes begin to soften.

SEED CLEANING AND STORAGE: Dry, then tumble or shake the fruit clusters to extract the seeds.

GERMINATION: Seeds require no pretreatment.

NURSERY PRACTICES: Sow fresh seeds in very well-drained soil in full sun. Take care not to disturb the roots too much during transplanting.

VITIS

Grape

V. labrusca (**fox grape**), *V. vinifera* (**vinifera grape**)

FAMILY: Vitaceae; grape family

FLOWERING: The flowers are perfect and borne in short panicles in May or June.

SEED COLLECTION: The fruit clusters usually have fewer than 20 berries, maturing August to October. Mature berries are green to dull black, depending on variety, and contain two to six brownish, angled seeds. Mature seeds have a dark brown seedcoat. Strip ripe berries from the vines or shake them onto sheets. Since the berries of purple varieties develop a purple color long before they're ripe, be sure to collect seeds from fruits that are dead-ripe to almost rotten.

SEED CLEANING AND STORAGE: Extract the seeds in a blender. The seeds are orthodox and store well for at least several years.

GERMINATION: Fox grape (*V. labrusca*) seeds exhibit dormancy that can be overcome by stratifying for 6 months. *Vinifera* seeds should be stratified for about 3 months before planting.

NURSERY PRACTICE: Spring-sow treated seeds, covering them with about ¼ inch of soil. The seedlings can be outplanted as 1+0 stock.

WISTERIA

Wisteria

W. floribunda (**Japanese wisteria**)

FAMILY: Fabaceae; legume family

FLOWERING: The perfect flowers open in mid-spring, just before the leaves appear.

SEED COLLECTION: The fruit is a fuzzy brown pod that ripens in October and persists into winter. Hand-harvest pods when they have turned fully brown and feel dry.

WISTERIA SEEDPODS

SEED CLEANING AND STORAGE: Separate the seeds from the pods, then store dry in a sealed container at a cold temperature.

GERMINATION: Seeds need no pretreatment, but if they are dry, soak them for a day in warm water before sowing.

NURSERY PRACTICES: Sow in moist, well-drained soil and full sun. Don't fertilize excessively because that will stimulate too much vegetative growth.

—————— WISTERIA ALERT ——————

Seedling wisteria plants have a long juvenile period — 10 to 15 years.

YUCCA

Yucca

Y. brevifolia (**Joshua tree**), *Y. elata* (**soaptree yucca**), *Y. glauca* (**great plains yucca**), *Y. schidigera* (**Mojave yucca**)

FAMILY: Agavaceae; century-plant family

FLOWERING: The perfect flowers are borne on terminal panicles from mid-May to mid-July and are pollinated by female moths.

SEED COLLECTION: The fruits are dehiscent capsules, each containing 120 to 150 shiny, flat, ovoid seeds. Hand-pick the fruits before the capsules dehisce, in mid-July to late September.

SEED CLEANING AND STORAGE: Dry the capsules, then pick out the seeds by hand. The seeds are orthodox and can be dried and stored at room temperature for a few years.

SEED TREATMENT AND GERMINATION: Seeds exhibit no dormancy or only very shallow hardseededness. Soak the seeds in water for 24 hours, or scarify them mechanically by nicking the seed coat near the hilum; then germinate at 86°F (30°C) for 3 weeks. Germinate the seeds of Joshua tree (*Y. brevifolia*) at 68 to 77°F (20 to 25°C).

NURSERY PRACTICES: Spring-sow the seeds after a 24-hour water soak. Germination begins after a week or two and may take 1 to 2 years for completion. Outplant the seedlings as 2+0 stock.

APPENDIX

ALCEA (HOLYHOCK) SEEDPODS

A HISTORY OF SEED SAVING
AND SELLING
IN NORTH AMERICA

EARLY SEED-SAVING AND SELLING VENTURES

1719 Evan Davies is among the first to sell "choicest" seeds as a side business in his Boston store.

1732 Samuel Everleigh advertises his sale of "diverse sorts of garden seeds" in the *South Carolina Gazette*.

1733 Justice Dudley of Massachusetts pens directions for raising onion seeds.

1747 First commercial sales of agricultural seeds have taken place. "Colli-flower," "long and smooth parsnips," and "Brown Dutch" cabbage are on the list.

1748 The Rev. Jared Eliot describes how to save clover seeds.

1763 Nathaniel Bird advertises English onion seeds for sale in the *Newport* (R.I.) *Mercury*.

1767 Charles Dunbar advertises seeds of cabbage, peas, beans, Strasburgh onions, and "orange carrots" for sale, along with some flower seeds.

1768 William Davidson of Boston offers seeds of 56 varieties of vegetables and herbs and one flower (carnation).

1770 Garden seeds are sold in New York City.

1772 Benjamin Franklin introduces rhubarb seed into the United States.

1773 Benjamin Franklin introduces kale seed into the United States.

1775 James Loughead offers "Colly-flower" seeds to customers.

FOR MILLENNIA, farmers and gardeners produced their own seeds or bought them from other local farmers who had extra to sell. If those seeds produced a good crop, the customers returned to those farmers to buy more, and in simple terms that's how production of seed evolved as a business venture. Seeds and seed production have played a role in the history of North America from the time of the Native Americans and the early settlements of the Europeans.

The first "European" seeds were brought into the New World by Columbus, and the first seed introduced from Spanish America was tobacco, in the Jamestown (Virginia) colony in 1611. Pilgrims to the Massachusetts colony brought seeds with them from England and Holland; the first extensive list of seeds appearing in a ship's manifest was an order placed by the Massachusetts colony in 1629. Seeds of most of our food and feed crops were thus introduced into the colonies by the end of the seventeenth century. These supplemented seeds of corn, squash, pumpkins, beans, and a very few other native crops that originated in the New World.

These first seed marketers were probably importers or single-crop vegetable growers who eventually merged to offer a greater variety of seeds. Boston was the center of the colonial seed business. Bean, pea, carrot, onion, cauliflower, and cabbage seeds were regularly traded before and during the Revolutionary War, but most of these were imported from Europe, especially England. It's said the tomato was first brought into the United States from Santo Domingo in 1798, but no one would eat it. It was not sold in Philadelphia markets until 1829. The Italian painter Michelo Corne introduced the fruit into Salem, Massachusetts, in 1802, but everyone there refused to eat it, too. Thomas Jefferson grew tomatoes at Monticello and encouraged his friends and neighbors to eat them, mostly to no avail. However, by 1812 tomatoes were regularly served at meals in New Orleans, and tomato seeds were first offered for sale in 1825 by J. B. Russell of Boston and appeared in seed catalogs of the 1830s.

Aggie Fairs and Farmers' Markets

Prior to the Revolutionary War, the colonial farmer had no easy way to communicate with other farmers about the latest in agricultural technology. "Market fairs," based on the medieval European model, were held in the colonies as early as 1686, but these were what today we might call farmers' markets. Their primary purpose was to sell produce, not to disseminate information. Fondness for gardening rapidly increased after the Revolutionary War, but

dissemination of improved varieties of plants and seeds was hampered by the difficulty in spreading the information among farmers and gardeners.

George Washington, Benjamin Franklin, Thomas Jefferson, and others recognized that the United States must gather and disseminate useful European agricultural information and plants to its farmers if it was to become a leading nation. Soon after the war, these leaders began to encourage the formation of agricultural societies. These societies facilitated seed exchanges among their members, but not everyone was allowed to join such societies, the privilege being reserved for the socially elite and wealthy estate owners.

Meanwhile, agricultural fairs had sprung up, one of the first started by George Washington Custis, step-grandson and adopted son of George Washington. Called the Arlington Sheepshearing, it began in 1803 at Custis's mansion in Arlington, Virginia. Other fairs were held in the Washington, D.C., area in 1804 and 1805, and annual agricultural fairs spread to other areas from there. At these fairs, common people could glean knowledge of the latest agricultural inventions and buy and trade new seed introductions from far-off lands or seeds especially prized by their own countrymen.

The Rise of Commercial Seed Companies

John Bartram's garden in Philadelphia was the best known of the gardens in North America in the last quarter of the eighteenth century. Even

Carl Von Linne — the Carolus Linnaeus of taxonomy fame — called Bartram the greatest naturalist in the world. Bartram specialized in growing North American plants, but many of the plants and seeds he raised were brought from Europe by Benjamin Franklin. Bartram grew the new plants and shared them with friends.

Many folks, mostly wealthy, dabbled unofficially in seed collecting and dissemination but did not sell their seeds commercially. Benjamin Franklin introduced Americans to two Scottish crops, rhubarb and kale. Thomas Jefferson, under penalty of death if caught, smuggled out of northern Italy seeds of an upland rice for South Carolina.

David Landreth was the first commercial American vegetable seedsman of note, establishing his business outside Philadelphia in 1784. It was Landreth who sold flower seeds to George Washington for his Mount Vernon gardens. Bernard McMahon (sometimes spelled M'Mahon) also established his seed business of sorts in that area, in 1800, and later became a well-known horticulturist. In the fall of 1805, Grant Thorburn began to sell seeds in New York and built up a substantial seed business. By 1850, 45 seed firms flourished in the United States, most of them in New England, New York, and Pennsylvania.

Seed "catalogs" were offered as early as 1805, but these were simply price lists. In 1822, Grant Thorburn published the first pamphlet-style seed catalog. The first mail-order seed company — B. K. Bliss of Springfield, Massachusetts — opened in 1853 and with it the first

mail-order catalog. With the rapid increase in seed growing after the Civil War, catalogs became prominent and included almanacs, cultural directions, and detailed descriptions of varieties. Most had few illustrations before 1867 but soon thereafter became heavily illustrated. After the 1880s, novelties were given special places in catalogs, starting a tradition that continues today.

Before the Civil War, most seed companies were owned by large landholders whose primary business was producing food — seeds were a sideline. That started to change in 1876, when 18-year-old W. Atlee Burpee founded his seed company in Philadelphia, with seed production as its primary goal. Young Burpee was already a poultry breeder of some note when he received a loan of $1,000 from his mother, with which he expanded his business to include breeding Border collies, hogs, and other animals. He then realized that breeding plants and selling their seeds was far less expensive and more profitable. That love of plant breeding was no doubt inspired or at least encouraged by his California cousin, the noted plant breeder Luther Burbank.

By the 1880s, W. Atlee Burpee's was the fastest growing mail-order company in the world and by 1890 the largest seed company in the world as well. Burpee carried on his vegetable-breeding experiments at the Fordhook farm he purchased near Doylestown, Pennsylvania, in 1888 and then went on to purchase more seed-breeding farms around the United States, especially when World War I caused a serious shortage of European vegetable seeds. It was Burpee's farms in Florida, Mexico, California, and more northern areas that supplied much of the garden seed for America's wartime Liberty Gardens. He "did his bit" to win the war.

The Import-Export Scene

By 1870, the United States could have become essentially self-supplying in seeds of all kinds, with the exception of a few specialty seeds still imported from Europe. Yet it was less expensive to grow seeds in Europe than in the United States. In the year ending June 1855, the United States had exported about $13,570 in seeds, but by 1863 that export figure had increased to $2,185,706, indicating that the Civil War had little effect on the seed-export business. In the year ending in June 1885, the United States exported a little over $2 million worth of seeds but in the same year imported about $4.5 million worth of seeds. There were somewhat over 2,000 acres in vegetable-seed production in the United States in 1867, a figure that had jumped to about 7,000 acres by 1878. But American industry could not yet supply the demand for garden seeds at a reasonable price, and in 1880 about half of America's vegetable seeds were still imported from Europe.

The American industry continued to grow, however, and by 1890 the United States had dedicated 170,000 acres to the production of vegetable and flower seeds. This blossoming of the industry was due in part to an increasing demand for garden seeds to satisfy the increasing population and in part to the expansion of a free-seeds program instituted by Congress.

Government Seed Program

Foreign seeds became more widely available in 1819, when the United States secretary of the treasury, William H. Crawford, began a formal program requesting that U.S. consuls and naval officers stationed in foreign lands collect and send to the United States seeds of improved varieties for American farmers. Congress appropriated no money for the work, but the Agricultural Society of South Carolina in 1823 gave $200 to naval officers to pay the costs of correspondence. President James Monroe installed a skilled botanist, Dr. William Baldwin, as ship's surgeon on a naval expedition to Rio de Janeiro, Montevideo, and Buenos Aires to collect seeds and plant material. The administration of President John Quincy Adams further encouraged the idea.

Continued poor farming techniques, which exhausted the land, and the Depression of 1837–38 made necessary the importation of several million dollars' worth of food and forced Congress to take heed of the need to promote domestic agriculture. In 1839, Henry Ellsworth, U.S. commissioner of patents, received $1,000 in congressional funds to collect and distribute cuttings and agricultural seeds, and established through the U.S. Postal Service what later became known as the Congressional Free Seed Distribution program. (Part of that initial allotment was for Ellsworth's office to gather agricultural data for the 1840 census.) By 1849, the U.S. Patent Office was sending out 66,000

packages of free seeds annually. By 1850, many United States farm papers and periodicals were advertising for sale packets of seeds of new varieties. The Patent Office began to make free seeds available to the public through members of Congress.

Congress judged this new program so successful, and no doubt politically expedient, that it increased the budget of the Patent Office from $4,500 in 1850 to $75,000 in 1856. In the spring of 1861, Congress distributed almost 2.5 million packages of seeds, mostly of vegetables (154 varieties) and flowers (230 varieties). Not only did this program successfully introduce improved varieties from Europe and elsewhere, it also surely promoted goodwill toward Congress and no doubt bought great numbers of votes in the next elections, a fact not lost on Commissioner of Patents David P. Holloway, who lamented that "we have no Secretary of Agriculture but only a subordinate bureau of the Patent Office designed mainly to furnish members of Congress cuttings and garden seeds to distribute among favored constituents, and tea plants, roses and other exotics for their greenhouses and pleasure grounds."

By this time, the growing population of city dwellers increased the demand for seeds, and a few companies began to specialize in selling vegetable seeds. When the Department of Agriculture was established in 1862, one of its primary missions was the propagation and distribution of new varieties of seeds formerly handled by the Patent Office. The free-seed program faded somewhat during the Civil War, but at the war's end, Congress demanded the program be revitalized. In 1866, nearly one million packages were distributed, and over 400,000 of these went to members of Congress. The rest were distributed to agricultural and horticultural societies and to "correspondents." By 1875, the distribution program (almost exclusively flower and vegetable seeds) accounted for a third of the Department of Agriculture's budget. In 1893–94, Congress distributed 9.5 million packages. The United States was then in the throes of a recession, but that didn't hamper the free-seed (read "vote-buying") program.

Seed companies struggled to turn a profit in the face of the overwhelming government competition. Secretary of Agriculture J. Sterling Morton, in his report of 1894, strongly urged Congress to curtail the "gratuitous, promiscuous" free-seed program and support the emerging seed companies. In that report, Morton indicated that about 1.8 million citizens received free seeds but only 940 acknowledged that they had. In 1895 a record number of free government seed packages were distributed, no doubt with the 1896 elections in mind. Commenting on the failure of Congress to heed his suggestions, Secretary Morton, in his 1896 report, told his readership that if all the seeds sent out in 1895 through "gratuitous" congressional distribution were planted in a row one rod wide, it would extend for 36,817 miles, or circle the globe one and a half times.

Still, in 1897 the free-seed program hit an all-time high (to that point) of over 22 million packages distributed. With each package containing five packets of different varieties, 1.1 billion packets of seeds were distributed free by friendly members of Congress.

By 1911, Secretary of Agriculture Wilson reported that 60 million packages had been distributed that spring. About a third of those seeds came from government surplus stores and the rest from a U.S. Department of Agriculture contract on competitive bids. Wilson's report for 1912 indicates that the distribution of vegetable and flower seeds for 1912–13 amounted to around 600 tons, in 61 million packages. By then, the U.S. Department of Agriculture was also increasing distribution of field seed for dryland farming in the face of the homestead boom in the West. In his 1913 report, Wilson mentions only distribution of field seed; there is no mention of vegetable- and flower-seed distribution in any subsequent USDA report. The program faded and finally ended in 1923.

The Impact of World War I

World War I revolutionized the U.S. seed industry. Until that time, 75 percent of vegetable seeds were imported from Europe, but by the war's end, the United States had become a net exporter of vegetable seeds. The inability at that time of the European countries to produce enough of their own seeds, let alone enough to satisfy U.S. imports, and the huge increase in the number of city dwellers, who had never had a garden but who now wanted to contribute to the war effort by planting a Liberty Garden, created a huge demand for seeds.

In 1916, European countries began placing large orders for seeds from the United States. U.S. growers sought out new areas for seed production and increased the acreage in older areas. They also planted many seed crops not previously grown in this country. All available acreage was put into production of food crops and their seeds. So great was the need and so high the prices that production spread to marginal lands. When prices fell after the war, those lands cost more to plant than the returns could cover, and both the falling prices and marginal production contributed to the agricultural depression of 1920–21.

Seedsmen reduced the number of varieties they grew, eliminated production of novelties, and produced mainly the old standard varieties that were tried-and-true producers, all in the spirit of conservation popular at the time. Because vegetable prices rose, many seed growers produced vegetables rather than seeds. That and other factors caused an increase in vegetable-seed prices. Between 1917 and 1918, the price of most seeds rose about 60 percent.

Prices settled down after the war, production of grass seeds increased, and the United States emerged as a major seed-producing country and a net exporter of fine vegetable seeds.

Land-grant universities began serious development of hybrid crops, probably stimulated by the tremendous increase in demand for food. These programs were in direct competition with those of seed companies, and eventually, most of the university programs were phased out, a move perhaps encouraged by those companies' political lobbies. But out of these university breeding programs came the 'Rutgers' and 'Early Summer Sunrise' tomatoes, 'Ponca' and 'Waltham' butternut squashes, and 'Rhode Island Red' and 'New Hampshire Midget' watermelons, to name only a few introductions.

After the war, only a very few seeds were imported, among them cauliflower seeds from Denmark and spinach seeds from Holland. Immediately following World War I, the agricultural depression of 1920–21 saw a slump in seed sales, but in general the market in the 1920s was strong. Through the Great Depression, with its Relief Gardens, followed by World War II and its Victory Gardens, seeds continued to be in great demand. In fact, during World War II, it was considered downright unpatriotic, even traitorous, not to plant a garden. The seed business boomed, with packets of standard, open-pollinated varieties selling for 10 cents each, though standard practice for most experienced gardeners was to save their own seeds for the next crop.

By the end of World War II, all vegetable seeds planted in the United States were grown in the United States. This tremendous production was spurred on in part by the more than 20 million Victory Gardens that produced more than 40 percent of American food by 1945.

Seed Production as Megabusiness

After World War II, there was a great upswing in production of hybrid seeds following the success of hybrid corn in the 1930s and 1940s. This in itself discouraged the home saving of seeds and eventually helped lead to its near-total demise. By 1998 U.S. vegetable-seed sales totaled $93,144,000; by 2007, $99,694,490.

Hybrid-seed-production technology spread from the United States to Japan, which began to export hybrid tomato and eggplant seeds to the United States in the 1950s. Hybrid-seed technology spread through the East, and by the mid-1970s, Taiwan was producing much of the world's hybrid tomato seed, as well as watermelon, melon, sweet pepper, eggplant, and cucumber seeds. Most of these hybrids were produced by hand-pollination, and as labor prices rose, the technology was exported again, to India and mainland China, the Philippines, Thailand, Chile, and even Baja California.

The raising of vegetable seeds has become a very big business indeed, both nationally and internationally.

GLOSSARY

A

abortive: Imperfectly developed.

abscission: Natural separation of leaves, flowers, and fruit usually associated with deterioration of a specialized layer of cells called the *separation layer* or *separation zone*.

achene: A small, dry, indehiscent, one-seeded fruit that has its seed attached to the ovary at only one point. Many species in the aster family produce achenes, and a sunflower "seed" is an achene.

afterripening: Biochemical and physical processes occurring after harvest that are often necessary for resumption of embryo growth and seed germination.

aggregate: Describing fruit formed from a group of ovaries with separate pistils in a single flower, as in *Rubus. Syncarp* is sometimes used synonymously for "aggregate fruit."

allele: One member of a pair of genes found at a specific position (locus) on a chromosome that controls the same trait.

androecious: Bearing staminate flowers.

andromonoecious: Bearing staminate and perfect flowers.

anemophily: Pollination by wind.

angiosperm: Member of the group of vascular flowering plants that mature their seeds inside a fruit.

anther: A sac producing and containing pollen.

anthesis: Shedding of pollen.

apetalous: Having no petals.

aril: Covering of appendage in seeds of some species that develops after fertilization as an outgrowth of the ovule, as in *Euonymus* and *Taxus*.

asexual reproduction: Reproduction without fertilization, as by stem cuttings, air-layering, and grafting. Also called "vegetative reproduction."

B

berry: A fleshy, indehiscent fruit developed from a single pistil, as in *Asparagus, Citrullus* (watermelon) and *Convallaria*.

bisexual: Having both sexes, as in a flower that has both male and female parts.

blocking: In hardening transplants in flats, cutting the medium into squares with one plant per square. This severs the roots, causing root branching and an increase in root surface area.

C

calyx: The inclusive term for separate or fused sepals.

carpel: A part of the ovary housing the ovules and, later, the seeds.

chiropterphily: Pollination by bats.

chlorosis: Yellowing of plant tissue due to a lack of light or a lack of certain mineral nutrients.

complete flower: A flower that has all parts present: that is, sepals, petals, stamens, and pistils.

cone: The dry, woody strobilus of a gymnosperm; a seed-bearing structure having a conical shape, as in *Liriodendron*.

corolla: Set of separate or fused petals that surrounds the carpels.

corymb: A flat-topped floral cluster as in *Rhododendron*.

cotyledons: Modified leaves formed by the embryo inside the seed. They may contain stored food for the initial growth of the seedling, as in *Quercus*, or they may become functional leaves after germination, as in *Pinus*.

cross-pollination: Exchange of pollen among different flowers from the same or different plants.

cuticle: The waxy coating on the surfaces of leaves, young stems, and some other plants.

cyme: Flower cluster having main and secondary axes, each terminating in a single flower.

D

damping-off: A rotting of the stem of very young seedlings at the soil line caused by some soil-borne fungi.

deciduous: Describing separation at the end of an event such as the growing season, as leaves, or at the end of bloom, such as flower petals.

dehiscence: Splitting open at maturity to discharge contents, as a capsule discharging seeds.

determinate flowering: A reproductive habit in which the terminal vegetative meristem differentiates into a reproductive meristem, essentially ending shoot growth and promoting flowering and fruit ripening.

dichogamy: Maturation of male and female organs on the same plant at different times, preventing self-pollination. If the staminate flowers ripen first, the plant is *protandrous*. If the pistillate flowers ripen first, the plant is *protogynous*.

dioecious: Having staminate flowers and pistillate flowers borne on different plants, as in *Acer*, *Asparagus*, *Ilex*, and *Spinacia* (spinach).

diploid: Having two complete sets of chromosomes per cell.

dormancy: A physiological state in which a plant or plant part is capable of growing but does not. *Primary dormancy* is a trait inherent in the genetics of the seed; *secondary dormancy* is usually induced by environmental or handling conditions after harvest.

double dormancy: Dormancy as a result of two or more primary factors. For example, some species have both embryo and seed-coat dormancy.

drupe: Fleshy, usually one-seeded, indehiscent fruit with seed enclosed in a hard, bony endocarp, as in *Prunus*. The flesh of the fruit is the mesocarp, the skin the pericarp, and the stone the endocarp.

E

embryo dormancy: Dormancy maintained by conditions within an embryo.

emergent: A very young seedling newly emerged above the soil.

endocarp: The hard, bony part (the stone) of the fruit of *Prunus* and some other genera.

endosperm: Nutritive tissue in a seed that maintains a food supply for the embryo.

entomophily: Pollination by insects.

epicotyl: Portion of the axis of an embryo or seedling between the cotyledons and the primary leaves.

epicotyl dormancy: A condition in which the radicle emerges and develops in the fall but the epicotyl remains dormant or only slightly emerges and becomes dormant again, then develops normally in the spring, as in *Aesculus* and *Quercus*.

evaporation pan: A widemouthed pan filled with water to a certain line and placed into a plant's environment. The amount of water that evaporates from the pan approximates the amount of water transpired by the plant.

exocarp: Outermost layer of pericarp; the skin on fleshy fruits like those of *Malus* (apple) and *Prunus*. Sometimes the synonym *epicarp* is used.

F

F₁: In a hybrid, the first filial generation. This is the first-generation offspring from a hybrid cross.

F₂: In a hybrid, the second filial generation produced by crossing or selfing F₁ individuals.

fertilization: Discharge of the sperm into the ovule followed by union of the male and female gametes. *Double fertilization* is said to occur when a second sperm fertilizes the two polar nuclei, forming the endosperm tissue.

filament: The tube that supports the anther.

flail: An ancient farm implement used to separate seeds from their plants. It was composed of two lengths of pole, each about an inch and a half in diameter.

follicle: Dry, dehiscent fruit, opening along one suture line.

frost-heaving: The condition wherein a seedling or plant is forced out of the ground by the expansion forces of frozen water in the soil.

full seeds: Those filled with tissue having a normal appearance.

G

gene: The smallest transmissible unit of genetic material consistently associated with a single primary genetic effect.

genetic diversity: The genetic variability within a population or a species.

genotype: An individual's hereditary constitution. It interacts with the environment to produce the individual's phenotype. *See* phenotype.

germinant: A seed that has just germinated.

germination: Resumption of growth in an embryo resulting in its emergence from the seed and the development of a viable seedling.

germinative capacity (germination percentage): Proportion of seeds, usually expressed as a percentage, that germinate normally during a period of time when germination is expected to be practically complete.

gibberellic acid (gibberellin): A class of plant-growth regulators or plant hormones that cause rapid cell elongation and can promote germination of some seeds.

glabrous: Smooth.

glaucous: Having a whitish or waxy coating that tends to rub off.

globose: Globular.

gymnosperm: A member of the plant subdivision having seeds not enclosed in an ovary (*naked seeds*) and borne on the scales of a cone, on the megasporophylls of other types of strobili, or singly with arils, as in *Taxus*.

gynoecious: Bearing female flowers only.

H

haploid: Having one complete set of chromosomes per cell.

hardening (hardening off): The process of adapting a greenhouse- or house-grown plant to its future outdoor environment by gradually exposing it to outdoor conditions over a period of weeks or months.

head: Densely packed cluster of stalkless flowers, as in *Cornus*.

hermaphrodite: Having both sexes; a term often used in place of *bisexual* and *perfect*.

heterozygous: Having one or more sets of unlike alleles: that is, having both the dominant and the recessive gene. A heterozygote

does not generally breed true. *See* homozygous.

hilum: Scar on a seed marking the point of attachment to the ovary in angiosperms or to the megasporophyll in gymnosperms.

hip: The ripened "false fruit" of *Rosa*, consisting of a fleshy receptacle that contains many achenes (one-seeded fruits).

homozygous: Having one or more sets of like alleles: that is, they are both dominant (AA) or both recessive (aa). A homozygote comes true when mated with the same genotype. Compare with *heterozygous*.

husk: Outside envelope of a fruit, as the involucre of *Carya*.

hybrid: Properly, the progeny of two inbred lines.

hydrophily: Pollination facilitated by water.

hypocotyl: That part of the axis between the cotyledons and the radicle or roots.

I

imperfect flower (unisexual): A flower that is missing the organs of one sex. Staminate and pistillate flowers are both imperfect by definition, one being male, the other female.

imbibition: The absorption of water by a seed prior to germination. Imbibition is also necessary during stratification and warm stratification.

inbred: Said of a plant that receives only gametes from self-pollination or from pollination by plants of the same variety.

inbreeding depression: A loss of vigor caused by repeated inbreeding.

incipient wilt: Initial stages of wilt that can be corrected by immediate watering.

incomplete flower: A flower that has one or more of its parts missing. If one or both of the sexual parts are missing, the flower becomes *imperfect* as well as *incomplete*.

indehiscent: Refers to dry fruits that normally do not split open at maturity.

indeterminate flowering: Flowers that open progressively from the base of an inflorescence. Compare with *determinate flowering*.

inflorescence: Floral axis with its appendages; a flower cluster.

internal dormancy: *See* embryo dormancy.

involucre: One or more whorls of bracts situated below and close to a flower cluster. It sometimes encloses the carpels, as in *Carya* and *Fagus*.

isolation: The act of separating varieties of plants by distance or by time of planting to reduce the chances of unwanted cross-pollination.

J

juvenile: Refers to a physiological phase of plant development during which a plant will not form flower buds and may display some characteristics that are not apparent in the adult phase, such as spines, varying leaf shapes, and a semi-evergreen habit.

L

landrace: Seed-propagated plants that have been selected over time to best adapt to a certain locale. Heirloom varieties are landraces.

legume (pod): Dry, one-celled fruit that usually dehisces along two suture lines, as in *Gleditsia* and *Lupinus*.

line: A group of seed-propagated plants of the same variety.

locule: A compartment of a fruit.

M

maceration: A process for removing the soft, pulpy tissue from fleshy fruits, usually by blending or grinding.

mericarp: One of the two halves of a schizocarp fruit, as in *Acer* (maple).

mesocarp: Middle layer of the pericarp; the pulp or flesh of drupes and berries.

micropyle: Minute opening in the integument of an ovule through which the pollen tube normally passes. It is usually closed in the mature seed to form a scar called the *hilum*.

monoecious: Having functional staminate and pistillate flowers on the same plant.

multiple fruit: Fruit formed from several flowers whose fused ovaries are inserted on a common receptacle, as in *Morus* (mulberry) and *Platanus* (sycamore).

N

nut: Dry, indehiscent, one-seeded fruit with a woody or leathery pericarp, as in *Quercus*, or a fruit partially or wholly encased in an involucre (husk), as in *Corylus*.

nutlet: Small nut, as in *Betula*, *Fagus*, and many ornamental flower species.

O

off-type: A plant or flower that is dissimilar to the normal species or variety type. For example, a white-rooted carrot among a row of orange-rooted carrots is an off-type.

open pollination: Pollination involving a mixture of related and unrelated pollen.

ornithophily: Pollination by birds.

orthodox: A term referring to storage behavior of some seeds. Most garden seeds are orthodox and are generally stored best in cool, dry, dark conditions.

outcrossing: Mating of unrelated individuals.

outplant: To plant a seedling to the field from the nursery bed.

ovary: In angiosperms, the basal portion of a pistil that bears the ovules.

ovule: Egg-containing structure in seed plants that develops into a seed after fertilization.

P

panicle: A branched flower cluster, as in *Aesculus* and *Fraxinus*.

papilionaceous: Resembling a butterfly, as in the flowers of many legumes.

pappus: A tuft of delicate fibers or bristles that form a feathery appendage of an achene (one-seeded fruit), as in *Asclepias* and some other members of the Asteraceae.

parthenocarpy: Development of fruit without fertilization.

parthenogenesis: Reproduction from an unfertilized ovule. The embryo may be haploid or diploid.

pedicel: Stalk of a single flower within a flower cluster.

peduncle: Stalk that bears a single flower or a flower cluster.

peloric: Refers to a flower type wherein the upper flower on an inflorescence is different in some way from the lower flowers.

perfect flower: *See* bisexual.

pericarp: The wall of a ripened ovary that may be homogeneous or composed of three distinct layers: exocarp, mesocarp, and endocarp; the fruit wall. In a peach, for example, the skin is the exocarp, the flesh is the mesocarp, and the stone is the endocarp.

phenology: Study of relations between climatic changes and biological phenomena such as dormancy, growth, flowering, and fruiting.

phenotype: The observed state, description, or degree of expression of a character; the product of the interaction of the genes of an organism with the environment. For example, a particular maple tree may have the genetic capacity for red fall leaf color controlled by genes (genotype), but that color will be much more intense if fall days are bright and sunny and nights cool but above freezing.

phytochrome: A photo-reversible plant pigment that operates between a red-light form and a far-red-light form.

pistil: A female organ of an angiosperm flower, composed of ovary, style, and stigma.

pistillate: Having pistils but lacking functional stamens.

plumule: The stem apex of the embryo from which the primary plant shoot develops.

pollination: Deposition of pollen on the receptive stigma or strobilus; the mechanical transfer of pollen from anthers to stigma.

polygamo-dioecious: Refers to species that are functionally dioecious but have a few bisexual flowers on some of the male and female plants.

polygamo-monoecious: Refers to species that are functionally monoecious but have a few bisexual flowers on some plants that also bear unisexual flowers of both sexes.

polygamo-trioecious: Refers to species that exhibit dioecious, monoecious, and bisexual flowering habits.

polygamous: Bearing both bisexual and unisexual flowers on the same plant or on different plants of the same species. The term pertains to species having mostly bisexual flowers.

pome: A fleshy fruit resulting from a compound ovary with seeds encased in a papery inner wall, as in *Crataegus* and *Malus*.

prechilling: Practice of exposing imbibed seeds to cool (40 to 50°F [4 to 10°C]) temperatures for a few days or months prior to germination. Also called *stratification*.

protandrous: *See* dichogamy.

protogynous: *See* dichogamy.

provenance: The original geographic source of seed.

purity: Percentage of clean, intact seed, by weight, in a seed lot.

pyrene: Individual seed of a drupe, as in *Ilex* and *Prunus*.

R

raceme: An unbranched inflorescence with flowers on stalks of equal length arising from a main axis, as in *Amelanchier* and *Prunus*.

radicle: The root of an embryo from which the primary root develops as the seedling emerges and grows.

recalcitrant: A term referring to the storage behavior of some seeds. Recalcitrant seeds generally store best in moist, dark, and cool conditions in temperatures between about 32° F and 50° F (0° C and 10° C).

rogueing: Systematic removal of undesirable individuals.

S

samara: Dry, indehiscent, winged fruit. It may be one-seeded, as in *Fraxinus*, or sometimes two samaras are fused, as in *Acer*.

scarification: Disruption of a seed coat by mechanical or chemical abrasion to increase its permeability to water and gases and to lower its mechanical resistance to radicle emergence.

schizocarp: A semi-dehiscent dry fruit derived from two or more carpels. Each carpel matures to contain a single seed.

seed coat: Protective outer layer of a seed. When two coats are present, the thick, tough outer coat is the *testa* and the thin, delicate inner coat is the *tegmen*.

seed-coat dormancy: Dormancy as a result of seed-coat conditions. These may involve impermeability to water or gas exchange, mechanical restrictions, or the presence of germination inhibitors in the seed coat, or any combination.

seed lot: A quantity of seeds having fairly uniform quality. It may be composed of seeds from a specific locale all collected within the same year, for example.

seed source: The locality where a seed lot was collected. *See* provenance.

serotinous: Flowering or fruiting late in the growing season. The term also pertains to cones that remain closed on a tree for several months or years after maturity and are therefore late in dispersing seeds.

silicle: A type of dry fruit.

silique: A type of dry fruit, similar to but larger than a silicle.

simple fruit: Fruit formed from a single ovary and sometimes including other flower parts. This is the most common type of fruit.

shatter: To break apart. A seed head from which the seeds fall freely as they ripen is said to "shatter."

species: Taxonomic category subdividing genera and comprising a group of similar interbreeding individuals having a common morphology, physiology, and reproductive process.

spike: Elongated inflorescence with sessile flowers on a main axis, as in *Amorpha*.

stamen: Pollen-bearing organ of a flower in angiosperms, consisting of a filament and an anther.

staminate: Having stamens but no pistils; a male flower.

steckling: A plantable rooted cutting; the root of a biennial root crop held over winter in storage and replanted the second spring.

stigma: The part of the pistil, usually the tip, often sticky, that receives the pollen during pollination.

stone: The hard, bony part of a drupe that consists of the seed within a hard endocarp, as in *Cornus* and *Prunus*.

strain: A group of organisms related by common descent but different in some respect from the main species; a subdivision of species.

stratification: Pregermination treatment to overcome dormancy in seeds and to promote rapid and uniform germination. It is done by keeping imbibed seeds in cold, moist conditions for a specified time. Also sometimes called *prechilling*.

strobile (plural: *strobiles*): Dry, conelike fruits that develop from pistillate catkins, as in *Alnus* and *Betula*.

strobilus (plural: *strobili*): Conelike male or female fruiting bodies of the conifers, composed of compact bracts or scales.

style: Neck of the pistil; the style connects the stigma with the ovary.

suture: The line of dehiscence on fruits that opens to disperse seeds.

syncarp: *See* aggregate.

T

testa: The outer, or only, seed coat. It is usually thick and tough.

thermodormancy: A type of dormancy imposed usually by subjecting seed to high temperatures, as in *Lactuca*.

thresh: To beat plants with a flail or against the sides of a container or other hard surface to separate seeds from fruits.

tilth: The condition of soil. Soil in good tilth is fluffy, well aerated, and fertile.

triploid: Having three times ($3n$) the haploid (n) number of chromosomes.

U

umbel: A flat-topped inflorescence with flower stalks arising from a common point, as in *Rhamnus caroliniana*. It is often compound, as in the paniculate umbels of *Aralia spinosa* (devil's walkingstick).

unisexual: Refers to flowers of one sex.

utricle: A bladdery, one-seeded, usually indehiscent fruit consisting of an achene surrounded by bracts, as in some ornamental flower species.

V

varietas: A botanical variety below the species taxon.

variety: A category usually intermediate between species (or subspecies) and forma, given a Latin name preceded by "var"; the older name for a selection within a species, now superseded by the term "cultivar" but popularly interchangeable with that term. For example, 'Big Boy' tomato is properly the cultivar but is commonly referred to as the variety. We use it in this sense.

vernalization: Subjection of a plant at a certain stage of growth to a period of cold to induce flowering.

viability: The state of being capable of germination and subsequent growth and development of a vigorous seedling.

vivipary: Germination of a seed within a fruit while still attached to the mother plant.

RESOURCES

Carolina Biological Supply Company
800-334-5551
www.carolina.com
Pollination bags

Fisher Scientific Inc.
Thermo Fisher Scientific Inc.
www.fishersci.com
Graduated cylinders

Sustainable Seed Company
877-620-7333
www.sustainableseedcompany.com
Pollination bags and seed-saving supplies

Seed-Saving Organizations

Association Kokopelli
+33-4-66-30-64-91
www.kokopelli-seeds.com

Henry Doubleday Research Association
Garden Organic
+44-0-24-7630-3517
www.gardenorganic.org.uk

Irish Seed Savers Association
+353-61-921866
www.irishseedsavers.ie

Native Seeds/SEARCH
866-622-5561
www.nativeseeds.org

Seed Savers Exchange
563-382-5990
www.seedsavers.org

Seed Savers Network
www.seedsavers.net

Seeds of Diversity
866-509-7333
www.seeds.ca

Recommended Reading

Ashworth, Suzanne. *Seed to Seed,* 2nd ed. Seed Savers Exchange, 2002.

Bienz, D. R. *The Why and How of Home Horticulture,* 2nd ed. W. H. Freeman, 1993.

Bubel, Nancy. *The New Seed-Starters Handbook.* Rodale, 1988.

Clothier, T. "Tom Clothier's Garden Walk and Talk." http://tom clothier.hort.net.

Dirr, Michael, and Charles W. Heuser. *The Reference Manual of Woody Plant Propagation,* 2nd ed. Varsity Press, 2006.

Gough, Robert E. *Encyclopedia of Small Fruit.* CRC Press, 2008.

Gough, Robert E., and Cheryl Moore-Gough. *Guide to Rocky Mountain Vegetable Gardening.* Cool Springs Press, 2009.

———. *Harvesting and Saving Garden Seeds,* MontGuide MT199905AG. Montana State University Extension, 2008.

Hartmann, Hudson Thomas, Dale Emmert Kester, and Fred T. Davies. *Plant Propagation,* 7th ed. Prentice Hall, 2002.

Hill, Lewis. *Secrets of Plant Propagation.* Storey Publishing, 1985.

Kumar, G. N. M., Fenton E. Larsen, and Kurt Anthony Schekel. *Propagating Plants from Seed,* PNW0170. Pacific Northwest Extension Publication, 2009.

Loewer, H. Peter. *Seeds.* Timber Press, 2005.

Rogers, Marc. *Saving Seeds.* Storey Publishing, 1990.

Turner, Carole B. *Seed Sowing and Saving.* Storey Publishing, 1998.

References

Anonymous. "American Seeds and Seedsmen." *American Agriculturist,* 45:117, 1886.

———. Testing Agricultural and Vegetable Seeds. *Agriculture Handbook 30.* U.S. Department of Agriculture, 1952.

———. *Woody-plant Seed Manual.* Forestry Service, USDA. Misc. Pub. 654, 1948.

Bailey, L. H. "The Seed Trade in America." In *Standard Cyclopedia of Horticulture,* 2nd ed. Vol. 6. Macmillan, 1917.

Bailey, L. H., and Ethel Zoe Bailey. *Hortus Third.* Macmillan, 1976.

Bonner, F. T., and R. P. Karrfalt, eds. "Woody Seed Plant Manual." In *Agricultural Handbook 727.* Forestry Service, USDA, 2008.

Boswell, V. R. "Flowering Habit and Production of Seeds." In *Yearbook of Agriculture.* U.S. Department of Agriculture, 1961.

Buskin, C. C. and J. M. Buskin. "Seed Dormancy in Wildflowers," in *Flower Seeds: Biology and Technology,* eds. M. B. McDonald and F. Y Kwon (CABI Publishing, 2005) 163–187.

Copeland, L. O., and M. B. McDonald. *Principles of Seed Science and Technology,* 2nd ed. Burgess, 1985.

Corbett, L. C., H. P. Gould, and W. R. Beattie. "Fruits and Vegetables." In *Yearbook of Agriculture*. U.S. Department of Agriculture, 1925.

Corbett, L. C., H. P. Gould, T. R. Robinson, G. M. Darrow, G. C. Husmann, C. A. Reed, D. N. Shoemaker, et al. "Fruit and Vegetable Production." In *Yearbook of Agriculture*. U.S. Department of Agriculture, 1925.

Emsweller, S. L., P. Brierley, D. V. Lumsden, and F. L. Mulford. "Improvement of Flowers by Breeding." in *Yearbook of Agriculture*. U.S. Department of Agriculture, 1937.

George, R. A. T. *Vegetable Seed Production*. Longman Group Limited, 1985.

Gough, R. E. *The Highbush Blueberry and Its Management*. Haworth Press, 1994.

Harrington, J. F. "Cleaning Vegetable and Flower Seeds." *Seed Technology* 20 (1977): 162–175.

Hawthorn, L. R. "Growing Vegetable Seeds for Sale." In *Yearbook of Agriculture*. U.S. Department of Agriculture, 1961.

Hawthorn, L. R., and L. H. Pollard. *Vegetable and Flower Seed Production*. Blakiston, 1954.

Justice, O. L., and L. N. Bass. "Principles and Practices of Seed Storage." In *Agricultural Handbook* 506. SEA-USDA, 1978.

Maynard, D., and G. Hochmuth. *Knott's Handbook for Vegetable Growers*. Wiley, 1997.

Save Our Seed Project. "The Seed Processing and Storage Guide." www.savingourseed.org/Survey/SeedProcessingandStorageSurvey.html.

McDonald, M. B., and F. Y. Kwong. *Flower Seeds: Biology and Technology*. CABI Publishing, 2005.

Pieters, A. J. "Seed Selling, Seed Growing, and Seed Testing." In *Yearbook of Agriculture*. U.S. Department of Agriculture, 1899.

Pollock, B., and V. K. Toole. "Afterripening, Rest Period, and Dormancy." In *Yearbook of Agriculture*. U.S. Department of Agriculture, 1961.

Quick, C. R. "How Long Can a Seed Remain Alive." In *Yearbook of Agriculture*. U.S. Department of Agriculture, 1961.

Rudolf, P. O. "What Do We Plant." In *Yearbook of Agriculture*. U.S. Department of Agriculture, 1949.

Russell, P. G., and A. F. Musil. "Plants Must Disperse Their Seeds." In *Yearbook of Agriculture*. U.S. Department of Agriculture, 1961.

Shurtleff, M. C. *How to Control Plant Diseases*, 2nd ed. Iowa State University Press, 1966.

Stanley, R. G., and W. L. Butler. "Life Processes of the Living Seed." In *Yearbook of Agriculture*. U.S. Department of Agriculture, 1961.

Wein, H. *Physiology of Vegetable Crops*. CAB International, 1997.

Whealy, K. *Seed Savers Exchange Yearbook*. Seed Savers Exchange, 2010.

ACKNOWLEDGMENTS

A BOOK AS COMPREHENSIVE AS THIS is not the product solely of its authors. It reflects the wisdom and experience of many people who have had a hand, directly and indirectly, in advising us over the years. Our earliest advisers were our parents and grandparents, who instilled in us a simple fascination for plants, from planting the seeds to enjoying the fruits, and then saving their seeds for the next year. Our professors taught us plant physiology, taxonomy, and anatomy and showed us how all of a plant's systems work together harmoniously to propagate the species. They also taught us patience and the scientific method.

Our combined experience adds up to nearly 100 years' worth of horticultural practice and experimentation. During much of that time, we were university academics, and participated in many professional conferences and published dozens of research reports. All of these experiences submerged us in the milieu of science, where we exchanged ideas and postulates with our colleagues and received their valuable help and support. But no matter how technical our research, we kept our feet on the ground and our fingers in the soil. We straddled two worlds.

The number of fine folks who inspired us along the way is legion; we cannot name them all. To colleagues Dr. Vladimir Shutak and Dr. Luke Albert, both formerly at the University of Rhode Island; Dr. Richard Stout and Dr. Rick Bates, both formerly at Montana State University; and Dr. Cathy Cripps at Montana State University, we owe our deep thanks for the vast opportunities they afforded for encouragement and collaborative research. To Dr. Norman Childers, formerly at Rutgers University and the University of Florida, we owe our gratitude for his tireless encouragement, promotion, and recognition of our horticultural achievements. To Dr. William Lamont of Pennsylvania State University and Dr. Paul Read of the University of Nebraska, both former presidents of the American Society for Horticultural Science (ASHS); to Dr. Barclay Poling of North Carolina State University and Dr. Larry Knerr of Shamrock Seed Co., both former vice presidents of ASHS; and to Mr. Dan Spurr, editor of *Zone 4* magazine, we owe our deepest gratitude for their helpful comments on improving the manuscript of this work. Finally, to the entire staff of Storey Books, and, most important, Gwen Steege, our heartfelt thanks for giving us the opportunity to write this book, and to Fern Marshall Bradley, much appreciation for her unexcelled Olympian efforts in editing the manuscript. Without help from all these wonderful people and from many others, professional and amateur, with whom we have crossed paths over the years, this book would not have been possible.

INDEX

Page references in **bold** indicate main entries; those in *italic* indicate photos or illustrations; those in ***bold italic*** indicate charts.

cardamom. See *Elettaria cardamomum*

Carex (sedge), **166**

carmel daisy. See *Scabiosa*

carnation pink. See *Dianthus*

Carolina basswood. See *Tilia*

carpel, 13

carrot. See *Daucus carota* var. *sativus*

Carum (caraway), **140–41**

Carya (hickory, pecan), **232–33**

Caryophyllaceae family, 84

caryopsis or grain, 23

Castanea (chestnut), *233*, **233–34**

Catalpa (catalpa), **234**

Catharanthus (periwinkle), **166**

catkin, 17, *17*

catmint. See *Nepeta cataria*

catnip. See *Nepeta cataria*

cauliflower. See *Brassica oleracea*

cedar. See *Thuja*

celeriac. See *Apium graveolens*

celery. See *Apium graveolens*

Celosia (cockscomb), **167**

Centaurea (bachelor's button, cornflower), 48, *167*, **167–68**

Cephalotaxus (plum-yew), **234**

Cerastium (mouse-ear chickweed), *168*, **168**

Cercis (redbud), **234–35**, *235*
  treatments for seed of, *235*

Cerinthe (honeywort), **168**

Chaenactis (pincushion flower), **169**

Chaenomeles (flowering quince), **235–36**

Chaenorhinum (dwarf snapdragon), **169**

Chamaecyparis (chamaecyparis), 33, **236**

Chamaemelum (chamomile), **141**

chamomile. See *Chamaemelum*

charming centaury. See *Centaurea*

chaste tree. See *Vitex*

checker lily/checkered lily. See *Fritillaria*

Cheiranthus (wallflower), **169–70**

Chenopodium (epazote, good King Henry, quinoa), **141**

cherry. See *Prunus*

chervil, salad. See *Anthriscus cerefolium*

chestnut. See *Castanea*

chia. See *Salvia*

chickpea. See *Cicer arietinum*

chili pepper. See *Capsicum* spp.

China aster. See *Callistephus*

Chinese cabbage. See *Brassica rapa*

Chinese forget-me-not. See *Cynoglossum*

Chinese parsley. See *Coriandrum sativum*

Chinese thornapple. See *Datura*

chiropterphily, 20

chives. See *Allium*

Chrysanthemum (chrysanthemum), **141–42**, **170**
  ×*superbum*, 152

Cicer arietinum (chickpea, Egyptian pea, garbanzo, gram), **106**

Cichorium endivia (endive & escarole), **112**

Cichorium intybus (chicory, Witloof), **107**

cigarflower. See *Cuphea*

cilantro. See *Coriandrum sativum*

citron. See *Citrus*

Citrullus lanatus (watermelon), **135**

Citrus (citrus), **236–37**

Clarkia (godetia), **170–71**

clary. See *Salvia*

clary sage. See *Salvia*

Claytonia perfoliata (miner's lettuce), **117**

cleaning seeds, 43–44, 47, *47*

Clematis (clematis), **237**

Cleome (spider flower), *171*, **171**

Clitoria ternatea (butterfly pea), **172**

clove pink. See *Dianthus*

cockscomb. See *Celosia*

coenocarp, 24

collards. See *Brassica oleracea*

collecting seeds. See harvesting and cleaning seeds

columbine. See *Aquilegia*

combined dormancy, 57

common cherrylaurel. See *Prunus*

common chokecherry. See *Prunus*

common hedge nettle. See *Stachys*

common pea. See *Pisum sativum*

common tansy. See *Tanacetum vulgare*

common winterberry. See *Ilex*

common yarrow. See *Achillea*

complete flower, 13, *13*, 14, *14*

compound fruits, 24

cone. See strobili/strobilus

coneflower. See *Echinacea*; *Rudbeckia*

conifers, 12

Consolida (larkspur), **172**

containers
  for seed storage, *54*, 54–55
  sowing seeds in, 72–73
  for starting seeds, 71–72, *72*

Convallaria majalis (lily of the valley), **172–73**, *173*

cool-tankard. See *Borago*

coral bells. See *Heuchera*

Coreopsis (calliopsis), **173**

coriander. See *Coriandrum sativum*

Coriandrum sativum (Chinese parsley, cilantro, coriander), **142**, *142*

corktree. See *Phellodendron*

corn. See *Zea mays*

drupe, 24, *24*, 262

dry fruits

cleaning seeds from, 47

types of, 23, *23*

dry fruits, seed extraction, 45–46

flailing, 45–46

heat treatment, 46

hulling, 45

threshing, 46, *46*

tumbling or shaking, 45

drying seeds, flower seed plants, 153

dusty miller. See *Artemisia*; *Centaurea*; *Chrysanthemum*

Dutchman's breeches. See *Dicentra*

dwarf snapdragon. See *Chaenorhinum*

Dyer's rocket. See *Reseda*

E

eastern redcedar. See *Juniperus*

eastern wahoo. See *Euonymus*

Echinacea (coneflower), 19, *19*, **180–81**, *181*

Echinops (globe thistle), **181**

edamame. See *Glycine max*

edema, 75

eggplant. See *Solanum malongena* var. *esculentum*

Egyptian pea. See *Cicer arietinum*

Elettaria cardamomum (cardamom), **143**

Ellsworth, Henry, 282

elm. See *Ulmus*

El Paso skyrocket. See *Gilia*

embryo dormancy, 57–58

endive. See *Cichorium endivia*

endosperm, 20

English daisy. See *Bellis*

English pea. See *Pisum sativum*

Enkianthus (enkianthus), **242**

entomophily, 19

epazote. See *Chenopodium*

eringo. See *Eryngium foetidum*

Erodium (heron's bill, stork's bill), **181**

Eruca vesicaria ssp. *sativa* (arugula), **94**

Eryngium foetidum (culantro, eringo, fitweed, shado beni), **143**

escarole. See *Cichorium endivia*

Eschscholzia (California poppy), **182**

Euonymus (euonymus), **242–43**

follicle of, 23, *23*

Eupatorium (Joe-pye weed), **182**

European centaury. See *Centaurea*

European cranberrybush. See *Viburnum*

eustoma. See *Lisianthus*

Eustoma (lisianthus), **182–83**

evening primrose. See *Oenothera*

evening stock. See *Matthiola*

evergreen orpine. See *Sedum*

everlasting. See *Ammobium*; *Helichrysum*; *Helipterum*

extracting/cleaning seeds, 43–47

extracting seeds, 43–46

dry fruits and, 45–46, *46*

fermentation process and, 45

fleshy fruits and, 44–45

float-test possibilities, 45

for flower seed plants, 153

from fruit, 44–46, *46*

F

F₁ hybrid varieties, 21, 22

Fabaceae family, 84

Fagus (beech), **243**

false indigo. See *Baptisia*

farewell-to-spring. See *Clarkia*

fascicle, 16, *16*

female flower. See pistillate flower

fennel. See *Ferula*; *Foeniculum*

fenugreek. See *Trigonella*

fermentation process, 45

fertilizing

seedlings, 77

transplants, 78

Ferula (giant fennel), **144**

field chickweed. See *Cerastium*

field marigold. See *Calendula*

filament, 13, *13*

finoccio. See *Foeniculum*

fir. See *Abies*

firecracker plant. See *Cuphea*

firethorn. See *Pyracantha*

fitweed. See *Eryngium foetidum*

flag. See *Iris*

flannel plant. See *Verbascum*

flax. See *Linum* spp.

fleshcolor pincushion. See *Chaenactis*

fleshy fruits

cleaning seeds from, 47

types of, 24

fleur-de-lis. See *Iris*

flint corn. See *Zea mays*

Florence fennel. See *Foeniculum*

flossflower. See *Ageratum houstonianum*

flower biology, 12–19

factors that affect flowering, 18–19, *19*

flower stalks, architecture of, *16*, 16–17, *17*

flower structure/development, 12–14, 16

juvenile period, 12

flower bud induction, 12

flowering, factors that affect, 18–19, *19*

light intensity, 18–19

nutrition, 19

photoperiod, 18

short- and long-day plants, 18

temperature, 18

flowering habit of plants, 91–92

flowering maple. See *Abutilon*

flowering quince. See *Chaenomeles*

flowering tobacco. See *Nicotiana*

flower seeds, saving, 152–54

    collecting seeds, 153

    drying/storage, 153

    extracting seeds, 153

    germination requirements, 153

    heat-treating seeds, 154

    inbreeding depression, 153

    pollination, controlling, 153

    raising transplants, 154

    seed viability, 154

    spacing plants and, 153

flower-seed yields, *43*

flower stalks, architecture of, *16,*
    16–17, *17*

flower structure/development,
    12–14, 16

    complete flower, parts of, 13, *13*

    "gender" of plants, 14

    grafting shortcut, 13

    imperfect flower, 14, *14*

    incomplete flower, 13

    perfect flower, 14, *14*

    pistillate flower, 14, *14*

    staminate flower, 14, *14*

Foeniculum (fennel), **144**

    vulgare (Florence fennel,
        finoccio, sweet fennel),
        112–13

        'Dulce', 144

    vulgare var. *azoricum* (anise,
        fennel, finocchio), 144

follicle, 23, *23*

food of the gods. See *Ferula*

forget-me-not. See *Myosotis*

four-o'clock. See *Mirabilis*

foxglove. See *Digitalis*

Franklin, Benjamin, 280

Fraxinus (ash), **243–44**

    americana (white ash), 57–58

    excelsior (European ash), 50

freezing point for seeds, 61

French sorrel. See *Rumex* spp.

French spinach. See *Atriplex*
    hortensis

Fritillaria (fritillary), **183**

fritillary. See *Fritillaria*

frost flower. See *Aster*

frost protection, 78

fruits, advanced seed saving,
    221–22

fruit/seed development, 22, *22*

fruit types, 23–24

    compound fruits, 24

    dry strobili, 24, *24*

    fleshy strobili, 24, *24*

    persistent, 40

    simple fruit, *23,* 23–24

G

Gaillardia (blanket flower), **183**

garbanzo. See *Cicer arietinum*

garden balsam. See *Impatiens*

garden pea. See *Pisum sativum*

garden pink. See *Dianthus*

garden sorrel. See *Rumex* spp.

garlic. See *Allium*

gas plant. See *Dictamnus*

gayfeather. See *Liatris*

Gazania (treasure flower), **184**

genetic diversity, 3, *3*

    preservation of, 4

Geraniaceae family, 84

geranium. See *Pelargonium*

Geranium (geranium, cranesbill),
    184, **184–85**

geraniumleaf heronbill. See
    Erodium

German chamomile. See
    Matricaria

germinants, 79

germination of seeds, 56–66

    dormancy, overcoming, 58–62

    dormancy, types of, 57–58

for flower seed plants, 153

    nature's logic, 56–57

    optimum, promoting, 62–65

    phytochrome physiology, 64

    waiting game, 65–66

    for woody plants, *65*

Geum (avens), **185**

giant hyssop. See *Agastache*

giant mallow. See *Hibiscus*

Gilia (gilia), **185**

gilliflower. See *Matthiola*

Ginkgo biloba (ginkgo), 12, *244,*
    244–45

Gladwin. See *Iris*

Gleditsia (locust, honeylocust), **245**

    triacanthos (honeylocust),
        58–59

globe amaranth. See *Gomphrena*

globeflower. See *Trollius*

globe thistle. See *Echinops*

Glycine max (soybean), **129**

gobo. See *Abelmoschus esculentus*

godetia. See *Clarkia*

golden buttons. See *Tanacetum*
    vulgare

goldenchaintree. See *Laburnum*

goldenraintree. See *Koelreuteria*

golden wave. See *Coreopsis*

gombo. See *Abelmoschus esculentus*

Gomphrena (globe amaranth), **186**

goober. See *Arachis hypogaea*

good King Henry. See
    Chenopodium

gousblom. See *Arctotis*

grafting shortcut, 13

grain or caryopsis, 23, *23*

gram. See *Cicer arietinum*

grape. See *Vitis*

grapefruit. See *Citrus*

grassy bell. See *Dierama*

green cockscomb. See *Sedum*

green pea. See *Pisum sativum*

I

Iberis (candytuft), **190**
ice plant. See *Mesembryanthemum*
Ilex (holly), **245–46**
    opaca (American holly), 58
imbibition, 57
immortelle. See *Helipterum*
Impatiens (impatiens), **190–91**
imperfect flower, 14, *14*
inbred parent, developing, 86–88
    EE, eE, or Ee, 88
    EE × EE, 87
inbreeding depression, 20, 88, 153
incomplete flower, 13
indehiscent fruits, 23, *23*
indeterminate growth habit, 16
Indian cress. See *Tropaeolum*
Indian-currant coralberry. See
    Symphoricarpos
indigobush. See *Amorpha*
indoor growing area, setting up, 73
inflorescence, 16
inkberry. See *Ilex*
internal dormancy, treatments,
    59–62
    germination-enhancement, 62
    light treatment, 61–62
    stratifying seeds, 60–61
    warm stratification, 61
invasiveness, 38
Ipomoea (morning-glory), **191**
    alba (moonflower), 59
Ipomoea batatas (sweet potatoes),
    92
Iridaceae family, 84
iris. See *Iris*
Iris (iris, flag, fleur-de-lis), **191–92**
isolating plants, pollination and,
    26–29
    bags and cages, 27–28, *28*
    distance and, 27
    planting time and, 28–29

J

jamberry. See *Physalis ixocarpa*
Japanese angelica tree. See *Aralia*
Japanese arborvitae. See *Thuja*
Japanese falsecypress. See
    Chamaecyparis
Japanese horsechestnut. See
    Aesculus
Japanese winterberry. See *Ilex*
Jefferson, Thomas, 280
Jerusalem artichoke. See
    Helianthus tuberosus
jewelweed. See *Impatiens*
Joe-pye weed. See *Eupatorium*
Johnny jump up. See *Viola*
"J-rooted" plants/position, 75, *75*
Juglans (walnut), **246–47**
juniper. See *Juniperus*
Juniperus (juniper), **247–48**
    occidentalis (Sierra juniper), 37
    pregermination treatments for,
    247
juvenile period/juvenility, 12

K

kale. See *Brassica oleracea*
Kalmia (mountain laurel), 74, **248**
kiss me quick. See *Portulaca*
Kniphofia (red hot poker, torch
    lily), **192**
Koelreuteria (goldenraintree), **249**
Korean arborvitae. See *Thuja*

L

Laburnum (goldenchaintree), **249**
Lactuca sativa (lettuce), *7*, 60, **115**
lady's finger. See *Abelmoschus
    esculentus*
Lagerstroemeia (crapemyrtle),
    249–50
lamb-kill. See *Kalmia*
lamb's ears. See *Stachys*

lamb's lettuce. See *Valerianella
    locusta*
Lamiaceae family, 84, 136
Lamium (deadnettle, spotted
    henbit), **192**
landraces, 81
Landreth, David, 280
Lantana (shrub verbena), **193**
larkspur. See *Consolida*;
    Delphinium
Lathyrus (sweet pea), 59, **193**
Lavandula (lavender), **250**
lavender. See *Lavandula*
leadplant. See *Amorpha*
leek. See *Allium*
legume, 23
lemon. See *Citrus*
lemon balm. See *Melissa officinalis*
lemon lily. See *Hemerocallis*
lemon mint. See *Melissa officinalis*;
    Monarda
Lens culinaris (lentil), **114**
lentil. See *Lens culinaris*
Lepidium sativum (cress, garden),
    110
lettuce. See *Lactuca sativa*
Leucanthemum (shasta daisy), **194**
Levisticum officinale (lovage), **145**
Lewisia (bitterroot, lewisia), **194**
Liatris (blazing star, gayfeather),
    191–92
Liberty Garden, 283
light treatment, internal dormancy,
    61–62
Ligustrum (privet), **250**
Liliaceae family, 84
lily-of-the-field. See *Anemone*
lily-of-the-Incas. See *Alstroemeria*
lily of the valley. See *Convallaria
    majalis*
lily seeds, 37
lime. See *Citrus*

starchy vs. oily seeds, 50
starflower pincushions. See
 Scabiosa
star-of-Bethlehem. See
 Ornithogalum spp.
starwort. See *Aster*
statice. See *Limonium*
stecklings, 35
stepping up, 74–75, *75*
stigma, 13, *13*
Stillman's tickseed. See *Coreopsis*
stinking Gladwin. See *Iris*
stock. See *Matthiola*
stonecrop. See *Sedum*
"stone fruit." See *Prunus*
storage of seeds, 48–55
 for flower seed plants, 153
 loss of vigor, 66, **66**
 moisture and, 50–53
 places to store seeds, 54–55
 recalcitrant seeds, 50
 summary of recommendations,
 55
 temperature and, **53**, 53–54
 viability of seeds, 48–50
stratifying seeds, 58, 60–61
strawberry bush. See *Euonymus*
strawberry tree. See *Arbutus*
strawflower. See *Helichrysum*;
 Helipterum
Striga asiatica (witchweed), 2
strobile, 23
strobili/strobilus, 12, 15, *15*, *259*
 color of ripe pinecones, **259**
 de-winging of seeds, 46
 dry strobili, 24, *24*
 fleshy strobili, 24, *24*
 immature/mature, 40, *40*
 specific gravity of ripe
 pinecones, **260**
style, 13, *13*
suborthodox seeds, 49, 53

summer radishes. See *Raphanus
 sativus*
sundrop. See *Oenothera*
sunflower. See *Helianthus*
sweet alyssum. See *Lobularia*
sweet chervil. See *Myrrhis odorata*
sweet cicely. See *Myrrhis odorata*
sweet corn. See *Zea mays*
sweet false chamomile. See
 Matricaria
sweetgum. See *Liquidambar*
sweet marjoram. See *Origanum*
sweet orange. See *Citrus*
sweet pea. See *Lathyrus*
sweet pepper. See *Capsicum* spp.
sweet potatoes. See *Ipomoea
 batatas*
sweet sultan. See *Centaurea*
sweet William. See *Dianthus*
Symphoricarpos (snowberry), **272**
synconium, 24

T
tabasco pepper. See *Capsicum* spp.
Tagetes (marigold), *215*, **215**
tagging, 31
tailwort. See *Borago*
tampala. See *Amaranthus*
Tanacetum vulgare (golden
 buttons, common tansy),
 150–51, *151*
tangelo. See *Citrus*
tangier scarlet pea. See *Lathyrus*
tansy. See *Tanacetum vulgare*
tarragon. See *Artemisia*
taurus chickweed. See *Cerastium*
taxonomists, 8
Taxus (yew), **272–73**
temperate-recalcitrant seeds, 49
temperature, seed storage and, **53**,
 53–54
 half life of seeds, 54
 orthodox seeds, 53

recalcitrant seeds, 54
terminology. *See* botanical
 terminology
Tetragonia tetragonioides (New
 Zealand spinach), **117–18**
Texas betony. See *Stachys*
Texas bluebonnet. See *Lupinus*
thermodormancy, 58
thistle sage. See *Salvia*
Thorburn, Grant, 280
thousandseal. See *Achillea*
threshing/threshing tools, 46, *46*
Thuja (cedar), **273–74**
thyme. See *Thymus*
Thymus (thyme), **151**
tickseed. See *Coreopsis*
Tilia (linden), *274*, **274–75**
 americana (American linden),
 61
timing the harvest, 37–43
 clues to ripening, 38–40
 seed-gathering techniques and,
 40, 42–43
toadflax. See *Linaria*
toad lily. See *Tricyrtis*
tomatillo. See *Physalis ixocarpa*
tomato. See *Lycopersicon
 esculentum*
torch flower. See *Kniphofia*
torch lily. See *Kniphofia*
toxicity, plants and seeds, 38
Trachymene (blue lace flower), **215**
Tragopogon porrifolius (salsify),
 126–27, *127*
transplanting, 80
transplants, raising, 70–80
 care after planting, 78
 containers for starting seeds,
 71–72, *72*
 fertilizing seedlings, 77
 flexibiity and, 71
 flower seed plants, 154
 germination of tiny seeds and, 71

winter radishes. See *Raphanus sativus*
Wisteria (wisteria), **277–78,** *278*
witchweed. See *Striga asiatica*
Witloof chicory. See *Cichorium intybus*
wolfbane. See *Aconitum*
woody plants
    ornamentals, advanced seed saving, 221–22
    seed-starting success, **65**
    self-fruitful or not, 85
    transplant hole for, 75
woolly betony. See *Stachys*
wormwood. See *Artemisia*

Y

yarrow. See *Achillea*
yellow buttons. See *Chrysanthemum*
yellow chaenactis. See *Chaenactis*
yellow flag. See *Iris*
yellow peavine. See *Lathyrus*
yellow sage. See *Lantana*; *Lathyrus*
yew. See *Taxus*
Yucca (yucca), **278**

Z

Zea mays (corn), **108–9,** *109*
    caryopsis or grain of, 23, *23*
    'Golden Bantam', 281
    hand-pollinating, *30,* 30–31
Zinnia (zinnia), **220**

Other Storey Titles You Will Enjoy

The Beginner's Guide to Edible Herbs, by Charles W. G. Smith.
A classic companion planting guide that shows how to use plants' natural partnerships to produce bigger and better harvests.
152 pages. Paper. ISBN 978-1-60342-528-5.

The Gardener's A–Z Guide to Growing Flowers from Seed to Bloom, by Eileen Powell.
An encyclopedic reference on choosing, sowing, transplanting, and caring for 576 annuals, perennials, and bulbs.
528 pages. Paper. ISBN 978-1-58017-517-3.

The Gardener's A–Z Guide to Growing Organic Food, by Tanya L. K. Denckla.
An invaluable resource for growing, harvesting, and storing 765 varieties of vegetables, fruits, herbs, and nuts.
496 pages. Paper. ISBN 978-1-58017-370-4.

The Homeowner's Complete Tree & Shrub Handbook, by Penny O'Sullivan.
The new bible of tree and shrub selection and care, showing hundreds of plant possibilities in full-color photographs.
416 pages. Paper. ISBN 978-1-58017-570-8.
Hardcover with jacket. ISBN 978-1-58017-571-5.

Landscaping with Fruit, by Lee Reich.
A complete, accessible guide to luscious landscaping — from alpine strawberry to lingonberry, mulberry to wintergreen.
192 pages. Paper. ISBN 978-1-60342-091-4.
Hardcover with jacket. ISBN 978-1-60342-096-9.

Secrets of Plant Propagation, by Lewis Hill.
Expert advice on techniques to grow beautiful, bountiful, healthy plants — and save money in the process!
176 pages. Paper. ISBN 978-0-88266-370-8.

Secrets to Great Soil, by Elizabeth P. Stell.
All the advice a gardener needs to have great soil — thus, a great garden!
224 pages. Paper. ISBN 978-1-58017-008-6.

The Vegetable Gardener's Bible, 2nd edition, by Edward C. Smith.
The 10th Anniversary Edition of the vegetable gardening classic, with expanded coverage of additional vegetables, fruits, and herbs.
352 pages. Paper. ISBN 978-1-60342-475-2.
Hardcover. ISBN 978-1-60342-476-9.

These and other books from Storey Publishing are available wherever quality books are sold or by calling 1-800-441-5700.
Visit us at *www.storey.com*.